Introduction to Climatic Geomorphology

J. Tricart
Director, Centre de Géographie Appliquée
University of Strasbourg

and

A. Cailleux
Professor, Faculty of Sciences,
University of Paris

Translated from the French by
Conrad J. Kiewiet de Jonge
Professor of Geography
California State University, San Diego

Longman

LONGMAN GROUP LIMITED
London

*Associated companies, branches and representatives
throughout the world*

English translation © Longman Group Limited 1972

Introduction à la Géomorphologie Climatique first published by the Société d'Édition
d'Enseignement Supérieur, Paris, 1965

English edition first published 1972

ISBN 0 582 48156 2

**Dedicated to Benita and Anna Maria Kiewiet de Jonge
for their love and forbearance.**

Set in 'Monophoto' Baskerville
and printed in Great Britain by William Clowes & Sons, Limited
London, Beccles and Colchester

Introduction to Climatic Geomorphology

Geographies for Advanced Study

Edited by Professor Stanley H. Beaver, M.A., F.R.G.S.

Contents

List of illustrations

Chart

Folding maps (*between pages* 274 *and* 275)

Foreword to the English edition

Geomorphology is the science whose object of study is the terrestrial relief and its evolution.

Geomorphology has an unusual history. An early branch of the discipline is discernible in the work of engineers such as Leonardo da Vinci (1452–1519) and Bernard Palissy (1510–1590). Later, at the end of the eighteenth century and at the beginning of the nineteenth, another branch developing out of geology made its appearance; it took form from the work of James Hutton (1726–97) and, especially, Charles Lyell (1797–1875).

Both branches remained almost wholly separated during the course of the nineteenth century. In France engineers such as Alexandre Surell (1813–1887) and L. A. Fabre created an applied geomorphology in connection with the control of torrents and the use of plants to control erosion. In Britain and America a geologic geomorphology developed, culminating in the work of William Morris Davis (1850–1934). One should not, however, generalise this historical development too much. For example, E. de Margerie (1862–1953), in France, originated structural geomorphology, while in the United States the eminent geologist G. K. Gilbert (1843–1918) combined the methods and preoccupations of the engineers with academic geomorphology.

It is true, nevertheless, that both branches have remained essentially ignorant of one another. In the United States, Britain, and France, academic geomorphology developed out of geology and was nourished with Davisian concepts. The theories of Davis fit right into the scale of geologic time. Such a phenomenon as the cycle of 'erosion' ending in peneplanation, or such phenomena as long periods of 'erosion' which succeed crustal uplifts that are short and violent enough to preclude any interaction between the internal and the external forces, can only take place during the course of millions of years. Because Davis's concepts are drawn from geology, they, by their very nature, embrace the order of magnitude of geologic time.

Even if we were to place ourselves on Davis's own ground, his concepts would be much open to criticism. Rather than a sudden crustal uplift succeeded by a long continued period of erosion, interaction between tectogenesis and morphogenesis is the rule. The too exclusive preoccupation with 'erosional' phenomena, neglecting everything that is depositional, gives a unilateral and incomplete view of the land-forming processes. These concepts are much too narrow and to a large extent inadequate. It is normal for such mistakes to occur in the evolution of scientific disciplines; it is only regrettable that it took such a long time for many geomorphologists to notice them. What was once progress has now become a handicap.

But there are even more serious defects. Although Davis had not forgotten the processes in his trilogy 'structure, process, and stage', they fared no better than Lepidus in the Triumvirate. They have practically been forgotten in Davisian geomorphology, having been disposed of in a convenient simplification: the theory of 'normal erosion'. The reason for such neglect is that the study of processes proceeds from the other branch of geomorphology started by engineers and to which an important part of the work of Gilbert belongs. Processes, however, cannot be studied on the scale of geologic time: the millions of years of its history are too long. The effects of the processes are measured on a different time scale involving shorter durations, evaluated in thousands of years. The concept of a spatiotemporal scale did not as yet exist in Davis's time, and an integration of phenomena belonging to different dimensional categories was impossible.

Even if Davis's views were correct as far as structural geomorphology is concerned, they would still be very incomplete because they are in fact incapable of integrating one of the three factors of the trilogy: the processes. His sketches never included a single plant or a single grass leaf. His landscapes look like deserts. Oddly enough, Davis used the word 'geographic' to describe his concepts, but his geomorphology was geologic rather than geographic. A geomorphology which gives due account of the morphogenic processes also takes account, in the highest degree, of the plant cover as influenced by the climate. The effects of a cloudburst are not the same on the bare frozen ground of a periglacial desert, below a temperate forest leafless in the spring, on a semiarid steppe, or in an evergreen tropical forest with sparse underbrush.

For these reasons the morphogenic processes, practically ignored by the Davisian school but considered essential by the German school, are the object of our investigations both within the framework of academic as well as applied research; this is the specific role of the Centre de Géographie Appliquée which I helped found and presently direct. Climatic geomorphology and the study of morphogenic processes are closely related. Investigations which lead to a better understanding of the processes are at the origin of a whole series of geomorphological applications. They permit the co-operation of geomorphologists and experts in soil conservation, of civil engineers, city planners, and specialists concerned with land utilisation and the problems of economic development. The remarkable work done in the United States since the 1930s to control soil erosion has been done by agronomists and pedologists who desired to protect the soils. They are the proof that yesterday's geomorphology has failed in its task, being exclusively preoccupied with physiographic descriptions.

Climatic geomorphology and the study of the morphogenic processes are therefore at the origin of a new approach to geomorphology. They also help to mitigate the disadvantages of a growing specialisation. Besides a more and more detailed analysis of the mechanisms, they create new links between different disciplines. Climatic geomorphology indeed must examine the interactions between all the factors influencing the processes: not only the geologic structure, which is a comparatively stable entity, but also the climate, which makes itself felt through its direct and, more important, its indirect influence on the vegetation and on the soils which, finally, are more immediately exposed to the external forces than the rocks. Climatic geomorphology cannot be satisfied with being a branch of geology: it is necessarily integrated into the entire field of physical geography.

We have elaborated these concepts during the course of a period of about twenty years, taking generous inspiration from the work of German geomorphologists who had remained outside Davis's influence, which is contrary to what happened to French geomorphologists. We have gradually enriched these concepts during the course of numerous investigations conducted in the different parts of the world, Antarctica excluded, usually in collaboration with engineers, agronomists, and pedologists to resolve the practical problems of land utilisation and economic development. A protracted collaboration with A. Cailleux has materially helped to clarify our ideas. The practice of teamwork and the enthusiasm and devotion of the group of workers who give their untiring efforts to the various aspects of physical geography at the Centre de Géographie Appliquée have contributed a large share of the elements on which our essay is based.

Since Christopher Columbus, many people of the Old World have been discovering America. Let us hope that the present work will help Americans discover Europe, whose research and publications are too often ignored by Americans. If this is true our friend Kiewiet de Jonge, to whom we owe the initiative of this translation, will deserve everybody's gratitude: a better mutual understanding will enable all of us to make more rapid progress.

J. TRICART
Strasbourg, March 1968

Introduction

Climatic geomorphology is the study of relief forms as determined by climate. This approach is widely accepted at the present time; it has even been contrasted to structural geomorphology. Thus contrasted geomorphology would be divided into two branches, as in a diptych.

We have supported the concept of climatic geomorphology with sufficient continuity so that it cannot be said that we are hostile to it. Nevertheless, we should question the objective value of such a dichotomy. Is it fully justified? Is it an unqualified progress? Does it not imply a theoretic and excessive oversimplification? Does it not reflect a tendency to split the unity of nature, to introduce divisions that are too rigorous and artificial? It is necessary to introduce divisions for the sake of exposition, but they should correspond to a reasonable reality and should not provide obstructions to the mind nor blinkers to the eyes. There is, then, a very real danger, and it is serious enough to receive careful treatment.

First, the structural geomorphology/climatic geomorphology diptych provides an incomplete image of the field of geomorphology. There are many processes that are no more closely and exclusively dependent on structural geology than they are on climatic factors. One cannot deny the originality of the littoral domain, which is a consequence of its very location at the zone of contact between the hydrosphere and the lithosphere, between marine or lacustrine waters and the land. To be sure, the distribution of continents and oceans is the result of structural phenomena, nevertheless one cannot explain coastal geomorphology solely in structural terms when dealing with individual coasts, even at the scale of 1 : 1 000 000. On all the world's coasts the fashioning of the littoral depends on the hydrodynamic mechanisms of the waves. True, climate introduces differences between polar coasts, locked in ice many months of the year, shielded from the action of swells, and subjected to frost-weathering in autumn, and in spring, and even in the middle of summer. Climate also introduces differences between desert coasts, where few waste products from the continents are accumulated; or between wet tropical littorals buried under mud liberally provided by rivers and fixed by mangroves. But such climatic influences on coastal landforms, however important they may be, should

I

not cause one to forget the unity of the processes set in motion by waves and marine currents. Climate only diversifies the actions of waves and currents, whose effects are reflected in the coastal landforms but do not basically modify them. The same is true of structural influences: rock resistance and tectonic deformations diversify the effects of wave action but do not change its laws.

An identical suprazonal and suprastructural unity is found in aeolian and fluvial processes. Water always flows according to the laws of gravity, whether it is at the Arctic Circle or at the tropics; whether the rocks are limestones or sands, basalts or gravels, the resistance the water of a stream encounters at the contact of an uneven bed produces turbulence, which, in turn, sets free mobile particles of a specific calibre. The wind is subject to analogous mechanisms, which, however, differ in relation to velocity and the relative density of the air. Nevertheless, the action of the wind as well as that of streams is influenced by climate: polar blizzards whip up mixtures of sand, silt, and snow, and these form deposits that are distinct from the unmixed sands of coastal dunes. The scorching winds of arid regions tear lumps of clay and silt, cemented by salt, from salt flats, and heap up the largest fragments into marginal ridges. In the forest, wind action is indirect: falling trees upset the earth caught between the roots. These are a few examples of the morphogenic effects of the wind as determined by specific climates. They are similar to the effects the wind has on irregularly cemented sandstones, delicately chiselling the weak beds, revealing the structure, and polishing the resistant beds. It is clear that in all these cases the wind acts according to its own laws; it is only the area affected by the wind that varies according to its composition or climate.

The climatic geomorphology/structural geomorphology diptych is therefore insufficient. It must be completed with a third major grouping of geomorphic phenomena: the *azonal processes* which include the littoral, aeolian, and fluvial processes. Each of these groups of processes is dependent on its own physical laws. Reduced to the schematic opposition climatic geomorphology/structural geomorphology, the science of geomorphology would be truncated or amputated from an important domain, precisely the one in which recent progress, stimulated by practical necessities, has advanced the science to quantitative concepts and the proposition of formulas, which, however insufficient they may be, bring our discipline closer to the more evolved physical sciences. Furthermore, by opposing climatic to structural geomorphology, one risks forgetting the mutual ties of interdependence which unite them. There is also the possibility that a truncated and insufficient concept is apt to distract the mind from observable realities and to lead it into erroneous theoretical speculations.

Every elementary form of relief results from the antagonism or the equilibrium between the wearing of the bedrock by a particular process and the resistance of the rock to this process. In this case the climate/structure diptych appears to be justified, subject to the objections which we have made above. An escarpment which breaks down reflects a relief determined by the geologic structure and the joint pattern. Climate only intervenes through the process of fragmentation: due either to frost-weathering[1] or to variations in moisture. The talus at the foot of an escarpment may be found in tropical deserts as well as in periglacial regions. But this is an exceptionally simple example which only affects a fraction of the world's relief. Usually things are more complex, and the two elementary processes (fragmentation and rockfall) are not the only ones operating. In most cases the rock is fragmented at the same time that it is weathered by the combined influence of physical, chemical, and biological agents, and its debris migrates under the joint action of various processes such as dissolution, solifluction, runoff, or creep. The respective role of these processes does not vary only according to climate or only according to lithology, but also according to their combination. The result is a double series of differences: limestone reliefs are not like granitic reliefs in the same climate, but the limestone landscapes of the tropics also differ from the limestone terrains of the periglacial and temperate zones. The same applies to granitic, shaly, or sandy reliefs. In one climate a rock is resistant, in another climate it is weak. Badlands, so typically sculptured by runoff, occur on marls and clays in semiarid climates and on sands in periglacial regions.

The relationships between structural and climatic factors also affect the relief differently according to scale. In the smaller hectometric or kilometric dimensions the lithologic influences predominate. A limestone slope differs from a granitic slope. In the larger dimensions this no longer holds true, and structural factors predominate; the regional relief of the continental plates contrasts with that of the orogenic belts, and the relief of sedimentary basins contrasts with that of ancient massifs. Morphoclimatic influences vary in the same way but in reverse order. In the smaller dimensions the contrast between a midlatitude argillaceous relief undulated by solifluction (during the Pleistocene), with irregular interfluves and poorly defined stream courses, and a semiarid argillaceous relief cut up into innumerable gullies, immediately strikes the eye. In the larger dimensions differences between tropical, temperate, or polar ancient massifs are less evident; the larger dimension implies an increased influence of surviving phenomena, particularly that of the structural evolution. The morphoclimatic influences which nevertheless exist are, as it were, subordinate in a more complex group of factors and, as such, less apparent and less dominant. Associated with the lithologic factors, they play a predominant role in elementary reliefs (slopes, plateaus) and in small taxonomic units. In larger physiographic units they gradually give way to structural influences.

In a general way, it is therefore possible to distinguish between topographies with a predominant morphoclimatic character and topographies

---

[1] *Frost-weathering* is the mechanism which subjects rocks to internal pressures caused by the increase in volume of freezing water contained in the rocks. Its effect is called *frost-wedging* (frost-riving or frost-splitting) (KdeJ).

3

with a predominant structural character. But to place them in opposition would do violence to the facts, deforming them to satisfy a too simple turn of mind. There is no exclusion nor dilemma (structural influence *or* morpho-climatic influence), but rather a combination in variable proportions: opposition as well as adjustment; action as well as reaction. In other words, there is *dialectic unity*. Neither of the two groups is uniform in these relation-ships between climatic influences and structural influences. Climatic influences range from purely meteorologic processes, such as frost action, to bioclimatic processes in which plants and animals appear, and to more or less secondary influences on the action of the azonal processes (littoral, aeolian, and fluvial). Structural influences include factors whose maximum intensity is reflected on different temporal and spatial scales: lithology, geologic structure (tectostatics), and tectogenesis (tectodynamics).

If the domain of climatic geomorphology is beyond question, the limits of its actions are nevertheless vague: morphoclimatic influences of decreasing relative intensity merge in one direction with the forces of the azonal pro-cesses, and in the other direction with the structural influences. The latter include rock types with which there is a real interaction, making climatic geomorphology, in essence, also a lithologic geomorphology. Geologic structure directly determines the relief in certain climatic zones, such as the arid zone; the cuestas of the Sahara are typical in this regard. In other zones, on the contrary, the geologic structure is poorly reflected in the relief: escarpments are exceptional on the face-slopes of the cuestas of the Paris Basin; the sandstone plateaus of the Fouta Djallon have a very confused relief in comparison to the sandstone plateaus of the Mauritanian Adrar of identical age. In other cases the geologic structure is so overpowering that the morphoclimatic influences only play a secondary role limited to the smaller relief forms. The Appalachians and the Australian cordillera, for example, have many more traits in common than differences, as do, in a more extreme form, the volcanic reliefs of Java and Kamchatka. Such similarities result from the fact that the group of factors commonly desig-nated as structural (which includes lithology) is less uniform as a whole than the zonal and azonal factors, which both belong to the group of external forces affecting the sculpturing of landforms. If tectonic deformations and their survivals, the geologic structure, are the result of the application of internal forces to the earth's surface, the same does not apply to lithology. The nature of the rocks is the result of a combination of internal and external influences: detrital rocks owe their facies as much to the action of the mor-phoclimatic systems on the landscapes as to the facies of the parent rocks. The internal forces, on the contrary, largely determine the facies of the igneous and metamorphic rocks, the kind of metamorphic environment playing only a subordinate role.

Lithology plays the role of a kind of hinge between the internal and the external forces. It is determined by both. There are important interactions between rock facies and morphogenic mechanisms: the rock facies influ-

4

ences the nature of the morphogenic processes, and these, in turn, partly determine the facies of the deposits.

There are also interactions between tectonics and geomorphology. On sufficiently large crustal units the transport of material by the morphogenic processes from predominantly eroded to depositional areas causes the functioning of isostatic compensations. Climate plays a role in this mechanism through its influence on the rate of erosion.

In short the analytical process should not distinguish two combinations (structural and climatic geomorphology) in the domain of geomorphology but rather three (structural, climatic, and azonal geomorphology). There is indeed a category of morphoclimatic facts on which the study of a part of geomorphology can be centred, but it cannot be isolated from the rest, and the unity of nature also must never be forgotten.

Having defined the concept under study, its principal aspects can now be stated. They will serve as an outline to this volume.

First we examine the concept of climatic geomorphology (Chapter 1), tracing its origin, comparing it with the concept of 'normal' and cyclic erosion and analysing an example in detail. We then define the morpho-climatic mechanisms, beginning with principles concerning geomorphic processes and the morphogenic system and indicating to what extent the development of relief is due to the direct actions of climate and to its indirect actions associated with the role of vegetation (Chapter 2). Thirdly, the principles of the morphoclimatic division of the earth are analysed by means of a confrontation with the climatic data and the azonal factors (Chapter 3). In Chapter 4 we deal with the problem of palaeoclimatic variations and show briefly what their influence has been on the evolution of landforms, before proceeding to discuss the fundamental question of the relationships between present and past causes. In the last chapter (5), we sketch a morphoclimatic division of the earth in broad zones and outline the distinctive characteristics of each one.

Bibliographical note

There does not exist at the present time, so far as we know, a single general treatise on climatic geomorphology, but only partial, more or less advanced, studies. Birot has posed the general problem of climatic geomorphology and has analysed some examples which include only certain climates and certain rocks. Cotton, student of Davis, has limited himself to a study of the concept of the cycle of erosion in glacial and arid environments. These attempts, therefore, come tightly within the framework of the debate on climatic geomorphology. They have a methodologic rather than a systematic interest. On all other matters one should consult the various chapters of this book. References to works by these writers will be found in the Consolidated bibliography (p. 256).

1

Normal erosion or climatic geomorphology?

To what extent does climate influence the sculpturing of landforms and impress its particular characteristics on the surface configuration of the land? This is the first question that we must pose. We begin by setting forth a point of view which does not admit a generalised climatic influence and which considers specific climatic influences as anomalies, as 'accidents', as freaks of nature. This is Davis's theory of 'normal erosion'. Next, we demonstrate the morphoclimatic evidence with specific examples by comparing reliefs formed by rocks that are similar but subject to different climates.

The concept of normal erosion

The concept of normal erosion appeared during the last years of the nineteenth century as the cornerstone of Davis's efforts to synthesise the scattered elements of the geomorphology of that time. Rather than a theory, it was originally a postulate destined to permit the construction by the imagination of the 'cycle of erosion'. In a way the concept of normal erosion was implicit before it was clearly defined; it only took form gradually as a result of discussion. Indeed, while Davis conceived cyclic geomorphology within the framework of the postulate of normal erosion, German geographers, for their part, insisted on differences of relief caused by climate. Because many American geologists had observed the fundamental differences between the geomorphology of the arid west and the humid east of the United States as well as the specific relief formed by Pleistocene glaciation, the postulate of normal erosion on which the cyclic theory was originally based became progressively less clear and susceptible to unanimous acceptance. Although Davis had himself observed these differences, it was then that he formulated the theory of normal erosion. He admitted the existence of glacial and arid 'accidents' caused by climatic oscillations in certain regions and at certain times in a general climatic evolution of a more stable type which, in fact, had served as postulate to his construction of the cyclic concept. 'Normal erosion' was thus defined by antithesis. The discovery of a number of facts bringing about new objections only further incited its promoters to systematise it progressively.

6

A critique of the theory of normal erosion requires two preliminaries: first, to give as complete a definition as possible of the concept of normal erosion; second, to give an outline of how this concept was elaborated.

Definition of the concept of normal erosion

The cyclic theory and the concept of normal erosion are indissolubly linked as a result of the very conditions of their elaboration. The one cannot be defined without the other.

Davis in his fundamental article concerning the cyclic theory, 'The geographical cycle' (1899), began by posing the necessity of a new geomorphologic vocabulary (explanatory rather than descriptive) based on the concept of the cycle. The major aspects of the cycle were then stated: development of streams and divides, relationships between stream capacity and solid load, adaptations of the drainage net to the lithology, formation of meanders, regularisation of slopes, and evolution from maturity to old age. It was only after this account of the characteristics of the cycle that Davis dealt with the 'interruptions of the ideal cycle'. The most important kind of interruption, according to Davis, is the one resulting from a tectonic uplift which modifies the relationships of the region with its base level. The 'accidental departures from the ideal cycle' followed later, almost at the end of the article. The morphogenic system was never defined in the whole work; the definition of normal erosion is strictly implicit. He wrote:

> The amount of change is limited, in the first place, by the altitude of a region above the sea; however long the time, the *normal destructive forces*[1] cannot wear a land surface below this ultimate base-level of their action, and glacial and marine forces cannot wear down a land mass indefinitely beneath sea-level (*Essays*, p. 251).

It was only later that he revealed other enlightening points, equally implicit:

> Besides the interruptions that involve movements of a land mass with respect to sea-level, there are two other classes of departure from the normal or ideal cycle that do not necessarily involve any such movements: these are changes of climate and volcanic eruptions, both of which occur so arbitrarily as to place and time that they may be called accidents. Changes of climate may vary from the normal toward the frigid or the arid, each change causing significant departures from normal geographical development. If a reverse change of climate brings back more normal conditions, the effects of the abnormal accident may last for some small part of a cycle's duration before they are obliterated. It is here that features of glacial origin, so common in northwestern Europe and northeastern North America, belong. Judging by the present analysis of glacial

[1] Our italics.

7

and interglacial epochs during Pleistocene time, or of humid and arid epochs in the Great Salt Lake region, it must be concluded that accidental changes may occur over and over again within a single cycle (p. 274).

The same concepts are revealed by the study of another fundamental article by Davis, 'Complications of the geographical cycle' (1904, also in *Essays*). Two paragraphs, under the headings 'Normal and special agencies' and 'Normal and accidental climatic changes', are devoted to the relationships between geomorphology and climate. The following are the main points with which we are concerned here:

> Thus far it has been tacitly implied that land sculpture is effected by the familiar processes of rain and rivers, of weather and water. It is certainly true that the greater part of the land surface has been carved by these agencies; but it is important to consider the peculiar work of other special agencies, namely ice and wind. It is not to be implied that any special agency ever works alone, but that it dominates in its time and place, as ice does in Greenland, as wind does in certain deserts, and as the rain and rivers do in better favoured lands; it is indeed important to recognise that the various agencies work to a certain extent in combination, for frosty weathering and the active washing of rainy thaws on the higher peaks and ridges is a characteristic accompaniment of glacial erosion in mountain regions; and even in deserts occasional cloudbursts may provide short-lived but strenuous streams that develop and maintain valley systems, with their well organised downhill lines, in defiance of the prevalent winds, which could never alone produce any such system of co-ordinated and ramifying slopes (p. 288).

The author then studies 'normal' climatic changes brought about by the gradual reduction of mountainous reliefs, especially the decrease in precipitation.

These quotations clearly indicate the essence of Davis's normal erosion. But normal erosion never was, on the part of its promoter, the object of an explicit definition. It was defined not in terms of facts arrived at by the analytical method but solely in terms of a theory: that of the cycle of erosion, whose end result is the peneplain. It was only defined as the antithesis of glacial erosion and 'desert' erosion supposedly dominated by the wind.

Normal erosion, then, appears as the combination of diverse mechanisms leading to the peneplanation of a continental relief under the predominant influence of the action of running water. Its essence is teleologic. This concept, moreover, was not peculiar to Davis; it was, in fact, widespread. More or less explicitly expressed it is found in A. de Lapparent and in E. de Martonne. It must be considered as characteristic of the entire cyclical school.

Thus in his textbook de Lapparent (1907) begins the fourth lesson, entitled 'Normal conditions of relief formed by running water', in the following manner:

All the agents working at the modification of the terrestrial relief are far from having the same importance. Thus, the actions of glaciers and wind are intense only on limited portions of the earth's surface. Chemical erosion only affects the external forms in certain areas where the lithology has a special composition. As to marine erosion, it is far from being as effective as it is bold. It only affects shores and limits itself to their retrogression without significantly influencing the actual relief of the continents. The real factor of this relief is running water. For any one who desires to acquire a comprehension of landforms it is therefore necessary to consider this agent first, before any other (3rd edn., p. 66).

As to de Martonne (1955), he writes in his *Traité de Géographie physique*:

This very reason would suffice to justify the determination to begin with the study of landforms of normal erosion and, with them, the analysis of geologic influences. But one should furthermore note that the extent of the relief formed by running water was at one time greater than it presently is. It preceded glacial relief and even alternated with it; it intervened during the Pleistocene and still occasionally intervenes in the deserts. It is really the normal relief (5th edn., p. 545).

Indirectly it appears that 'normal erosion' is that which is determined by running water. The evolution of watercourses exclusively determines the genesis of landforms. True, they fix the base-level of slope development, but, as we will see, they only influence it very indirectly; the form of the slopes results from a whole series of processes much more varied than running water alone. The combination of processes are precisely those which give slopes, conditioned by lithology and climate, their peculiar forms. But these processes are neglected in the concept of normal erosion. They are only considered in the sum total and reduced to a sterile uniformity. It is because of their having been studied in detail that the insufficiency of the concept of normal erosion was revealed.

Having thus defined the concept of normal erosion, it is now necessary to examine how it came about and what has been its evolution up to the present.

Evolution of the concept of normal erosion

Of the various disciplines of the natural sciences geomorphology is one of the last to appear. Its development, like that of geology, to which it is intimately related, presupposed an awareness by man of a time scale beyond the reach of his senses and his daily experience. Did not Voltaire, the enlightened and progressive mind of eighteenth-century France, affirm without joking that sea shells observed in the Alps were put there by some pilgrims? For his contemporaries, and even those living at the beginning of the nineteenth century, the concept of a valley was a most confusing one: every depression was labelled as such, and for some even the Atlantic was a

9

valley. It is clear, therefore, that the connection between morphologic action, running water, and the existence of valleys remained extremely vague. Many educated men still considered 'valleys' to be pre-existent to rivers; they believed them to be the result of a subsidence of the earth's crust caused by some unspecified cataclysm and to have been occupied by streams from the moment of their creation. An eminent engineer, E. Belgrand, still explained the Paris Basin and the origin of the Seine in an analogous manner in the middle of the nineteenth century. He imagined the centre of the Paris Basin to have formed a depression into which water had accumulated until, by taking advantage of a fissure that, when widened, produced the lower Seine valley, it overflowed into the English Channel. We find, then, at the origin of geomorphology the same error that we find at the origin of tectonics: a cataclysmic interpretation of natural phenomena which had its origin in and was propagated by certain traditions such as the biblical Flood.

In spite of the obstinacy of some, other brilliant minds already had more exact scientific concepts. Leonardo da Vinci, who put a keen intelligence at the service of practical ends, arrived at the beginning of a theory of fluvial erosion. He even wrote, in the part of his notebooks which, unfortunately, remained unpublished until 1797: 'Each valley has been cut by its river, and the size of valleys is related to that of streams.' The modifications of a riverbed could not have escaped the attention of an engineer who had built canals, as it did the armchair geographers of the time. H. Baulig (1950, pp. 6–8) has shown how the concept of fluvial erosion was born at the end of the eighteenth century at the cost of numerous attempts at demonstration, and how Hutton and Playfair confirmed and formulated it.

The first half of the nineteenth century records a definite step forward which again was made thanks to a mass of observations gathered by a competent engineer during practical work. A. Surell, who was put in charge of the maintenance and construction of roads in the southern French Alps, was faced with the misdeeds of torrents. In his contest against the natural calamities he drew a valid theory of fluvial erosion and, it should be noted, arrived at the base-level concept, later developed and formulated by J. W. Powell. The introduction and spread of these concepts into the domain of theoretical research was mainly the work of Britons. Hutton and Playfair developed the main outlines of the drainage net concept, pointing to the hierarchy of streams, the gradual alluviation of lakes, and the slow but continuous erosion of rocksteps. The concept of geologic time was clearly introduced into the domain of geomorphology in spite of the positions taken by the catastrophists, a fundamental progress which was due to a Scot, A. Geikie. In a textbook of physical geography, which ran into several editions and was translated into French, Geikie showed how rocks slowly disintegrate in geologic time owing to frost action, variations in temperature, and the contact of waters containing carbon dioxide; and how the debris is subject either to chemical weathering or to transport by mechanical agents. The work of streams was clearly described in its main outlines: formation of potholes, meanders, terraces, and deltas. The relief was said to evolve under the effect of running water which transports the products of disintegration formed by the contact of the atmosphere and the rocks or by a process called weathering. But Geikie refused to go into adventurous extrapolations: he did not co-ordinate all the facts into a system with a universal pretension nor formulate laws on the evolution of relief. His description of the development of riverbeds, however, was not as well executed as Surell's. Nevertheless, there was a definite progress in relation to a large number of his contemporaries. Leaning on Hutton and Playfair, Geikie gave a correct view of the general mechanisms of fluvial erosion and granted them a proper place in the evolution of the landforms of the temperate lands, the only ones that he knew and the only ones about which he wrote. Lyell, his contemporary and compatriot, still believed in the Deluge and ascribed erratic boulders to icebergs and certain valleys to the action of the sea and other water bodies of the diluvial period.

The second half of the nineteenth century witnessed the application of Surell's discoveries to theoretical geomorphology. Another French engineer, M. F. B. Dausse, arrived at the concept of the profile of equilibrium, already noted by Surell, and applied it to lowland streams basing his concept on observations gathered as he was working on artificial embankments and canalisations. De la Noé and de Margerie, in a textbook of geomorphology in 1888, clearly set forth a coherent and detailed theory of fluvial erosion based on the concept of the profile of equilibrium. Simultaneously, the most brilliant field geologists in the American West elaborated an analogous doctrine. Powell, in 1875, analysed the magnificent example of fluvial dissection of the Colorado River, and G. K. Gilbert described the mechanism of fluvial erosion.

By 1890, shortly after Davis began to write, some of the foundations of geomorphology were already laid. The ideas of the destruction of relief during the course of geologic time, of the slow co-ordinated work of streams in networks, and of the peneplain had been clearly elaborated and defined. This knowledge was based on careful observations, most of which were the result of practical necessities. It became valuable to theoretical geomorphology through prudent yet sufficiently thorough generalisations which gave the observations a meaningful framework. But their authors refused to separate this knowledge from the observed facts and to venture into purely theoretical speculations. Davis broke with this prudence and linked together the hitherto separate data; he associated them into a logical system. He did not contribute any new facts, nor did he demonstrate the existence of a link between peneplains and the work of streams in a temperate environment. He only created a logical scheme, the fruit of his imagination, and set himself against those who refused to yield to theoretical speculations and be carried away by a well co-ordinated exposition of successive hypotheses. In fact, he went as far as to imagine and to draw hundreds of possible theoretical cases and then to eliminate certain specific factors because the case he had himself imagined to be explained by them did not exist in nature. The only *proof* which justifies the cyclic theory is its logic and its internal cohesion. The Davisian theory of the cycle of normal erosion is an illustration of the dangers of Bergson's *creative imagination* applied to science.

There are several postulates at the base of the theory of the cycle of normal erosion. Two of them concern us here:

1. The uniformity of the geomorphic mechanisms on the earth's surface outside the glaciated regions and the arid regions subject to the predominant action of the wind.

2. The sudden nature of tectonic deformations (in relation to geomorphic evolution) which triggers the initiation of the cycle, a position which, after all, is catastrophist.[1]

The two postulates correspond to a certain stage in our knowledge of the earth; but both are as insufficient from the point of view of the nature and the speed of tectonic processes as from that of the mechanisms of erosion.

The oversimplified character of the concept of normal erosion became evident as early as the first years of the twentieth century. Scientific explorers described the characteristics of the development of relief in non-temperate lands that are neither glaciated nor arid. Two fundamental works appeared simultaneously in 1916: H. M. Eakin elaborated the theory of altiplanation of periglacial regions, a theory irreducible to that of normal erosion, and W. Salomon demonstrated the capital importance of Pleistocene gelifluction processes in the elaboration of the mountainous relief of middle Germany. At the same time knowledge of the humid tropics progressed. A relief very different from that of the temperate zone was found there in spite of the existence of an integrated drainage net which should have imposed a morphology conformable to the schemes of the type of erosion said to be normal. If the fundamental work of Passarge on the Kalahari did not deny 'normal erosion' because it concerned a desert, a type of region excluded by hypothesis from the theory, the same cannot be said of Sapper's study on the landforms of the humid tropics nor of the works of Freise. Other discoveries dealt with semiarid lands where inselbergs were described (Bornhardt, beginning in 1900).

Like the famous legend of the *peau de chagrin*, the domain in which the Davisian theory of the cycle of normal erosion could still be defended shrank at every step as the geomorphological knowledge of the world progressed; it shrank to such a degree that its advocates had to abandon the realm of facts for that of subjective justifications. Typical in this regard was Fourmarier's apology in his preface to Macar's textbook (1946):

In this book, which he has just finished, P. Macar limits his domain to the study of *normal geomorphology*. Some, perhaps, will think that this expression is not appropriate, as it is justified only from the point of view of the

---

[1] This position is discussed in vol. XI of the *Treatise*, dealing with orogenic belts.

man accustomed to living in the most favoured lands for the development of civilisation, i.e. under temperate climates. For the Bedouin of the desert the landforms of the Sahara are normal and those of our countries are, as far as he is concerned, contrary to all the ideas he could have had on the physiognomy of the landscape. If the land surfaces are due to identical causes in all parts of the earth, it is nevertheless true that a great part of the relief of the continents is governed, from the point of view of sculptural forms, by the same agents of erosion. The expression used by the author can thus be justified without difficulty (p. 7).

But during the course of the same period the theory of the cycle of normal erosion was undergoing internal criticism. The study of processes, neglected by Davis himself as by most of his students, revealed the fallacious character of the relationships admitted in the theory between series of different facts. The progress of geology led to the demonstration that ancient peneplains, regarded as the end result of the evolution of relief sculptured by 'normal erosion', were formed under very different conditions: those of the biological deserts of the early Palaeozoic or under the arid and semiarid climates of more recent geologic epochs. The influence of vegetation on the processes of mechanical erosion was gradually admitted by geomorphologists who made use of exceptionally early studies by foresters. The plant cover withholds an important part of the water, facilitates infiltration of another part, and vigorously opposes the gullying of slopes by runoff. On the other hand it favours chemical weathering. The evolution of relief in densely vegetated regions is not the same as the evolution of relief in sparsely vegetated regions. The morphogenic system of forested lands and of grasslands is not identical to that of sparsely vegetated steppes. But the Davisian theory of normal erosion associates the gullies of the densely dissected hills of the American arid and semiarid West with the soil and forest covered hills of the eastern United States, just as de Martonne does for the gullies of the sparsely vegetated southern French Alps and the forested slopes of the Paris Basin. The gullies would represent the stage of youth, and the slopes, protected by vegetation, the stage of maturity. But we know that the gullies of the Department of Basses-Alpes are due to overgrazing and imprudent farming methods which have destroyed the natural vegetation; they are a manifestation of anthropic erosion. Nobody can say when the slopes will have been sufficiently reduced to allow the return of the vegetation so as to look like the hills of Argonne or the Côtes de Meuse, but there are numerous texts to prove, with the help of field observations, that the now gullied slopes of the southern Alps still harboured a continuous plant cover one or two centuries ago, or even less, and that gullies were then the exception or even non-existent.

Furthermore, the success of Davis's theories discouraged research, and research has forced his theories to be discarded. The seduction was great indeed: the qualities of exposition, the scientific activity, even the poetic talents of the author were compounded by the attractiveness of the fine logical construction. It was thought that a decisive step forward had been made, and that a veritable revolution had been accomplished in the domain of geomorphology. The magic of words provided the illusion of the explanation: is it not more attractive to speak of a relief that is in a youthful or old age stage of dissection, in submaturity, or in rejuvenation, than it is to undertake patient analyses of processes, numerous slope measurements, or careful age determinations by means of correlative deposits? The contemplation of the landscape essentially served to stimulate the imagination which, referring to the theoretical schemes, recreated it in a more or less brilliant synthesis depending on the qualities of the author. For some Davis's contribution was mainly reduced to a terminology, for others he went beyond and offered both a method and a doctrine. The excessive ease of explanation in the latter case was harmful to the establishment of facts, and observation gave in too rapidly to the paraphrase of a theory. Progress was impeded. The German school of geomorphology, however, was not much influenced by Davis, and although the excesses of theoretical speculation detached from the facts (or, in a word, idealism) did not spare it, it is there that the first elements of climatic geomorphology developed the most decisively. At the very moment that Davis elaborated his theory, F. von Richthofen, in his guidebook for scientific explorers, gave equal importance to the geologic and the climatic factors in the development of relief. This tradition was not lost, and Davis's success remained minor in Germany. The great travel facilities accorded German university professors, the education of geomorphologists as naturalists, and the considerable efforts at climatic classification, as those of W. Köppen and R. Geiger, explain this orientation toward climatic geomorphology in terms of the presently existing landforms as well as in terms of the influences of palaeoclimates.

Davis's influence was much greater in France, probably largely because the very education of French geographers turns them away from the natural sciences as well as from practical applications.[1] Even though the work of French foresters in the Aigoual (Cévennes) and elsewhere demonstrated the marked influence of the vegetation on the morphogenic system as early as the middle of the nineteenth century, it was necessary to wait until the middle of the twentieth for this knowledge to be incorporated into theoretical geomorphology. The fact that Davis was an extremely able teacher facilitated the introduction of his theories in France. They became the basis of the most widely read textbooks, as those of de Lapparent and, later, de Martonne, and as such influenced the initial orientation of a whole generation of French scholars. De Martonne, however, remained hesitant. He became interested in the influence of climate on relief as early as 1913 but without ever leaving the framework of the Davisian concepts. He also had some reservations on Davis's outlook, which he stated in his *Traité*:

[1] Geography in French universities is taught in the Faculty of Letters and Human Sciences (formerly the Faculty of Letters) (KdeJ).

The most vigorous effort to systematise the ideas on the evolution of land-forms was made during the last years of the nineteenth century. It was the work of a somewhat abstract mind, constructing rather than observing. In a series of memoirs published between 1889 and 1900, W. M. Davis described all the essential concepts, created words which have since become accepted (e.g. peneplain, cycle of erosion), and devised a fault nomenclature related to structure. He worked hard at giving the ideas developed by other observers the appearance of a body of theory. As he gradually widened the bounds of his activities, he tried to do the same for landforms which are not the result of normal erosion but, it seems, with less success (5th edn., p. 546).

It was only in 1940 and 1946 that de Martonne took a stand in favour of a zonal geomorphology in two articles published in the *Annales de Géographie*. But we still had to await the 1950s to witness a rupture *vis-à-vis* the theory of normal erosion and a clearly deliberate orientation toward climatic influences on landforms with the works of, notably, Cholley, Dresch, Birot, Cailleux, and Tricart, while Baulig, who had as systematic a mind as Davis, essentially preserved the theory of the cycle of normal erosion and directed his criticism on other points of the American professor.

Davis's teachings were generally adopted in the English-speaking countries, but they coincided with a general decline in geomorphological research. The coincidence, as a result of the dangers we have decried above, was probably not fortuitous. The main trend in American geomorphology became the making of an inventory of the relief of the United States within the framework of Davis's theories. N. M. Fenneman, who had only minor reservations about them (he even preserved the term 'normal erosion' although he found it open to criticism), compiled the physiography of the whole of the United States. Another trend pursued the study of certain types of semiarid landforms, notably the pediments of the American southwest, an undertaking which cannot be dissociated from the work of K. Bryan and which led to a considerable departure from the Davisian concepts but without a categorical rejection of the cyclic theories within the framework of which Davis endeavoured to encompass the new results. C. A. Cotton, of New Zealand, has preserved the essentials of the Davisian doctrine in his manuals, compiling two parallel volumes, one on normal geomorphology and the other on climatic accidents and producing a third on volcanoes as landscape forms. L. C. King, of South Africa, going further, attacks the link between the theory of normal erosion and the development of peneplains and makes of the latter the end result of the process of pedimentation in semiarid climates.

It seems that two different aspects must be considered in the concept of normal erosion:

1. The co-ordinating role of streams in the evolution of relief, a basic factor in the cyclic concept, must first be considered. This role exists although in

reality important reservations must be made. In dry lands, in certain peri-glacial regions, the overabundant slope waste is not removed by the streams and begins to bury the landscape. It is not the streams which then command the evolution of the relief, but the evolution of the relief which commands that of the streams! In the wet tropics the incision of watercourses is impeded by the resistance of rockbars to chemical weathering and by the lack of a coarse bedload in such a way that the long-profiles of streams become stepped. The general evolution of streams in the tropical zone is altogether different from that in the temperate zone.

2. The second aspect that must be considered in the concept of normal erosion is the uniformity of the geomorphic processes within the cyclic framework. The only differences between the relief forms are due to the different stages in their evolution: some are 'older' than others. This concept is totally wrong. The evolution of a hillside depends, it is true, on the rate of incision of the watercourse located at its foot and on the condition, moreover, that the stream is close enough, but the evolution depends even more on the geomorphic processes and the lithology, which, in turn, are affected by the climate and the vegetation.

The influence of climate on landforms was never denied, even by Davis, but it was considered to be minor, both in time and in space. The part attributed to palaeoclimates was most limited, and wind and glaciers were only granted a secondary role as accidents. Frost, chemical weathering, and the role of vegetation were hardly ever mentioned. It is therefore necessary to demonstrate the full importance of the morphoclimatic influences before studying them in detail.

The morphoclimatic evidence

According to the cyclic theory the influence of geologic structure on relief is most marked during the stage of maturity. In a youthful relief the streams gradually penetrate into the surface of the original peneplain and seek their way. Erosion, progressively increasing in intensity, does not yet adapt itself to the geologic structure. It is only later, once the principal rivers are sufficiently incised and their profile of equilibrium is realised or close to realisation, that adjustments to the geologic structure become increasingly important. These adjustments to lithology and structure occur in a sup-posedly immobile terrain. In this way the structural influence, admitted by Davisian cyclical geomorphology, essentially takes the form of differential erosion; the only consideration necessary is the role of the lithologic factor, which is the only factor deemed to be in direct relation to climate. According to this theory, lithologic differences are reflected in the rate of slope develop-ment (i.e. in the evolution toward final peneplanation). Hillsides composed of 'soft' rocks (soft because evolution proceeds relatively fast on them) erode rapidly. The transition from maturity to old age (i.e. to a relief of subdued

16

interfluves separated by wide open valleys) is relatively rapid although the rate of erosion decreases progressively as the final peneplain is approached. On 'hard' rocks (hard because evolution proceeds relatively slowly on them)[1] the cycle proceeds in exactly the same way; it only takes longer. The ineluctable end, imposed by the teleologic essence of the theory, is also a peneplain, rigorously similar to the one developed on soft rocks. In this way the evolution from maturity to old age implies decreasing lithologic influences as well as the progressive destruction of the relief caused by the geologic structure.

If we want to demonstrate the insufficiency, even the falsehood, of the Davisian cyclical theory, we should put ourselves on its own ground. We shall therefore consider several reliefs that have arrived approximately at identical stages of evolution; the stage of maturity proves best for examination as it is the one for which structural influences are admitted by the theory. We will demonstrate that the ultimate evolution does not necessarily end in the wearing away of the lithologic influences, contrary to what Davis maintained. Our method will compare several regions that are composed, as far as possible, of identical rocks which have all endured a long-protracted evolution. These regions have theoretically reached the stage of maturity in which is reflected a similar geologic structure (a platform structure, for greater simplicity), but which are all located in different climates. In this way the specific features that characterise each one of them can be attributed only to differences in climate, whether past or present.

Example comparing the morphology of similar highland regions in different climates

During the Palaeozoic, at the beginning of the Mesozoic, and to a lesser degree in the Cenozoic, the deposition of sandstones was relatively important on the earth's surface. A considerable part of the lower Palaeozoic beds of the continental plates are composed of shales and sandstones, series which have been explained by the hypothesis of biological deserts; these sandstones are often aeolian and are terminated by a final fluvial deposit. Palaeontology tells us that terrestrial plants did not as yet exist. During the Devonian, and perhaps even down to the Permian, they seem to have been restricted to swamps. As a consequence, fluvial Palaeozoic beds have peculiar characteristics, including more or less well cemented sandstones ranging from sedimentary quartzites to poorly consolidated sandstones alternating with shales. Very resistant beds are exceptional; beds with a poor cohesion, inclined to fragmentation by frost, variations of temperature, or weathering

---

[1] We may note, in passing, the verbalism of this explanation, however classic: it is a magnificent example of a vicious circle. Never has Davisian geomorphology taken the trouble of giving an objective explanation of the notions of hard and soft rocks. It has failed to discern mechanical or chemical resistance although these have been experimentally determined by engineers and pedologists.

of the argillaceous-ferruginous cement (a common cement) are the rule. In nearly all climates such sedimentary rocks are of mediocre resistance although sufficiently well consolidated to be able to produce relatively abrupt slopes, steeper than the angle of repose of sliderock. Minor differences in resistance to fragmentation, regardless of the weathering agents, would be favourable to a perfect uniformity of relief in the various morphoclimatic zones, and therefore these rocks should offer a particularly demonstrative example of the Davisian theory of normal erosion. For this reason we will study them first (cf. Fig. 1.1).

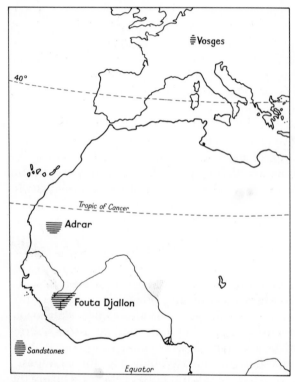

FIG. 1.1. Location sketch map of the sandstone highlands examined in this chapter

Basing our thesis on our own observations, we will compare three highlands located in very different climates: the Fouta Djallon of Guinea, the Adrar of Mauritania, and the Sandstone Vosges of Alsace. They are bold massifs with reliefs consisting of plateaus deeply dissected by valleys reaching a depth of 300 m (1 000 ft) on the average. The earliest dissection goes back to the Stampian (middle Oligocene) or Burdigalian (lower Miocene) in the case of the Vosges (date of the last peneplanation), and perhaps even further in the case of the two other massifs (Cretaceous?). Evolution is sufficiently protracted and the relief sufficiently bold so that all three highlands may be classified in Davis's stage of maturity. The comparison is therefore valid.

THE FOUTA DJALLON

In its western extremity the Guinea Arch is composed of a sandstone cover more than 1 000 m (3 300 ft) thick lying nonconformably on the Pre-cambrian basement. Ranging in age from the Cambro-Ordovician to the Devonian, the beds are almost exclusively sandstones and shales including several very resistant, well cemented beds of a thickness up to 50 or 100 m (165 to 330 ft). The proximity of the ocean (less than 50 km = 30 miles from the southwest flank) and the very gentle gradients of the lower Senegal River and the Gambia (both on the northwest flank) assure a deep dissection which has been in progress for a very long period of time. The prominence of the relief precludes it from being considered as old age.

The entire highland is composed of a system of stepped plateaus at elevations between 700 and 1 500 m (2 300 to 5 000 ft) cut by deep divergent valleys. The erosional relief is characterised by the following features:

1. A coexistence of steep slopes and very wide valleys exists. The plateau breaks up at its margin into a series of narrow ridges occasionally interrupted by depressions that may be quite deep and may widen into embayments several kilometres wide. The relief is thus very much cut up and deeply hollowed out; the measure of erosion may be judged from the mass removed. At first sight the hollowed out volume of the valleys and embayments is approximately equal to the rock volume of the plateaus, promontories, and alignment of buttes which still separate them. Such erosion implies a long evolution, confirmed by what is known about the regional geomorphic chronology. Nevertheless, the slopes are steep even though they have re-treated quite a distance. Their average gradient is usually more than 25° and not infrequently even 35°. Abrupt escarpments are common; they frequently show up through the vegetation, adding a picturesque touch to the landscape. Abrupt slopes are not only located on cuesta fronts: they exist on the flanks of innumerable valleys in the most diverse structural positions.

2. Another characteristic is a concentric areal zonation in the entire massif. It is particularly evident in the northwest where there is a regular outward dip of the beds. To the south and the southwest a number of faults, which cause the sudden drop of the arch toward the coastal plain, eliminate part of the zonation. Where the zonation does exist, it is characterised by the succession of three types of topography:

(*a*) At the outer margin, for example near Gaoual and Youkounkoun, on the crystalline basement, there is a low, slightly undulating plateau incised by shallow valleys with rolling forms. It is dominated by abrupt sandstone buttes which stand without any transition above the low plateau surface like veritable inselbergs. Their flanks, sloping 30 to 40°, join, over a distance of a few metres, the billowing surface found at their base and whose slopes are never steeper than 4 or 5°. Their upper surface is usually tabular and sometimes preserves remnants of former erosion surfaces. The volume of the buttes is very limited in comparison to the mass that has been eroded away: 5 per cent, perhaps even less.

FIG. 1.2. An embayment in the sandstone plateaus of the Fouta Djallon (SW of Pita), elevation about 1 000 m (3 300 ft) (*drawing by D. Tricart from photo by J. Tricart*)
In the foreground: the embayment, wide open, its floor slightly undulating and dominated by steep slopes. *Beyond the foreground:* a deep valley where the embayment ends by way of an escarpment over which the waters drop in waterfalls without incising it; note the difference with Fig. 1.3. *In the left foreground:* beginning of bowalisation.

(*b*) A second ring is formed by the region of embayments (described above) which breaches a highly dissected plateau.
(*c*) The centre, around Labé, forms a massive plateau lacking any incision but having a slightly undulating surface of broad basins and convex hills without much relief. Its margin consists of several more or less wide and unevenly dissected stepped benches. The long-profiles of the watercourses are very irregular here; slightly inclined bottomlands, where the water wanders out of the channel in a dense vegetation, alternate with waterfalls where the stream spills across the edge of a massive ledge. There is therefore no evidence of regressive erosion into the central plateau of the Fouta Djallon. The abundant waters which leave it fall into the embayments surrounding it and, later, drop to successively lower embayments, indicating that the levels of the floors of the embayments are at different elevations.
3. The details of the morphology are also original. The plateau margins are hardly incised by secondary streams. The waters, which are highly abundant during the rainy season, rush down the margins of the embayments or the benches of the central plateau without cutting any valleys. The streams descend the slopes without incising them; their average gradient is thus about the same as that of the slopes. The water runs swiftly over the flats which separate the escarpments and drops in waterfalls at every outcrop. When at the beginning of the dry season the torrents dry up, their beds are only visible because of the absence of vegetation and because of a darker

brown colour of the rocks; the streams do not leave a trace in the topography. The same type of falls over which the waters pour without leaving an incision are also found on large rivers whose long-profiles are as irregular as the width of their beds.

FIG. 1.3. The Fétoré valley (Fouta Djallon)
Note the width of the valley and the dense forest. *In the background*, between the trees, on the opposite slope two affluent waterfalls are visible. In spite of their copious flow the streams have not incised valleys in the hillside. The waters rush down a plateau bench without dissecting it.

The Fouta Djallon, in the humid tropics,[1] belongs to the morphoclimatic zone of the wet–dry tropics. The dry season extends from November to April, but it alternates with a very rainy season which drops between 1 500 and 2 000 mm (60 and 80 in.) of water on the massif and permits the growth of an often very dense forest, at least where it has not been despoiled by man. Runoff to the ocean is always assured although most secondary streams dry up near the end of the dry season. The region is therefore a part of the realm of 'normal erosion' in the implicit Davisian sense. We are therefore justified in comparing its relief with what it ought to be in terms of the Davisian cyclical theory:

1. The coexistence of steep slopes and an advanced stage of erosion is incompatible with the theory of the progressive proportional reduction of

[1] The term 'humid' in geography is normally opposed to the term 'dry'. The critical limit between humid and dry regions is generally considered to be the zone where annual precipitation equals evaporation (or potential evapotranspiration). The dry regions are further commonly divided into arid and semiarid regions. On the same basis the humid regions may be divided into perhumid and semihumid or, better, wet and wet–dry regions. These terms, however, are most often applied to the humid tropics (*zone chaude*). The above scheme has been adopted for the English edition of J. Tricart's works (KdeJ).

FIG. 1.4. Souma Falls, near Kindia, in the Fouta Djallon (*drawing by D. Tricart from photo by J. Tricart*)
The river drops into a cleaned out fractured zone without indenting the rim.

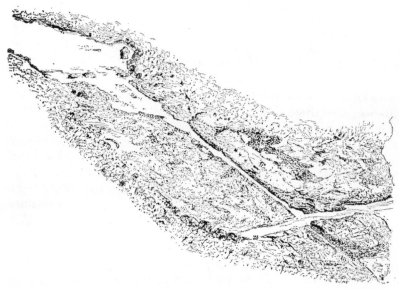

FIG. 1.5. Rapids on the middle Bafing, western Mali (*from an aerial photograph taken in January 1954, MAS collection*)
Jointed pavement of Ordovician sandstones truncated by a surface of erosion. The river narrows down into a fractured zone and follows a right-angled course. High waters flood the sandstone pavement. There are no pebbles in the narrow channel, only large scattered boulders on the sandstone pavement.

slopes as the cycle advances. In the case of the Fouta Djallon one must admit that the slopes have retreated considerably without having simultaneously become noticeably gentler. In the outer zone the existence of buttes shows that steep slopes persist to the end. The volume of the residual reliefs has shrunk without a concomitant noticeable decrease in slope. The level surfaces that have developed between the buttes cannot be assimilated to the theoretical Davisian peneplain because the evolution during which they appeared is completely different. They are the result of a slope retreat parallel to itself, not the result of a progressive lowering of the relief.

2. The stream profiles and the relationships between the stream lines and the type of erosional relief confute the law of the profile of equilibrium. The profile of the watercourses is quite dependent on the geologic structure in spite of a very long evolution. Water runs on the surface of a resistant bed, drops into a fractured zone, and descends a pre-existent slope. It does not seem to reduce the irregularities of its bed. Moreover, the absence of pebbles, the rareness of sands, and the insignificant quantity of transported silts is a sure indication of the weakness of mechanical abrasion in the bed.

The Davisian scheme, then, is completely in default in this region; the relief of the Fouta Djallon simply cannot be accommodated to it.

The Mauritanian Adrar

In the southwest of Mauritania the Precambrian basement disappears below a sedimentary cover which, here also, is of Cambrian to Devonian age and is composed of the same type of rocks as in the Fouta Djallon. The main difference consists in the intercalation of a layer of dolomitic limestones in the Cambrian. But it only outcrops in a small area and its topographic role, at the foot of the sandstone cuesta of the Grand Dhar, is limited. The highest points of the Adrar reach 800 m (2 600 ft), whereas the plain stretching on Precambrian igneous and metamorphic rocks at its foot does not exceed 200 m (650 ft) above sea-level. The total relief, although less than in the Fouta Djallon, is nevertheless important and permits the existence of deeply incised valleys like that of the Wadi Séguélil which follows the foot of the Grand Dhar rising 400 or 500 m (1 300 or 1 600 ft) above it. The relief, here again, is in Davis's stage of maturity.

The Adrar is formed by a system of prominent cuestas, the front of which is generally exposed towards the west, and the back of which slopes gently in an easterly direction. The principal cuesta, the Grand Dhar, which overlooks Atar, marks the watershed between the Atlantic and the Sahara. The waters do not reach the ocean at the present time, but during the Tertiary an important erosional plain, covered here and there by rubble all the way to the coast, was formed at the foot of the Adrar. Exterior drainage must therefore have existed during part of the very long evolution of this relief of plateaus and cuestas.

The erosional relief is characterised by the following features, in which

23

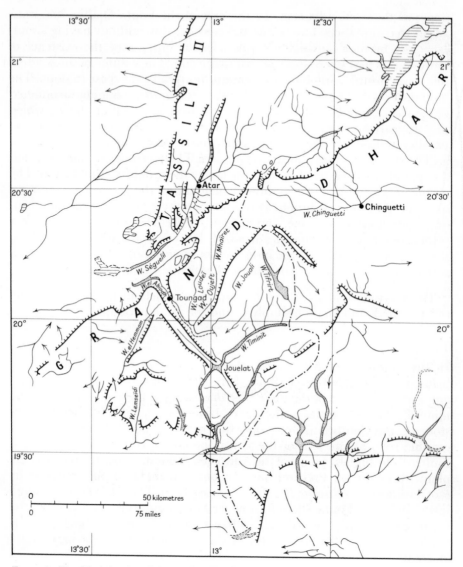

FIG. 1.6. The Mauritanian Adrar, after Th. Monod

structural influences clearly predominate. The drainage net is strongly controlled by structure and lithology. The Grand Dhar forms the drainage
divide between Wadi Séguélil, which occupies a subsequent position at its
base, and the resequent wadis, which wither away in an easterly direction
and slightly incise the homoclinal plateau that forms the backslope of the
cuesta. In the faceslope there is not a breach, or it is very short (a few kilometres) and does not provide easy communication between the subsequent
depression and the backslope. A similar dependence of the linear features on

24

the structural relief also exists on the other cuestas, particularly the Tassili
II located west of the Grand Dhar and only slightly less high although still
prominent. Wadi Séguélil is located between the two cuestas and breaks
through the Tassili II at the locus of the only consequent water gap in the
area, a gap which is due to an important fracture almost perpendicular to
the front of the cuesta. Obsequent wadis in the faceslopes of the cuestas
themselves are minor, and the indentations which they produce hardly
scallop the cuesta front. The longest do not exceed a length of about 10 km
(6 miles); they are exceptional and are located on minor fractures. Obsequent wadis, such as the one going through Toungad, do not even link up
with Wadi Séguélil although they flow at the foot of the Grand Dhar, which

FIG. 1.7. A. The Tassili II cuesta at Gara Tamzak (view toward the east). B. The
Grand Dhar cuesta at the outlet of the valley of Wadi Terjit (view toward the
south) (*drawing by D. Tricart from photograph by J. Tricart*)

Two different cuestas of the Mauritanian Adrar. Desert vegetation colonises the
alluvial aprons at the foot of the highlands. Gullies are located in the weak rocks
of the faceslopes, and sheer escarpments crumble in more or less large masses
depending on rock cohesion (note the overhangs on the Gara Tamzak escarpment).

In the foreground of Fig. 1.7A is Wadi Séguélil, in a subsequent position at the
base of the cuesta. In Fig. 1.7B it is entrenched 100 m (330 ft) into the substratum
of the alluvial apron, at the right, beyond the bounds of the sketch. This considerable difference in base-levels is without effect on the dissection of the cuesta.
The concept of base-level is inadequate here.

25

forms a clear escarpment. The same wadi (though Toungad) cuts a gorge about 100 m (330 ft) deep in the backslope of Tassili II right at the foot and west of the Grand Dhar. The affluent wadis do not dissect the walls of this gorge but fall in dry cascades.

The Adrar is also characterised by the combination of a whole series of presently related morphogenic processes. The resistant escarpment producing strata are frequently sculptured into abrupt, even sheer, capping rocks and retreat through the double action of massive landslides and detailed fragmentation. The more shaly subjacent strata are etched and carved by runoff into shallow gullies which are deep enough (if the escarpment is sufficiently high) to evacuate the debris that falls down. The waste spreads out at the foot of the cuesta, burying under a piedmont slope the eastern extremity of the backslope of the next cuesta to the west. This piedmont slope is uninterrupted where the faceslope is not indented by important wadis. When the escarpment is more scalloped the uninterrupted piedmont slope gives way to a succession of rocky ridges and alluvial fans. The wind reworks the important sandy fraction of the alluvium of the wadi beds burying some areas under dunes.

The Adrar, which receives a mean annual rainfall somewhere between 100 and 200 mm (4 to 8 in), is located within a desert region. Vegetation is extremely sparse and nowhere forms a continuous cover. On the beds of washes or on alluvial aprons where the presence of subterranean water reserves permits the vegetation to reach a maximum development, it is always composed of bushes and tufts of high, widely spaced grasses (at intervals of about 10 m: 33 ft). Smaller plants which sprout after rains, although they are more densely spaced, only play a temporary, episodic role. Because of the existence of an important runoff during heavy rains, the Adrar, although it is in an arid climatic region, really has a semiarid morphology.

The role of the wind is here subordinate to that of running water; the latter spreads the debris that the wind later reworks. The Adrar therefore does not fit within the regions to which is applied the arid cycle of erosion imagined by Davis. Furthermore, because runoff is insufficient to ensure a regular drainage of waters toward the ocean, the area cannot be classified within the regions of 'normal erosion'. Davis's theory is not in contradiction to the facts here; it is in default simply because it did not take into account nor did it provide for such a case.

THE SANDSTONE VOSGES

The Hercynian crystalline basement of the Vosges plunges toward the north and west under a Mesozoic sedimentary cover the first elements of which are composed of nonconformable Permian, forming a local basin, and of lower Triassic sandstones. The latter, reaching a thickness of 500 m (1 600 ft), are formed by a succession of more or less well cemented beds ranging from sand poorly consolidated by iron oxide to conglomeratic

Fig. 1.8. A. The Zorn valley above Saverne: Sandstone Vosges (*drawing by D. Tricart, from a photograph*)

Horst of the Sandstone Vosges above the Rhine graben. The land was subjected to intense Plio-Quaternary erosion caused by tectonic uplift. Note the contrast between the slightly undulated surface of the plateau, the steepness of the slopes fashioned by periglacial Pleistocene scree, and the flat floor of the valley of variable width.

    Forest vegetation. On the right, at the base of the slope, a forest clearing has caused accelerated erosion.

    The Marne-to-Rhine Canal follows the bottom of the valley.

B. Humid midlatitude sandstone relief: the Taintrux valley in the Vosges (*drawing by D. Tricart, from a photograph by J. Tricart*)

The steepness of the hillsides and the width of the valley bottoms might suggest U-shaped glaciated valleys (foreground), which they are not.

quartzitic sandstone. The discontinuity of the most resistant beds and the existence of numerous faults, most of them beheaded, cause such an irregular stratigraphy as to preclude a systematic development of a cuesta topography. The relief is that of a plateau cut up by valleys 200 to 300m deep (600 to 1 000ft).

The characteristics of the erosional relief of the plateaus of the Sandstone Vosges are the following:

Important regional differences exist due to tectonic influences. In the west, in Lorraine, the beds dip gently below more recent series of the Paris Basin. The rivers traverse long distances before reaching the sea, so that their valley floors are relatively high and not more incised than in the northeast of the massif in spite of the higher elevation of the plateaus of the west. Recent tectonic deformations have caused a slow, half domical tilting toward the west. In the northeast the Sandstone Vosges form an asymmetric horst on the margin of the Rhine graben. The massif suddenly drops down to the Alsatian Plain by way of a series of fault scarps which have continued to function right down to the Quaternary. Despite the lesser elevation of the plateaus here, their dissection has been quite active as a result of the proximity of the local base-level of the downfaulted Alsatian rift. One would therefore expect to find a marked contrast between a mature area in the west and a youthful area in the northeast. But the difference is not so clearcut. While the average slope is steeper in the northeast, it is only the result of a different frequency curve: steep slopes (over 25°) are more frequent in the northeast although not absent in the west.

One can also observe quite a contrast between the forms of the hillsides and the long-profiles of the streams. In the entire Sandstone Vosges the long-profiles of the permanent streams are regular; exceptions are few and unimportant. A number of rapids are observed only where small streams clear particularly resistant ledges. The drainage net is evenly ramified, and the affluents have incised the hillsides and scalloped the margin of the plateau. The influence of the geologic structure is not reflected in a marked fashion. One may speak, in a Davisian manner, of a mature drainage net. But slopes are often steep. If abrupt escarpments are seldom developed and always limited to conglomerate beds a few metres thick and never very long, hillsides with average slopes of 20 to 30° are not exceptional. Their profile, which is often rectilinear, is hardly interrupted by rock benches due to outcrops of resistant beds. They merge, however, into flat valley bottoms often extended into wide depressions. They have therefore retreated without a marked loss in gradient. To be sure, there are gentle slopes with billowy, irregular forms, but they are scattered in distribution and coexist with steep slopes. In the west such gentle slopes are commonly found at the bottom of depressions; in the northeast at the head of small valleys.

The topography of the Sandstone Vosges results from the alternation of temperate and periglacial climates during the Pleistocene Epoch. Erosion during the Holocene has been insignificant. The part played by palaeoclimates is therefore predominant in the elaboration of the relief. But there has been no disorganisation of the drainage net due to aridity during the Quaternary; streams functioned continuously and even deposited much more alluvium during the cold than during the temperate stages. Thus the 'system of normal erosion' has continuously functioned in the massif for a very long time. It should therefore have impressed upon it the standard type of topographic dissection. We note, however, two major contradictions in the Davisian scheme:

1. An important retreat of some slopes without concomitant loss in gradient.
2. The coexistence of a drainage net with a typically 'mature' profile of equilibrium and generally steep slopes typical of the stage of 'youth'.

Further examples extended to other rock types and conclusion

Is the case of sandstone which we have just analysed in some detail a particular case? Can it be generalised to other rocks before drawing conclusions about it concerning the scientific value of the Davisian theory of the cycle of normal erosion?

CRYSTALLINE ROCKS AND LIMESTONES

Let us take the case of crystalline rocks and limestones, two groups of rocks particularly widespread on the earth's surface, and examine their geomorphic evolution under different climates, as we have done for sandstone regions.

In the wet tropics crystalline rocks (such as granites, gneisses, and schists) produce a topography which in many respects is much like that produced by sandstones in the same morphoclimatic environment. Advanced erosion generally coexists with very steep slopes. Valleys broaden into wide depressions with flat or slightly undulating floors, forming bottomlands where waters stagnate in the midst of tall grasses and arborescent thickets. The width of such lowlands may reach several kilometres even on divides. Their surfaces of faint relief are surrounded by slopes averaging 20°, but sometimes as much as 30 or 35°. Their forms are varied and betray structural and lithologic influences. In metamorphic rocks asymmetric, cuesta-like ridges are common. In intrusive rocks differences in petrology or variations in jointing may cause convex bedrock hills to be eroded out in bold relief, as in the case of sugarloaves or monolithic domes with abrupt or precipitous slopes. In other cases the relief is characterised by a sharp contrast between a sea of deeply weathered convex ridges (*mar do mauros*) and abrupt reliefs of sugarloaves or broader monolithic domes. In some regions, for example at the head of Guanabara Bay (Brazil), steepsided, deeply weathered convex hills, appropriately named 'half oranges' (*meias laranjas*), stand in the midst of flats formed by alluviation during the Flandrian.[1] This alluvial

---

[1] The late-Glacial rise in sea-level (KdeJ).

veneer further increases the uniformity of the relief which was already flat. Here, too, steep slopes and a long evolution are associated. The stream beds of such regions are far from having regular profiles even if the stream flows on a level or slightly undulating surface which would well merit the name of peneplain provided the inselbergs which dominate it were left out of account; the water frequently forms rapids, sometimes real falls. Quiet reaches, almost without current at low water, normally alternate with short, steep reaches where the river clears jumbles of large, practically unweathered crystalline boulders or ledges of quartzite. Such rapids are frequently found at the lower outlet of wide depressions.

In an arid environment a topography developed on crystalline rocks is also characterised by an association of steep slopes and almost level lowlands. When the volume of the removed material is not too important, the relief is characterised by the formation of wide depressions dominated by steep uplands with structural outlines, a disposition which agrees with joint texture and degree of metamorphism. On the uplands the texture of dissection is quite dense with innumerable, slightly incised ravines or gullies. The spacing of the gullies may be as little as 10 m (33 ft) with streams barely incising the bedrock between slightly convex ridges formed by slowly

Fig. 1.9. Granito-gneissic topography in the wet tropics
A. The Paraguaçu valley above Cachoeira, Bahia, Brazil.

The relief is produced by dissection of a Neogene erosion surface visible in the background. The valley is poorly graded, with stretches of flat bottomlands. The river branches on an outcrop of more resistant granite, forming rapids. The plateau is broken up into steepsided hills which are later reduced to deeply weathered convex hills (*meias laranjas*), as in the vicinity of the Paraguaçu River.

B. Erosional relief near Cabo, Pernambuco, Brazil.
The evolution of the relief is strictly similar to that shown in Fig. 1.9A. We again find steep slopes here (up to 40°) which plunge down to a flat bottomland where a small river meanders in fine sediments.

In the wet tropics granite is subjected to intense weathering. This liberates mainly clays and sands that are protected from mechanical erosion by the plant cover.

weathering outcrops. Important sand masses forming coalescing fans are spread at the mouths of the ravines in the depressions. A more advanced erosion ends in the formation of inselbergs and plains which may be immense. The unbroken plain of Inchiri (near Akjoujt, Mauritania) extends over a width of more than 100 km (60 miles). The inselbergs dominating such plains generally have conical or truncated conical forms rather than convex shapes such as sugarloaves. The peneplain is realised but only through a combination of processes (shattering, flaking, thunderstorms) very different from those of the Davisian scheme.

The relief of the ancient crystalline massifs of temperate lands should be more familiar to most of our European readers. It is characterised by a dense dendritic drainage net with long-profiles approaching the theoretical profile of equilibrium, although generally broken by breaks in slope attributed to the succession of erosion cycles. But even the most irregular stream courses have a less broken long-profile than that of intertropical streams; it is composed of successive concave sectors which connect with one another by short convexities rather than by a series of steps. Interfluves are characterised by a dissection into ridges, convex at the top, concave at the base and becoming more ramified as erosion is complete. It is furthermore necessary

Fig. 1.10. Granito-gneissic topography in the wet–dry tropics
A. Pediplain with inselbergs near Ipara, Bahia, Brazil.

Savanna vegetation. Rainfall: approximately 900 mm (36 in). Large pediplains have developed and coalesced. They are dominated by steep isolated hills or inselbergs.

B. Pediplain, closed depression, and inselberg, João Amaro, Bahia, Brazil.

In the foreground there is a small closed depression where water stagnates, and the soil remains bare. Fine debris accumulates here between rock knobs. The depression is caused by more intense weathering, promoted by moisture, than in the surrounding area. Such depressions are characteristic of savannas and semiarid climates in granito-gneissic environments. In the background there are well individualised inselbergs with steep slopes.

C. Broken inselberg relief near João Amaro, Bahia, Brazil

Rainfall: approximately 600 mm (24 in). The granite is jointed, which would facilitate weathering if the climate were sufficiently humid. But weathering being slight the rock remains exposed and is washed by runoff, producing a broken relief which contrasts with the hills of the wet tropics padded with the products of weathering (*meias laranjas*).

FIG. 1.11. Relief carved from gneisses in the wet climate of tropical mountains: Santo Domingo Valley, 15 km (9 miles) above Altamira, Venezuelan Andes

Altitude 1 800 m (6 000 ft); rainfall over 3 000 mm (120 in). The natural vegetation is a rainforest. The torrent causes intense dissection. The gneisses weather rapidly, but the weathered material is almost immediately evacuated by periodic landslides, such as the one clearly visible in the foreground. Numerous slightly incised torrents mark the side slopes.

33

FIG. 1.12A. Relief carved from granites in an arid environment, Chosica, Peru

Vegetation is nonexistent except in the oasis at the bottom of the valley. Although the disintegration of the granite is intense because of variations in temperature and humidity (dew), some rock knobs outcrop on the slopes. The slopes are covered with a veneer of debris incised by the watercourses which are washed every now and then by a freshet.

FIG. 1.12B. Disintegration of granite, Chosica, Peru

The view shows a detail of the granitic slopes shown in Fig. 1.12A. The granite disintegrates on the surface into a coarse sand which slides on a slope with an average gradient of about 45°. Somewhat more resistant nuclei (corestones) eventually produce boulders.

Fig. 1.12C. Granitic hills and pediment in the Peruvian littoral desert at Río Seco, north of Lima
In the foreground: a pediment covered by debris, which is spread apronlike by the waters which debouch from the hills. Slopes are steep (about 40°) and form a characteristic angle ('nick') with the pediment. The slopes weather through granular disintegration (as the slopes in Fig. 1.12B), and the debris is evacuated by sliding and runoff in the ravines during sporadic cloudbursts.

to emphasise that the relief of these Hercynian massifs was influenced to a very high degree by tropical Tertiary and periglacial Quaternary phases, which tend to obliterate the zonal characteristics corresponding to the temperate climates.

The relief of ancient crystalline massifs again reveals different characteristics in the cold zone. Besides glacial actions, which produce cirques and troughs, periglacial processes produce slopes, convex at the top, concave at the base, rendered uneven by irregularities such as solifluction lobes. Minor rock outcrops are broken up by frost-weathering. Depressions are filled with debris which form alluvial aprons or concave valley fills. Sometimes the flank of a mountain is composed of successive steps, or altiplanation terraces, formed by a vertical zonation of slightly inclined benches and steep slopes (20 to 25°).

Limestones also produce quite different reliefs depending on the type of climate. They were the concern in the first instance of German geomorphologists, and then of others.

Two typical types of karstic landforms develop in sufficiently humid tropical climates: 'haystack' or 'needle' karsts and 'cockpit' karsts, as in Jamaica. The first are a highly evolved type characterised by level plains interspersed by abrupt hills, rounded at the top and excavated at the base.

35

FIG. 1.13. South-facing slopes on porphyroidal granite in the Aigoual
(Cévennes Mts), near Arphy, southern France

An open chestnut forest with occasional cultivated terraces covers the slopes.
Frost action and chemical weathering produce rubble and sand. At the
present runoff is rather important. It remains unconcentrated on the ridges
and becomes concentrated in the rather widely spaced watercourses (compare
with preceding views). During the cold stages of the Pleistocene frost-wedging
produced important quantities of rubble which slid down the slopes and
covered them with an uneven mantle, which restricts present runoff.

The second represent an early stage of evolution in which plateaus are
pitted with deep funnel or V-shaped sinks which alternate with abrupt
pinnacles. An intermediate type displays small bottomlands enclosed by
contiguous uplands. Notable is a certain convergence of forms between
karstic haystacks and the sugarloaves of crystalline areas of identical
climates.

Under arid and semiarid climates limestones in tablelands produce a
relief of stony plateaus cut by steepsided valleys (20 to 35° on the average)
with benches corresponding to the more resistant strata. Dissolution forms
are mainly represented by lapiés and scattered sinks. There is a substantial
contrast with the karst of the wet tropics.

Under cold periglacial climates limestone slopes develop concavities at
the base and convexities at the top. Ridges alternate with wide, flat, or
concave valleys. The topography of the calcareous plateaus of central
Europe which still displays fresh evidence of Pleistocene periglacial actions
offers an excellent example. The steepness and the regularity of the hillsides
depend on the nature of the carbonates. The least frost-susceptible limestones produce benches or scattered outcrops. In dolomitic limestones

flatbottomed valleys are dominated by jagged escarpments whose feet are buried in talus.

The geomorphic facts noted in the three examples of sandstone regions are therefore not isolated. Analogous facts can also be found in regions of crystalline rocks and of limestones (i.e. in all three categories of well consolidated rocks that are most widespread on the face of the earth). Furthermore it should be noted that the morphologic expressions of different rocks in one and the same climate bear a certain family likeness. The family likeness between limestone haystacks and crystalline sugarloaves is even greater than that between the limestone plateaus (Causses) and the granitic topography of the Massif Central. There are also other convergences: closed depressions resembling the sinks of midlatitude karsts can also be found in the crystalline rocks of savanna regions. The *dayas* of deserts, which form similar depressions (which are cultivated), are also far from being limited to limestone outcrops. The interplay of climatic influences and lithology, therefore, produces very diverse morphologic combinations which contradict the rigidity of Davis's theory of normal erosion.

CONCLUSION: INADEQUACY OF THE THEORY OF THE CYCLE OF NORMAL EROSION

The geomorphic facts briefly analysed above contradict the theoretical scheme of the cycle of normal erosion imagined by Davis.

The incompatibility of the theory with the facts stands out particularly on two fundamental points:

1. The profile of stream courses is the first significant point. The profile of equilibrium of streams, which enables a regular evacuation of all the debris of mechanical erosion to the general base-level of the ocean, is rapidly realised only under temperate and certain cold climates. As soon as there is a tendency towards aridity, the characteristics of fluvial morphology change: in the semiarid tropics there is a sharp juxtaposition of areas of wash slopes and ravines maintaining a finely dissected and steep relief, and areas of piedmont slopes[1] littered with debris abandoned at a lesser or farther distance, depending on the importance of each flood. In the wet tropics streams do not appear to develop a profile of equilibrium of the kind found in the temperate zone. Everywhere quiet reaches alternate with rapids, which are the dominant characteristic of *all* intertropical streams, in whatever degree of dissection the land they drain, and whatever length of time they have taken to form their courses.

The Davisian scheme of the evolution of stream courses during the cycle of erosion is therefore not susceptible to general application. Its aspects are valid only in the humid temperate zone, which includes less than 10 per cent of the earth's surface. But even in this zone it includes serious errors. It

[1] The term 'piedmont slope' in this work is used in a non-committal sense and may be either a pediment or a bajada or both (KdeJ).

37

associates gullies or ravines and the quiet reaches of streams: the ravines would characterise the stage of youth in the upstream area, and the quiet reaches the downstream area in the stage of maturity. It puts end to end a form of dissection peculiar to semiarid or anthropic erosion and a drainage regime characteristic of regions with a dense plant cover where spring fed streams predominate. As long as man has not excessively despoiled the

FIG. 1.14. Morphoclimatic evidence in a limestone environment of two distinct climatic regions

Both views (14A and B) deal with a highly dissected limestone relief with steep slopes sculptured in horizontal beds.

A. At St Chéli (Causses) periglacial morphogenic processes of the cold stages of the Pleistocene have played the major role in the present relief. While the more friable beds were then reduced to rubble by frost-splitting, the more massive rocks resisted and produced escarpments. The rubble which fell or slid down the slopes now veneers the ground and has become stabilised.

B. Evolution is exclusively chemical at Bom Jesus da
Lapa (Bahia, Brazil) which is situated in the wet–dry
tropics and receives an annual rainfall of 1 000 mm
(40 in). Corrosion sculptures pinnacles from thick
bedded limestone. Flutings intersecting in sharp crests
gradually develop because of rainwash. Dissolution
exploits stratification joints and occasionally produces
overhangs and carves out blocks which lose their
support and collapse. Because of past and present
climates evolution is completely different from that of
the Causses and ends in an utterly distinct relief.

forest and the natural grasslands, the steepest reliefs of the humid temperate
zone never show erosion gashes nor general gullying. This can be observed simply by taking a walk through the northern French Alps, the
Massif Central, or the Ardennes. Even when non-consolidated rock formations are accidentally cleared, gullying does not develop, and the vegetation
if left to itself recolonises the momentarily lost space. Gullies have a morphoclimatic significance: they only develop in regions having a marked dry
season with a semiarid tendency. They are antinomic to the spring fed
stream regime humid temperate lands.

Depending on what regions of the world one considers, the scheme of the fluvial cycle of erosion, which is the cornerstone of the Davisian system of normal erosion, is therefore not only inadequate but false.

2. The other weak point is the supposed link between the progress of erosion and the evolution of slopes. The Davisian theory of the cycle of normal

FIG. 1.15. A. Tropical limestone relief: the needle karst (or haystacks) of southeast China (*drawing from photograph by von Wissman*).

B. Periglacial limestone relief of the midlatitudes: the village of Germaine on the Langres Plateau (*drawing by D. Tricart from photograph by J. Tricart*)

A plateau topography dissected at least since the Pliocene. Very subdued forms mostly developed by Pleistocene periglacial processes. Contrast with 1.15A.

erosion asserts that slopes become progressively gentler as evolution proceeds during the cycle. But a fair observation of the *entire* face of the earth
reveals, on the contrary, that the maintenance of steep slopes is the most
general case. From the equator to the Mediterranean zone, in the most
varied rocks, one is struck by the presence of precipitous reliefs which look
down on intensely worn plains. The inselberg is associated with the most
advanced planations and stands witness to an enormous amount of denudation. At least in this part of the earth, the continuation of the work of erosion
results in an increased degradation but not in a reduction of slope angles
below certain values which are rather high (20 to 30°). Rather than the idea
of a generalised reduction of slopes, it is the concept of the parallel retreat of
slopes associated to the concept of the minimum slope[1] which is the more
realistic. It has led Lester C. King to formulate his theory of *pediplanation* in
opposition to that of Davisian peneplanation. In King's theory the development of the remarkable planations under semiarid conditions, those which
lead to the development of pediments, is closely associated with the retreat
of slopes parallel to themselves. The end result is the formation of inselbergs
and pediplains, often stepped, separated from one another by steep slopes
which retreat slowly. With this theory King has interpreted the relief of
southern Africa and Brazil. Its application, however, runs into many difficulties: more detailed studies show that it is somewhat rudimentary because
it fails to take into account tectonic deformations. Even so, its application is
more valid than Davisian peneplanation. Indeed, in those places where the
finest present or recent planations are to be found the facts rebuff the
Davisian theory. It is precisely in the humid temperate zone where the
theory of the cycle of normal erosion might have applied that recent peneplains developed under the present climate are wanting.

The only conclusion, then, is that the Davisian theory of the cycle of
normal erosion is inadequate; the facts prove it. The conclusion the facts
impose is that of the diversity of relief types as a function of the climatic
zones: a diversity which maintains itself over a very long period of time in
spite of a considerable amount of denudation. The reliefs formed by the
various dominant categories of well consolidated rocks display among themselves important analogies, a family likeness, within the same morphoclimatic zone. In the wet tropics closed depressions develop on limestone
plateaus as well as on plateaus underlain by sandstones or crystalline rocks.
The reduction of the relief produces monolithic domes or 'half oranges'
(*meias laranjas*), and the outline of limestone haystacks does not differ funda-
mentally from that of crystalline sugarloaves. In a semiarid environment
inselbergs and pediplains develop regardless of the rocks. An obvious morphoclimatic factor may be deduced from these facts: the postulate of

[1] The concept of the minimum slope is a fundamental concept in geomorphology. It is
based on the fact that a specific morphogenic process cannot function below a certain slope
angle under certain conditions of lithology, climate, and vegetation; refer to Tricart,
'L'Evolution des versants', *Inf. Géogr.*, **21** (1957), 109–10 (KdeJ).

41

'normal erosion' must be replaced by a law of the dependence of relief on climate. We must therefore examine the effects of climate on relief.

But, first, it would be well to draw a conclusion on method from this discussion of the concept of normal erosion. It is worth while to analyse the reasons which led so eminent a research worker, so active a scholar, and so enthusiastic a professor as Davis to be thus mistaken and to elaborate and later propagate a gravely erroneous theory. Such an analysis is indeed indispensable to all those who desire to continue the untiring research of a correct and ever deeper understanding of the facts of nature. It will help us avoid similar errors.

In 'The geographical cycle' (*Essays*, pp. 251–2) Davis devotes himself to a rapid statement on method:

> It is evident that a scheme of geographical classification that is founded on structure, process, and time must be deductive in a high degree. This is intentionally and avowedly the case in the present instance. As a consequence, the scheme gains a very 'theoretical' flavour that is not relished by some geographers whose work implies that geography, unlike all other sciences, should be developed by the use of certain of the mental faculties only: chiefly observation, description, and generalisations. But nothing seems to me clearer than that geography has already suffered too long from the disuse of imagination, invention, deduction, and the various other mental faculties that contribute towards the attainment of a well-tested explanation.

Davis himself makes recourse to the imagination the peculiar characteristic of his method. And, in fact, the cycle of normal erosion is a product of the faculty of imagination whose merits he praises and whose increased use he advocates. The theory is not based on new methods of observation nor on a mass of newly established facts. It only re-uses a pre-existing common fund of knowledge based on observation and often on practical experience (especially the behaviour of stream courses). Between these data it establishes supposed links of succession, of cause and effect, and of co-ordination into a logical structure. It is symptomatic in Davis's life work that this synthesis was one of his first geomorphological works, which was not preceded by a co-ordination of methods or new techniques nor by numerous detailed and varied investigations. It is also significant that this theory immediately took an exclusive turn, that it aimed at a total satisfaction of the mind, that it left no room for interrogation, and that it made no new appeal to further research. That is why, once formulated, it restricted its author and part of those who followed him. It forced Davis to defend it foot by foot as knowledge of the globe became progressively more specific and new data became established. The artisan became the prisoner of his tool and all there was left for him to do was to try to fit the new data into the system. Thus he attempted the elaboration of a cycle of arid erosion and also a cycle of glacial erosion, both of which are based on a still more limited number of facts than the theory of the cycle of normal erosion. Davis immediately adopted the myth of the creative imagination and rejected, by this very fact, the perpetual dialectical oscillation between observation and hypothesis. In his work the establishment of facts holds a too limited space while hypotheses abound, one producing another, and postulates arise every time observed data are wanting. At the same time, as a cause as well as an effect, Davisian geomorphology became separated from practice. The collaboration between methodic and theoretical thought and the work of engineers, which had provided the very bases of our discipline in the nineteenth century, was totally abandoned; the facts established by engineers were embarrassing to those who clung to a theory which had tried to grasp everything before its time, and the theoretical schemes which it presented were utterly devoid of interest to the engineer because its mental constructions did not permit him to act on the facts, on which the constructions were not based to begin with.

Contemporary geomorphology suffers from this divorce as much in its theoretical as in its applied form. The division sets back the progress of our knowledge of nature and, in its ineluctable corollary, limits our action upon it. In the realm of slope development, for example, the theoretical utilisation of all the material assembled by engineers would be a source of rapid progress. In the same way an effort of confrontation and synthesis, of methodologic elucidation, would materially help engineers.

For this reason our method will consist in having recourse, every time it is possible, to this link with practice which is so cruelly wanting in our discipline. For this reason, too, we will make every effort by means of new methods to probe deeper into the analysis of facts and to obtain a greater knowledge of the secrets of nature before proposing new hypotheses and theories. It is necessary, in order to respect the indispensable oscillation between fact and hypothesis, that the formulated hypothesis does not reach too far in advance of the facts. It is also necessary, whenever a complete knowledge of reality is not attained, not to adopt overly exclusive schemes of causality without leaving the door open to other hypotheses which tomorrow the facts may suggest. Hypotheses should open up perspectives, not shut them out.

To be sure, this method will perhaps be less satisfactory to those who possess the absolute turn of mind of Davis's rigid constructions. In each scientific truth there is a relative element. The sparkling flash of firework is the appanage of the poet. As for the naturalist, he must patiently investigate the earth step by step in order to progress in the knowledge of its intimate nature. He does not carry its truth *within himself*, but he must wrest the truth *from nature* bit by bit.

43

Bibliographic orientation

Works on 'normal geomorphology'

We will limit ourselves to some of the better known and more accessible works which deal with geomorphology according to the Davisian concept of the cycle of normal erosion:

DAVIS, W. M. (1899) 'The geographical cycle', *Geogr. J.* **14**, 481–504 (reproduced in *Geographical Essays*, Boston, Ginn, 1909 and New York, Dover, 1954, pp. 249–78).

DAVIS, W. M. (1904) 'Complications of the geographical cycle', *8th Int. Geogr. Congr.*, pp. 150–63 (reproduced in *Geographical Essays*, pp. 279–95).

LAPPARENT, A. DE (1907) *Leçons de géographie physique*, 3rd edn., Paris, Masson, 728 p.

DAVIS, W. M. (1912) *Die Erklärende Beschreibung der Landformen*, Berlin, Teubner, 565 p.

MARTONNE, E. DE (1935) *Traité de géographie physique*, vol. 2, *Le Relief du sol*, 5th edn., Paris, Colin, pp. 499–1057.

MACAR, P. (1946) *Principes de géomorphologie normale, étude des formes du terrain des régions à climat humide*, Liège, Masson, 304 p.

COTTON, C. A. (1948) *Landscape as Developed by the Processes of Normal Erosion*, 2nd edn., Christchurch, New Zealand, Whitcombe & Tombs, 509 p.

Historical development of the concepts of 'normal geomorphology' and climatic geomorphology

Research on the evolution of geomorphological concepts is meagre. The best work up to but not including Davis is that by Chorley, Dunn, and Beckinsale. A good analysis of the beginnings of geomorphology and a review of Davis's work is found in Baulig's *Essais*. De Martonne devotes a short paragraph to Davis in his *Traité* (pp. 545–7).

BAULIG, H. (1948) 'L'oeuvre de William Morris Davis', *Inf. Géogr.*, pp. 101–8 (reproduced in *Essais*, pp. 13–25).

BAULIG, H. (1950) *Essais de géomorphologie*, Strasbourg, Publ. de la Fac. Lettres, no. 114, article entitled: 'La philosophie géomorphologique de James Hutton et de John Playfair', pp. 1–11.
   We refer to the references of this work as concerns the pioneers of geomorphology.

BAULIG, H. (1956) 'La géomorphologie en France jusqu'en 1940', in *La Géographie française au milieu du XXe siècle*, Paris, Baillère, pp. 27–36.

SCHMITTHENNER, H. (1956) 'Die Entstehung der Geomorphologie als geographische Disziplin', *Petermanns Geogr. Mitt.*, C, pp. 257–68.
   Shows the orientation of various authors but without methodologic perspectives. Useful references to which we refer.

LOUIS, H. (1957) 'The Davisian cycle of erosion and climatic geomorphology', *Proc. Int. Geogr. Union*, Regional Conf., Japan, pp. 164–6.

JAUDEL, L. and TRICART, J. (1958) 'Les précurseurs anglo-saxons de la notion davisienne de cycle d'érosion', *Rev. Gén. Sc.* **65**, 237–51.

CHORLEY, R. J., DUNN, A. J., and BECKINSALE, R. P. (1964) *The History of the Study of Landforms or the Development of Geomorphology.* vol. 1, *Geomorphology before Davis*, London, Methuen; New York, Wiley, 678 p.

CHORLEY, R. J. (1965) 'A re-evaluation of the geomorphologic system of W. M. Davis', in R. J. Chorley and P. Haggett, eds., *Frontiers in Geographical Teaching*, London, Methuen, pp. 21–38.

A good understanding of pre-Davisian geomorphology may be obtained from a consultation of the following works:

LYELL, C. (1833) *Principles of Geology.* Edinburgh, 3 vols.

SURELL, A. (1841) *Etudes sur les torrents des Hautes-Alpes*, Paris.

RAMSAY, A. C. (1863) *The Physical Geology and Geography of Great Britain*, London, 145 p.

RUTIMEYER, P. (1869) *Über Thal- und Seebildung*, Basel, 95 p.

GEIKIE, A. (1875) *Physical Geography*.

POWELL, J. W. (1875) *Exploration of the Colorado River of the West and its Tributaries*, Washington (especially pp. 149–214). Reprint by Univ. of Chicago and Cambridge University Press, 1957.

GILBERT, G. K. (1877) *Geology of the Henry Mountains*, Washington, 160 p.

DE LA NOÉ, G. D. and DE MARGERIE, E. (1888) *Les Formes du terrain*, Paris, 205 p. together with atlas.

The major steps which have led to climatic geomorphology are marked by the following works, which, like the preceding references, only represent a choice which is voluntarily reduced to a minimum:

RICHTHOFEN, F. VON (1886) *Führer für Forschungsreisende*, Hannover, 734 p.

PENCK, A. (1894) *Morphologie der Erdoberfläche*, Stuttgart, 2 vol.

WALTHER, J. (1900) *Das Gesetz der Wüstenbildung*, Berlin, 175 p.

BORNHARDT, W. (1900) *Zur Oberflächengestaltung und Geologie Deutsch Ostafrikas*, collection *Deutsch Ostafrika*, vol. 7, Berlin, 595 p.

PASSARGE, S. (1904) *Die Kalahari*, Berlin, 822 p.

PENCK, A. (1905) 'Climatic features in the land surface', *Amer. J. Sc.* **169**, 165–74.

MARTONNE, E. DE (1913) 'Le climat facteur du relief', *Scientia*, pp. 339–55.

SAPPER, K. (1914) 'Über Abtragungsvorgänge in den regenfeuchten Tropen', *Geogr. Z.*, pp. 5–18 and 81–92.

SALOMON, W. (1916) 'Die Bedeutung der Solifluktion für die Erklärung der Deutschen Landschafts- und Bodenformen', *Geol. Rundsch.* **7**.

EAKIN, H. M. (1916) 'The Yukon Keokuk Region (Alaska)', *USGS*, no. 631, 88 p.

THORBECKE, F. (1927) 'Morphologie der Klimazonen, herausgegeben von . . .,' Selection of conferences by various German authors, Breslau, Düsseldorfer Vorträge, 100 p.

MARTONNE, E. DE (1940) 'Problèmes morphologiques du Brésil tropical atlantique', *Ann. Géogr.* **49**, 1–27 and 106–29.

DRESCH, J. (1941) *Recherches sur l'évolution du relief dans le Massif Central du Grand Atlas, le Haouz et le Sous*, Tours, Arrault, 703 p., 206 fig.

MARTONNE, E. DE (1946) 'Géographie zonale: la zone tropicale', *Ann. Géogr.* **55**, 1–18.

BÜDEL, J. (1948) 'Das System der klimatische Morphologie, Beitrage zur Geomorphologie der Klimazonen und Vorzeitklimate', *Deut. Geographentag*, Munich, 1950, pp. 65–100.

LEUZINGER, V. R. (1948) *Controversias geomorfológicas*, Rio de Janeiro, 209 p. Excellent critique of Davisian geomorphology.

BIROT, P. (1949) *Essai sur quelques problèmes de morphologie générale*, Lisbon, Instituto para a Alta Cultura, Centro de Estudios Geográficos, 176 p.

TRICART, J. (1950) *Le Modelé des Régions périglaciaires*, Paris, CDU; 3rd edn., SEDES, 1967, 512 p.

CHOLLEY, A. (1950) 'Morphologie structurale et morphologie climatique', *Ann. Géogr.* **59**, 321–35.

TRICART, J. (1952) 'Climat, végétation, sols et morphologie', *Jubilaire 50e Anniv. Labo. Géogr. Rennes*, pp. 225–39.

TRICART, J. (1953) 'Climat et géomorphologie', *Cahiers Inf. Géogr.*, no. 2, 39–51.

TRICART, J. (1955) *Introduction à la Géomorphologie climatique*, Paris, CDU, 2nd edn, Paris, SEDES, 1965, 306 p.

TRICART, J. (1955) *Le Modelé glaciaire et nival*, Paris, CDU, 2nd edn., SEDES, 1962, 508 p.

TRICART, J. (1960–61) *Le Modelé des Régions sèches*, Paris, CDU, 2 vols, 129 and 179 p. (mimeographed).

45

Examples of the influence of climate on relief

The main attempt to compare reliefs formed by identical rocks under different climates is that of Birot. Unfortunately, schemes often marked with a theoretical bend take much more space than descriptions and analyses. One can also refer to Thorbecke (1927), but this collection of articles by different authors is too heterogeneous to provide the necessary material adequately. Whoever desires to acquire a personal opinion will have to have recourse to numerous regional monographs. We will list some of them as examples:

FREISE, F. W. (1933) 'Brasilianische Zuckerhutberge', *Z. Geomorph.* **8**, 49–66.

SAPPER, K. (1935) *Geomorphologie der feuchten Tropen*, Geogr. Schriften, herausgegeben von A. Hettner, Leipzig, Teubner, 150 p.

JESSEN, O. (1936) *Reisen und Forschungen in Angola*, Berlin, Reimer.

LAUTENSACH, H. (1950) 'Granitische Abtragungformen auf der iberischen Halbinsel und in Korea, ein Vergleich', *Peterm. Geogr. Mitt.*, pp. 87–196.

BIROT, P. and JOLY, F. (1952) 'Observations sur les glacis d'érosion et les reliefs granitiques au Maroc', *Mém. Centre Doc. Cartogr.*, *CNRS*, **3**, 9–55. Some excellent photographs.

AUBERT DE LA RUE, E. (1954) *Reconnaissance géologique de la Guyane française méridionale*, Paris, Larose, 128 p.

LEHMAN, H. (1954) 'Bericht von der Arbeitstagung der internationaler Karstkomision', *Erdkunde*, **8**, 112–21, and the articles which follow this paper, in the same issue, especially those of Corbel, Lasserre, and von Wissmann.
     An attempt at classification is made by P. Birot.

HEMPEL, L. (1955) 'Konvergenzen von Oberflächenformen unter dem Einfluss verschiedener klimatischer Kräfte', *Deutsche Geogr. Blätter*, **47**, 188–200.

BIROT, P. (1960) *Géographie physique générale de la zone intertropicale* (à l'exclusion des déserts), Paris, CDU, 1960, 1965, 244 p. (mimeographed).

ROUGERIE, G. (1960) *Le Façonnement actuel des modelés en Côte d'Ivoire forestière*, *Mém. I.FAN*, no. 58, Dakar, 542 p.

RAYNAL, R. (1961) *Plaines et piedmonts du Bassin de la Moulouya (Maroc oriental), étude géomorphologique*, Rabat, Imframar, 619 p.

TRICART, J. (1961) 'Les caractéristiques fondamentales des systèmes morphogénétiques des pays tropicaux humides', *Inf. Géogr.*, pp. 155–69.

COQUE, R. (1962) *La Tunisie présaharienne, étude géomorphologique*, Paris, Colin, 476 p.

JOLY, F. (1962) *Etudes sur le relief du Sud-Est marocain*, Rabat, Institut scientifique chérifien, 578 p.

BATTISTINI, R. (1964) *L'Extrême sud de Madagascar, étude géomorphologique*, Paris, Editions Cujas, 636 p.

DOLLFUS, O. (1965) *Les Andes centrales du Pérou et leurs piedmonts (entre Lima et le Péréné), étude géomorphologique*, Lima, Inst. Franç. d'Etudes Andines, 404 p.

BORDE, J. (1966) *Les Andes de Santiago et leur avant pays, étude de géomorphologie*, Fac. Lettres, Bordeaux, 559 p.
     This list should, of course, be completed with de Martonne's articles and Dresch's *Recherches* (1941) mentioned above, and with the volumes of Tricart's *Treatise* and their bibliographies.

2
Morphoclimatic mechanisms

Because the postulate of normal erosion is false, it is necessary to replace it by an analysis of the effects of climate on relief. Which climatic factors cause differences in topographic forms? This is the question that must be answered in this chapter. First we define and classify the morphogenic processes which were insufficiently investigated by the theory of the cycle of normal erosion. Later we study the direct influence of climate, particularly those processes that are limited to a specific climatic zone. Lastly, we discuss the indirect influence of climate as it is felt through plants and the soil or the biosphere, and the soil-forming processes.

Basic definitions

The Davisian cyclic theory, which was a premature synthesis, was not concerned with a thorough analysis of the various mechanisms of relief development. A number of its adepts, such as Baulig (1950, p. 24), have explicitly recognised this and regretted it. A successful analysis of the morphoclimatic mechanisms must be preceded by an understanding of the basic principles which affect them. In this way we hope to avoid Davis's grave methodological error which implicitly defined 'normal erosion' by way of the theory of the 'cycle of erosion'.

Cholley (1950) has clearly analysed the concepts dealing with the morphogenic mechanisms in an article whose basic outlines we will follow here.

Criticising the concept of 'climatic accidents' of the Davisian school, Cholley poses the problem in the following way:

In reality there are not two morphologies but one, and its genesis is related to the kind of erosion imposed by climate. But we are wrong in thinking that it is the sole action of a single agent which produces a whole topography. . . . in the example of glacial erosion, and also in that of desert erosion, it is a whole complex of agents, a veritable system of erosion, which each climate sets in motion. Given the structure of the atmosphere and the nature of climates, how could it be otherwise: has one ever observed an element of climate act by itself on the earth's surface? It is, therefore, a concept more conformable to reality to envisage the action of complexes or of a combination of factors. They should be called *systems of erosion* because they are capable of producing a morphology whose elements belong together and which are also systematically interdependent. . . .

Let us consider fluvial erosion as an example: it implies, consecutively, the disintegration of the rocks (mechanical or chemical erosion), the setting in motion of the rock-waste (by overland flow and creep), and the concentration of the overland flow into channels (stream beds) permitting the evacuation of the materials; operations which result in the form of the valley, the slopes, and the hill crests (pp. 323–4).

The problem is well posed: the development of the relief is due to a hierarchy and a succession of mechanisms intimately linked together, the action of which is co-ordinated into a system. This hierarchy may be described with the help of a few terms which will help to clarify it, but one should never lose sight of the fact that the development of the relief is due to a series of interrelated factors in which there is no complete break. The new terms to be introduced are of necessity artificial. Our intellect requires them, and they should be considered as a convenience. They are markers, not barriers, used to guide reasoning and observation, not to occlude them.

If we analyse the mechanisms of erosion we arrive at relatively simple actions caused by a limited number of factors. A rockfall from a precipice towards the foot of the slope on which the rock rolls is an example. This mechanism, oft repeated, gives birth to a minor decametric or hectometric landform; the talus or scree slope. Its construction develops through two kinds of successive actions: fragmentation of the bedrock and free fall of the rocks under the effect of gravity. The example is ideally simple and constitutes an elementary landform. In the same way, a porous waterlogged rock subjected to frost action experiences high internal tensions because of the increase in volume of the water that has become frozen. If the stresses go beyond the mechanical resistance of the material they cause a fragmentation. The mechanism involved is frost-weathering, another simple case, which presupposes only the freezing of water in the pores, fissures, or joints of the rock and the consequent increase in pressure inherent in the force of crystallisation above a certain limit, variable according to the rocks.

We designate such simple mechanisms by the expression *elementary processes*. They include, for example, the formation of potholes by swirling grinders; the formation of rock pavements by the differential action of deflation or the wash of running water (wind and water remove all particles smaller than a certain dimension, leaving the remainder on the ground); other examples are the formation of calcium bicarbonate from the contact of limestones with water containing carbon dioxide; the undermining of the foot of waterfalls and their consequent retreat; and the drifting of pebbles and sand on the shore with each breaking wave.

48

Other morphoclimatic mechanisms are more complex and result from the combination of several elementary processes. For example, slope development by solifluction in a periglacial environment: there is, first, the elementary process of frost-weathering which not only produces the rock-waste but continues to act on it as it flows downslope. Then there are the more complicated mechanisms of the pasty flow of waterlogged materials that contain enough clayey matter to have mechanical properties which permit a mass movement. There are also variations in the volume of water due to the effects of alternating frost and thaw; these variations cause movement of particles and effects of sorting between the muddy matrix and the rubble. The whole process is complex because it implies the combination of several different mechanisms and of several elementary processes; their number, however, remain restricted to a few units (3, 4, or 5). We call such combinations *complex processes*.

Let us give some other examples of complex processes in order to make the concept absolutely clear. The formation of a bajada in a semi-arid climate proceeds according to a mechanism which requires the combination of several elementary processes: rock fragmentation of a highland, overland and stream flow and the acquisition and transportation of debris, and, finally, the spreading of the waters as soon as the slope decreases, producing unstable rills and the deposition of the materials. The formation of meanders in a fluvial channel presents a no less complex association of elementary processes: the undermining of the concave bank, the formation of fissures due to desiccation and frost, slumps, the washing of the slumped material by the current, the reduced stream velocity of the convex bank due to lesser depth, and the accumulation of successive point-bars, which deflecting the current toward the concave bank, allow the migration of the meander. The same is also true of a sea cliff evolving under the joint action of subaerial processes (frost-weathering, variations of temperature and of humidity, etc.), marine processes (effects of pressure in cavities under the impact of waves, dissolution, mechanical impact of pebbles and boulders, etc.), and the process of shore drifting which occurs when the materials have become sufficiently triturated. This reduction in size itself results from the grinding of sand and finer clastics on boulders and, especially, from processes of fragmentation by dissolution, chemical weathering in contact with water, and subaerial processes of fragmentation (crystallisation of salt, frost-weathering, etc.) between two high tides. The accumulation of a terminal moraine is brought about by the combination of the abandonment of the materials contained in the ice under the effect of its fusion and processes of sorting and reworking by meltwaters. The formation of a dune is also an example of a complex process: the sand is repeatedly deposited and moved by the wind, depending on its strength and direction, in such a way that a dune results from a series of complex actions of the elementary aeolian process. These actions interfere more or less with the resistance offered by the vegetation or by the relief itself, if it is bare.

49

A much greater number of factors constitute what Cholley has called a *system of erosion*. Such a system is a whole ensemble of complex processes which are associated together and which determine the evolution of the relief: processes of slope development, which vary according to the nature of the rocks and the degree of slope; processes affecting stream beds, which are determined by their attributes (unconsolidated or consolidated rocks); the stream gradient; the regime of stream flow; the bedload; and the suspended load. We should also include the mesoclimatic influences which are by no means negligible in the case of a high intermontane basin. The very concept of a system of erosion implies a rather vast region in which processes of erosion as well as of deposition are simultaneously taking place at a given stage in the evolution of the relief. The link between the processes is evident: the kinds of sediments accumulated and the kind of relief which they form are determined by the denudational and transportational processes in the highlands. But this obvious link, which is at the origin of the profitable concept of *correlative deposits*, does not appear in the expression *system of erosion*, which places the emphasis on erosion only and, as such, betrays the concept on which it is based. It seems that it was forged with the Davisian theory of the cycle of erosion in mind; systems of erosion and cycle of erosion although different in concept are two expressions with similar construction. For this reason we prefer the more adequate term *morphogenic system* rather than the expression 'system of erosion'.

Elementary processes, complex processes and morphogenic systems only constitute steps or convenient references in the description of a series of interrelated natural phenomena. As is the case with every scientific concept, they should not be given an absolute and formal value. It is nonetheless true that they are more than convenient terms: each represents an aspect of natural reality, fits into a taxonomic classification of geomorphic facts, and corresponds to a definite spatiotemporal dimension.

A *morphogenic system* operates over a vast region that may be as large as an entire morphoclimatic zone. There is a morphogenic system of midlatitude deserts and a morphogenic system of tropical deserts. They differ because the elementary or complex processes linked to frost and snow operate only in the midlatitude deserts. There is a morphogenic system of the maritime regions and another of the continental regions of the forested midlatitude zone; they differ by the degree of intensity of the morphoclimatic processes. Physiographic divisions may also support different morphogenic systems. For example, the morphogenic system of an important mountain range with considerable relief is not the same as that of a much lower plateau with modest relief in the same morphoclimatic zone. One can compare and contrast the morphogenic system of the Paris Basin with that of the central or northern Alps. The morphogenic system is a dimensional unit of the second order of magnitude in the taxonomic classification of external forces and comes immediately after the morphoclimatic system. It extends over several tens or hundreds of thousands of square kilometres [an area ranging from the size, for example, of the Black Hills of South Dakota to that of Amazonia].

The *complex process* affects a much smaller area. It functions in areas which must be rather homogeneous in geologic structure within the framework of a given morphogenic system. For example, in the morphogenic system of the Paris Basin the complex processes are not the same on the Loire, with braided channels between sand deposits, as on the Seine, with a well incised channel describing meanders. They also differ depending on whether the terrain is argillaceous or marly and subject to sliding and solifluction, or calcareous and evolving principally under the effect of dissolution. Lithologic factors strongly influence the complex processes which is not the case for the morphogenic systems on which in general they only have a subordinate influence. The complex process functions on areal units ranging from a few square kilometres to several hundreds of square kilometres. They are usually characterised by certain types of landforms: moraine, hillside, meander, or cliff; a form repeated a greater or lesser number of times in a certain region. The morphogenic system is on the other hand characterised by a combination of forms which may reach a high degree of variety; for example, the troughs, cirques, riegels and morainic systems which are found in glaciated valleys.

The *elementary process* characteristically applies to a still smaller part of the domain of a given morphogenic system. Along sea cliffs, for example, frost-weathering occurs only in areas where there are rocks susceptible to frost. On the topographic form of the cliff the elementary process is reflected in variables affecting secondary characteristics, as, for example, in the profile of the cliff. It produces a kind of *geomorphic facies* which is closely dependent on local, principally lithologic factors. It is associated with a specific facies of the detrital products which correlates with the process of erosion: a muddy shore at the foot of an argillaceous cliff, a gravelly shore at the base of a cliff composed of solid rock, or a sandy shore along a granitic cliff subject to granular disintegration. The elementary process produces elementary forms, such as an escarpment overlooking a scree slope. The area which it affects, depending on the kind of terrain, extends over small and discontinuous surfaces of a specific slope, often forming narrow strips of land, such as a riverbank, the crest of a ridge, or the foot of a slope.

The taxonomic classification of the external forces which we have just outlined includes a much smaller number of units than that which we have proposed for the actions of the internal forces:[1] three instead of seven orders of magnitude. The reason is that the uniformity of the climatic factors is much greater than that of the structural factors on the earth's surface. The nature of the rocks and the geologic structure both vary from place to place much more than the climate. The physiographic divisions determined by structural geomorphology are generally smaller than the morphoclimatic

---

[1] Concerning the taxonomic classification of structural geomorphology refer to Tricart's Treatise, vol. 11, on orogenic belts or his *Principes et Méthodes de la Géomorphologie*, 1965.

zones. For this reason there is a progressively increasing influence of the structural factors on the morphogenic processes as one proceeds from larger to smaller units. At the scale of the Sahara the relief is characterised by a particular morphogenic system due to the predominating influence of the climate. At the scale of the Saharan *gour* (butte) the relief, on the contrary, is characterised by the lithologic factor and by the structure, both of which determine the nature and the intensity of the elementary process within the climatic framework.

The direct influence of climate

When, under the effect of denudation, a rock comes into contact with the atmosphere, it is subject to a series of alterations which are expressed by the term *weathering*. Weathering includes mechanical, physical, and chemical changes which a rock undergoes at the hand of the agents of the weather; it may be one of the agents of pedogenesis, which consists in the formation of a soil at the expense of the rock. It is first of all a physical and chemical process; by definition (by convention), intervention of organisms is excluded from it. The shattering of a rock due to frost is a typical process of weathering. In the soil-forming process, however, not only the atmosphere but also such organisms as plants and animals (unicellular or multicellular), bacteria, fungi, insects, worms, and other burrowing animals almost always intervene. One way in which this intervention is reflected is by an introduction of organic matter which produces biochemical processes that are always much more complex than those of simple weathering.

The classical distinction drawn above is justified. It is helpful in understanding the natural phenomena, for it brings to light an aspect of reality; but, as in every natural law, it only contains part of the whole truth. There are no natural environments from which life is totally excluded; our planet is not aseptic, and organisms, no matter how simple or rare they are, nevertheless exert some action. Water close to freezing contains bacteria: in Greenland ponds formed by summer thaw often take on a rusty colour due to the presence of ferruginous bacteria. The coldest infiltration waters contain unicellular organisms which produce a small quantity of carbon dioxide; this carbon dioxide, no matter how little, is effective in the chemical mobilisation of limestone by subterranean waters. One cannot, and one should not, therefore, adopt a categorical attitude: weathering necessarily includes some biological action, but it is very small in certain cases and does not play a determinant or important role. Chemical analyses only reveal 'traces' of organic matter in the products of weathering, whereas they find a determinable proportion in the soil. There are, then, greater or smaller quantities of organic matter of one kind or another in the series of interrelated natural phenomena; they may therefore be studied as such. It is only necessary, as always, not to forget that none of these phenomena should be artificially extracted from nature.

Weathering thus understood is simpler than pedogenesis because it implies one factor less, the biological. In nature weathering is often the first act of pedogenesis; it is only when a rock outcrop has been sufficiently attacked by weathering and forms a layer of loose debris that the higher plants can occupy the area and participate, in turn, in the evolution of the soil.

Mechanisms under the direct influence of climate

The direct influence of climate not only affects weathering but also, to a lesser degree, the processes that transport and deposit the debris. It affects the various stages of the *morphogenic processes*: erosion, transportation, and deposition.[1] The direct influence of climate on the morphogenic processes is reflected in two different ways: in the nature of the processes (qualitative manner) and in their intensity (quantitative manner).

QUALITATIVE INFLUENCES

Certain morphoclimatic mechanisms are under the direct influence of climate and are characteristic of certain climates.

This is, for example, the case of frost. The critical temperature of 0°C (32°F) is of major geomorphic importance. It sets in motion particular zonal processes and considerably modifies the course of the azonal processes.[2]

Frost and thaw alternations are dependent on the climatic factors of atmospheric temperature and solar radiation. These alternations produce considerable variations (of the order of 10 per cent) in the volume of water contained in the fissures and pores of the rocks, often enough to cause solid rocks to break up (frost-weathering). In nonconsolidated formations such frost–thaw alternations create important variations of volume as soon as

[1] Traditional geomorphological vocabulary unfortunately lacks an expression to designate this sequence of events, which helps to obscure it to the mind. The emphasis is usually on erosion alone, as in the expressions 'cycle of erosion' and 'forces of erosion'. The expression 'littoral erosion' to designate all three processes is also deficient; a better term would be 'littoral morphology'. The same applies to the expression 'fluvial erosion' which hides the fact that streams are primarily agents of transportation. (In English we have the term 'gradation' [Chamberlin and Salisbury, 1904] which includes both degradation and aggradation. Apart from other objections, which may be found in textbooks, this term implies that the relief is evolving from higher to lower relative elevation, which, of course, is not necessarily so. The self-explanatory term 'morphogenic' is therefore to be preferred. 'Morphogenic' is shorter than 'morphogenetic' and has a construction similar to that of pedogenic and tectogenic. (KdeJ.)

[2] Following Troll (1944), processes or mechanisms proper to a given climatic zone are called *zonal*. *Azonal*, on the contrary, are those which take place indifferently in other climatic zones. *Extra-zonal* are those phenomena proper to a given climatic zone but which occur locally, for exceptional reasons, outside of this zone; for example, particular modern polygonal soils around lakes in the forest zone of the Alps. For phenomena which are common to several climatic zones we propose the term *polyzonal*. This is the case, in a strict sense, of stream actions which are operative in the cold, temperate, and tropical zones but which do not exist in glacial and areic environments.

clayey particles exceed 2 per cent of the total. On slopes they produce displacements which play a role in solifluction. Frost considerably softens up the deposits, which lose their cohesion during thaw. The tiny frost-needles which form on the ground (by virtue of the cold wall principle)[1] cause a mixing in the superficial deposits, making them more friable and more apt to removal by the *azonal* agents of wind and running water: the wind easily picks up the silts, which once deposited produce loess; meltwaters are often turbid when they have washed the bare ground. Frost, too, seems to affect colloids, disturbing their internal cohesion during thaw.

Frost also intervenes in the regulation of runoff and underground drainage. Sufficiently prolonged, it withholds stream water and directly influences its regime. In frost climates the flow of water takes on a seasonal aspect, dependent not on rainfall but on temperature: low in the winter, flooded during spring or summer. It may increase the irregularity of stream flow and, through it, the intensity of morphogenic work. *Permafrost* (perennially frozen subsoil) forms when frost is sufficiently intense and prolonged. It is rigorously impermeable when it is continuous; meltwaters only infiltrate the thin thawed out superficial layer (the *active layer*). It causes a high runoff which corresponds to the absence of springs, creating a certain analogy with the bare terrains of semiarid regions. A particular river of gentle gradient in a basin of minor relief may transport a relatively coarse load because the meltwater floods are devastating. The modification is especially pronounced in this case: the polyzonal processes of the fluvial environment are not modified in themselves, but they may be. In the northwest of Greenland Malaurie has described spring floods on top of a still frozen stream bed. There is then so much less friction that the transport mechanism of the particles is different from that of midlatitude streams. That may explain why pebbles of small periglacial vales in the east of the Paris Basin are aligned parallel rather than normal to the current as is the case in ordinary streams (Tricart). In this case frost introduces a qualitative change in the polyzonal processes of stream flow.

Frost also modifies the action of the wind. Snow which covers the ground and is blown by the wind modifies the geomorphic action of the wind. Blown in blizzards, it may be hardened by frost, and its aptitude to act as an abrasive is presently being discussed by specialists. According to Fristrup it is not capable of producing an aeolian polish on rocks, of pitting them, or of producing ventifacts as does sand. Mixed loads of sand (or silt) and snow carried by the wind produce deposits that are very different from those of pure aeolian sands. The material, which is not as well sorted, forms irregular accumulations rather than dunes. Since 1946 the Belgians and Dutch have designated them by the term *niveo-eolian deposits*.

Frost acts on the littoral processes in two ways: first, in a quantitative

---

[1] Ice crystals prise rock fragments apart by tensile fracture, as they grow in the direction of maximum heat loss (i.e. normal to the 'cold wall') even though this may be against some mechanical confining pressure. (Editor's note).

manner by shielding the shore from wave action with pack ice during part of the year; second, by the frost-weathering of the cliffs. Although, as it appears now, sea ice riding the strand is incapable of exerting any notable erosion, Nansen has attributed it with the formation of an abrasion platform along the Norwegian coast, the so-called *Strandflat*. Sea ice as well as river ice is, however, an active agent of transportation; both may even carry debris of a higher calibre than can be transported by running water. Secondly, icebergs, which are detached fragments of glaciers, seed the sea bottom as they drift with large erratics, sands, or clays. Lyell explained in this way the erratic boulders of northern Europe which we now know to be morainic.

Frost, provided it is sufficiently intense, may even completely modify the drainage conditions on the earth's surface as soon as glacial ice starts to form. It is needless to emphasise the geomorphic differences which result from glacier action; they are classic, and all textbooks contrast glacial and fluvial reliefs. Conditions of velocity, viscosity, pressure and turbulence are completely different, resulting in entirely dissimilar morphogenic processes and landforms.

Frost, which is a climatic phenomenon, has an enormous influence on the morphogenic processes. It not only produces specific processes but also modifies the azonal processes. It creates on the earth's surface a zone characterised by the predominance of the solid state of water over the liquid state, whether above the ground (glacial zone) or below the ground in the form of permafrost (periglacial zone). Of all the morphoclimatic agents it is the one which has the most intense qualitative influence.

Temperature variations without frost also have their effect on rocks. Differences of superficial heating between night and day may reach more than 50°C (90°F) in arid regions. The expansion coefficient of rocks is far from negligible: that of granite reaches two-thirds that of iron. There are, consequently, mechanical forces on rock surfaces. Processes of fragmentation such as the granular disintegration of granite and flaking are attributed to them; perhaps they intervene in the creation of fissures and joints. Granite is composed of crystals of a highly different colour which absorb heat unequally; their expansion is therefore also unequal and produces tiny forces at the contact of the various crystals, tending to separate them. In fact, it has been observed that the granular disintegration of granite is more complete where the crystals are larger. Flaking would then result from the difference of heating at the rock surface and the area immediately below it. The low thermal conductivity of ordinary rocks only allows a very weak diurnal penetration of the solar radiation, not exceeding a few centimetres. The difference in expansion between the surface of the heated rocks and the mass unaffected by the diurnal thermal flux produces forces which if repeated often enough would cause the flaking of the rock parallel to the heated surface.

Wet–dry alternations, like freeze–thaw alternations, are also of direct climatic origin. They depend on the rainfall regime and are an important factor in weathering because they occasionally produce powerful mechanical forces in the superficial part of the rocks. Thus clays contract under the effect of desiccation, and their uniform surface is transformed into a net of polygons. Montmorillonite is particularly sensitive to moisture variations; water may cause it to swell as much as 10 or 20 per cent. Intense pumping of the ground water under Mexico City has caused a general subsidence of the ground of as much as 2 m (6·6 ft) in a period of about fifty years. The density and width of desiccation cracks which result from the variations in volume of the sediments affect the circulation of rainwater. The latter, after a period of drought, penetrates the cracks to a depth which may reach 2 or 3 m (6 to 10 ft) and accumulates at this level where it encounters less dry and consequently less permeable rock. This level may occasionally be transformed into a sliding plane and cause earthflows, as in Mediterranean regions when heavy rains succeed a protracted drought. Other clays, such as kaolinite, only have a very small contraction coefficient. Alternations of drought and humidity act in a different way on their surface: raindrops pack it and drought later hardens it into a crust, so that the impermeability of the rock is reinforced, thereby facilitating runoff.

Wet–dry alternations also produce displacements of matter in the superficial part of the rocks. Certain soluble products are dissolved during the wetting phase and later progressively crystallise during the drying phase. Such is the generally admitted origin of desert varnish, a black or brown coating on the surface of certain rocks; as rainwater penetrates the rocks it dissolves the iron and manganese and later, as it rises to the surface, it evaporates and precipitates the metals in the form of oxides (in reality it seems that the process is not quite that simple but necessitates the intervention of bacteria and algae fixing the iron and the manganese). Alternations of wetting and drying also strongly affect biotite, causing it to cleave into flakes, weakening the rock and sometimes beginning its granular disintegration. They also seem to favour chemical processes of hydration. Lastly, they contribute to the formation of calcareous and saline crusts and tropical cuirasses, but these are related to a pedogenic process rather than to a direct effect of climate.

The influence of the variations of humidity and temperature (above freezing) on rocks is, unfortunately, still poorly known because of the want of laboratory experimentation and lack of sufficient observations in the field. The mechanisms which they produce, and which we have just reviewed, therefore remain hypothetical.

QUANTITATIVE INFLUENCES

The elements of the weather also influence the intensity of the azonal processes. These quantitative influences may be strong enough to produce qualitative changes; for this reason one should be careful to avoid an overly formal distinction between the two influences.

56

The instability of the atmosphere at sea affects the intensity of the littoral processes. Frequent cyclonic storms moving over the midlatitude oceans have geomorphic effects on remote shores. The shores of Mauritania and Senegal, for example, are subjected to an important north–south drifting of sand due to powerful northwestern swells originating in the North Atlantic between Newfoundland and Florida. Local swells, much weaker because of the different character of the intertropical atmospheric circulation, only play a minor role in the sculpturing of this littoral. A similar situation exists in the South Atlantic where enormous austral swells set in motion shore drifting in distant Cameroon. The atmospheric instability of the midlatitudes tends to increase the intensity of the littoral process in that zone, where there are various types of swells, ranging in intensity (height, wavelength) as well as in direction. Storms, which subject coastal regions to intense effects and often brutal modifications, alternate with periods of calm during which activity is reduced and is sometimes of a different nature. The large exotic swells of tropical shores, on the other hand, are much more constant and have a more continuous and steady geomorphic effect. We have here a good example of a quantitative modification added to an azonal mechanism by atmospheric agents.

Atmospheric factors also affect the intensity of the fluvial processes. The seasonal distribution of precipitation is one of the factors in the regime of a stream, and its influence is reflected in the intensity of stream flow. The runoff of tropical deserts is closely patterned on the rainfall, for there is no frost and snow withholding effect. Washes only fill when there is enough rain, and the importance of the flood depends on the importance of the storm.

Stream discharge and its effect on the stream bed operate according to physical mechanisms which obey hydrodynamic laws that are valid over the entire earth. The geomorphic work produced, however, varies directly as a function of the regime; abundant flow is nearly always reflected by increased velocity; the increase depends on the type of stream bed and is much more rapid when the water is clearly canalised as, for example, in a rocky gorge or in a channel incised between high banks. Velocity is less, even negligible, when the water spreads and splits into a multitude of uncertain channels. The geomorphic activity of running water is also a function of stream velocity; according to Du Boys approximately of its sixth power. Large floods with rapid flow, therefore, do most of the geomorphic work. Some exceptional floods, as that of the Guil (French Alps), in June 1957, can even permanently change the morphology of the valley bottom and the dynamics of the stream.[1]

[1] See Tricart (1959), 'L'évolution du lit du Guil au cours de la crue de juin 1957'. *Bull. Comité Trav. Hist. Scient., Sect. Géogr.* (Ministère de l'Education Nationale), **72** (1960), 169–403; summary in Tricart (1959), 'Les modalités de la morphogenèse dans le lit du Guil au cours de la crue de la mi-juin 1957', *Int. Ass. Sci. Hydrol.*, Publ. no. 53, Commission d'Erosion Continentale, 1959, pp. 65–73 (KdeJ).

Stream regimes are thus of great geomorphic importance. In humid regions they are only in part a function of climate; plant cover and lithology also play an important role. In dry regions they are more directly dependent on precipitation alone. Large floods sweep the channels and transfer the debris further downstream; small floods abandon the material at the foot of the highlands in which the arroyos originate. The irregularity of stream flow imparts special characteristics to the drainage net of dry regions, introducing qualitative differences with the streams of humid regions. Because of the scarcity of springs, the hydrologic regime is as irregular as the rainfall regime, and the stream flow is intermittent. Washes remain dry during long periods permitting their degradation; the wind, for example, reworks the fine fractions of the alluvium into dunes, which often constitute an obstacle to the next floods and intensify divagations. General drought causes heavy losses to streams by evaporation and infiltration so that they rapidly decrease in size and soon wither away. All this results in different ways of accumulation in comparison to those taking place in lands of perennial stream flow although they are all determined by the same hydrodynamic laws. The quantitative difference due to climate produces qualitative consequences on the geomorphic plane.

Wind action has an identical logic. Aeolian relief is dependent on variations in the regime of the wind (intensity and direction). Dubief (1952) has shown that the most unstable winds are the most favourable to deflation, but that the role of steady strong winds is predominant in the development of depositional landforms. The transportation of sand is indeed proportional to the cube of the wind velocity. The shape of dunes does not reflect the resultant of all the winds but corresponds to the direction of the strongest winds. Capot-Rey (1945) has also demonstrated that the predominant action of storm winds imparts its general characteristics to the relief and that the normal winds only affect details. The concept of a critical limit is therefore clear as regards the wind and also, moreover, of all the other factors: stream floods, marine storms, etc.

The direct influence of climate is also felt on chemical weathering. It normally increases with temperature and humidity: abundance of water favours dissolution; and heat activates all the reactions. The well known law of Van 't Hof indicates that a temperature increase of $10°C$ ($18°F$) multiplies the rate of reversible reactions by about $2\cdot5$. Certain reactions having a geomorphic importance are influenced by temperature in a different, indirect way: in the dissolution of limestone by water containing carbon dioxide the reaction is greatest at low temperatures (carbon dioxide is then more soluble). This explains the importance of limestone dissolution around snow patches producing lapiés and the existence of rather well developed karsts at relatively high latitudes (Lapland, notably). The direct climatic mechanism is not the only cause, however. The CO_2 content of the water also intervenes and is often higher in water seeping from the soil of the wet tropics where the carbon dioxide is provided by the rapid mineralisation of a considerable mass of plant debris. On the whole, there is an acceleration of chemical weathering with temperature on the earth's surface. It explains the thickness of the regolith and the considerable development of such processes as the leaching of silica or of ferrallitisation that occur in the wet tropics. Even processes of karstification are more intense here than in high latitudes, as is shown by the enormous amount of denudation suffered by limestone regions; instead of plateaus pitted with caverns there is a development of plains with scattered pinnacles ('haystacks') pierced by solution cavities. One of the reasons is that in addition to processes dependent on mineral chemistry there are biotic activities which rapidly increase with temperature and humidity.

Before showing their importance, however, it is necessary to go into some detail concerning the climatic data which geomorphology requires.

CLIMATIC DATA REQUIRED BY GEOMORPHOLOGY

Climatic geomorphology needs certain basic climatic data which are indispensable to the study of the intensity of the processes and their distribution over the face of the earth. In order to understand the effect of temperature on rocks we need to know the temperature range which affects them and the frequency with which certain critical limits are passed, such as the freezing point of water. The usual meteorological data concerning temperature are, unfortunately, not obtained in a manner that satisfies our needs. We are provided with monthly averages (average temperature, average maxima and minima), rather than with frequential distributions. Unless one undertakes time consuming analyses of the data of weather stations it is impossible to know, for example, how many days per month the daily range has reached or exceeded 10, 20 or 30°C. Neither is it possible to know the number of freeze–thaw alternations which have occurred within a specific period of time. In too many stations observations are made at predetermined times rather than in a continuous manner by thermographs. In order to prepare morphoclimatic data it is necessary to have recourse to the minutes of the observations and to sort the data personally, which implies lengthy stays at the stations.

But there are even greater problems: some fundamental data are not recorded in current observations. For example, the temperatures generally provided are those of the air, under shelter, at 2 m (6 ft) above the ground. Temperature measurements at the ground are extremely rare. They do not interest climatologists who study the *air* and who work at the scale of the entire world. They are extremely variable, depending on the nature of the rocks and the plant cover. But it is particularly important to know such differences of heating in order to evaluate the influence of thermal variations. Geomorphologists need to know the propagation of thermal fluxes in the ground but very few stations make such recordings. The only recourse left is to use general formulas to calculate the data we lack; to use, for example, actual air temperatures at the ground from temperatures obtained

59

under cover. Such extrapolations are always vitiated by a considerable margin of error in such a way that they can only be attempted with a reasonable chance of success in areas whose characteristics differ little from those of the station where the parameters used in the computation have been determined. Most of the time one is restricted to sporadic and dispersed data, often, moreover, of difficult access. A large number of research workers belonging to different disciplines are interested in such data: zoologists, because soil temperature determines the presence of burrowing animals and numerous insects; botanists, because it influences the germination of seeds; and pedologists, because it strongly influences the formation of soils. One must therefore gather scattered observations, generally of short duration, which have been made by all kinds of investigators and are recorded in extremely diverse publications. We have not as yet reached the stage of methodic description, of systematic mapping of such and such a morphoclimatic fact, but only that of sporadic sampling, disposing over data established in large part by unco-ordinated research, and, on account of it, often difficult of comparison and irregularly representative. Some ecologists have refused to use such data and have themselves undertaken systematic measurements. Such was the case with Cachan, in Ivory Coast, in satisfying the needs of a UNESCO project concerned with the wet tropical zone. A tower was established in the midst of the forest, with the least disturbance to the vegetation. Recording instruments were placed at various levels in the different tiers of the forest, recording temperatures, rainfall, wind, humidity, insolation, and light intensity. Such ecologic climatology is exactly what is needed by the geomorphologist (but remains cruelly absent as data of this type are exceptional). We must try to establish contact with ecologists and together attempt to multiply such observations in the most diverse environments.

In the case of frost there is a relatively satisfactory documentation on the number of freeze–thaw alternations which occur in the various parts of the world. The laws of frost penetration into the ground are also fairly well known. However, very little is known about the quantities of solar radiation received by the surface of glaciers, or about the superficial temperature of the rocks which fringe them or litter their surface. And little is known about the temperature variations of rocks in the tropics.

Data on rainfall are, in general, also deficient. The number of days with snowfall and the proportion of precipitation in solid form are more or less known; but we ignore almost everything about the intensity of rain showers and about the duration of rainless periods between showers. Weather stations everywhere record only the total precipitation falling within twenty-four hours. But the character of the runoff is determined both by the intensity of the showers and by their duration. A very short but violent shower may cause more havoc to cultivated fields than a fine rain of long duration. For this reason hydrologists are more and more interested in the characteristics of showers. It is indispensable to know them, for example, in order to apply the method of the unitary hydrogram. But such pluviometric observations are only obtained in particular cases: in experimental basins and by agricultural experiment stations, where erosion is measured on experimental plots. In any case it pays geomorphologists to take them into account and, also in this domain, to try to associate themselves with other specialists in order to establish a common documentation.

The mode of occurrence of showers is also very important. A shower which falls after a period of drought causes much more damage, with equal intensity, than a shower which is interspersed within a rainy season. It is also common knowledge that a shower accompanied by thaw causes more serious erosion than when thaw is absent, because the soil structure is destroyed and the still frozen subsoil opposes infiltration and favours runoff. The erosion potential of runoff depends on all these factors, which can be rendered fairly accurately by formulas. The Soil Conservation Service of the United States Department of Agriculture, for example, has proposed the following formula, valid in the area of the upper Mississippi:

$2 \cdot 5 \, A + 8 \cdot 5 \, (B - C)$, which gives the intensity of erosion, in which:
A represents the total precipitation in inches;
B represents the average intensity during 30 minutes in inches per hour;
C represents the average intensity during 5 minutes in inches per hour.

The value of such formulas is evident: they permit geomorphology to pass progressively from the qualitative to the quantitative stage. Unfortunately their formulation and use are exceptional: adequate climatic data are lacking. It is important, not only for geomorphology but also for a number of other sciences studying nature, that the observations of weather stations be oriented in this direction and that responsible stations equipped with recording rain gauges and thermographs be multiplied. It is also desirable that a sorting of existing data, based on the concept of frequency and not on that of averages, which is insufficient, be carried out every time it is possible. The evolution of meteorology should lead towards a more flexible utilisation of the observations rather than to the establishment of averages; curves of temperature frequency, temperature ranges, and showers would be complementary elements to the 'type of weather' concept. Important progress should therefore be realised in this domain by efforts with which it would pay geomorphology to be associated.

Morphoclimatic regions subject to the direct influence of climate

The concept of the direct influence of climate on relief only examines one aspect of the development of landforms. Before examining the other aspect (that of the indirect influence of climate) we must inquire in what regions of the earth the direct influence of climate makes itself felt, and to what degree it affects the morphogenic processes of those regions.

The deserts and the glacial regions are the two major morphoclimatic realms on the earth's surface in which the plant kingdom is most reduced or completely absent.

In very cold climates the surface of glaciers is completely devoid of vegetation (except for extremely rare snow plankton). The case of forests growing on moraines which rest on top of glacier ice is exceptional; it is characteristic of decrepit glaciers and extremities of ice tongues which have advanced very far beyond the snowline. This vegetation, furthermore, has a very restricted geomorphic role: its only influence, together with the moraine which supports it, is in retarding the melting of the subjacent dead ice; it does not noticeably modify its flow nor its geomorphic action. In the vicinity of glaciers and in some very cold regions without glaciers, the periglacial zone also includes vast expanses that are either bare or occupied by rare isolated plants (the so-called barren grounds). The living matter is so reduced there that its influence is insignificant. The relief is subject to the direct impact of the climate. Frost-weathering, solifluction, and the wind operate without hindrance. But as soon as tundra covers the soil a difference appears: organic matter accumulates in bogs, soil forms and plants exert an antagonistic influence on the mechanism of solifluction; this interference produces specific but minor relief forms such as solifluction lobes, tundra hummocks and the wrinkles of string bogs. The vegetation also favours the accumulation of aeolian deposits.

A similar paucity of vegetation exists in the landscapes of arid and semi-arid climates where bare rock outcrops over vast surfaces, and the vegetation is restricted to alluvial deposits in which a little infiltrated water provides a meagre reserve which the plants utilise to the maximum. But even here the plant cover is deficient: it is composed of bushes and isolated tufts of grass with extensive bare spaces between them. Small plants which form a mat after a shower are too ephemeral to pose an effective obstacle to the weather; it exerts its action almost without hindrance. The principal obstacle is the tufts of vegetation which stop the wind driven sand, producing small hillocks, *rebdou* or *nebkas*. The rainwater flows exactly according to rock permeability, steepness of slopes and intensity of showers. Variations of air temperature and insolation are reflected on rock surfaces whose temperature rises according to their exposure and colour. The mechanisms are ideally simple, and some investigators have attempted a mathematical analysis of the aeolian processes with as much chance of success as that of hydrologic problems. Moreover, the disintegration of rocks through variations of temperature or frost-weathering can be studied experimentally, if the conditions of the laboratory permit the duplication of the essential natural factors.

In other climates analogous conditions are exceptional; they are only realised where the bedrock is exposed. They most often occur on the precipitous slopes of mountains: the limestone escarpments of the Vercors (Dauphiné Alps) are sculptured before our very eyes by processes of frost-weathering and chemical dissolution. The wind, too, freely exerts its action on beaches or on certain alluvial bars too often swept by floods to be occupied by plants. In dry climates, such as those of the Magreb or Senegal, the concentration of salts by evaporation may be sufficient to eliminate the vegetation and at the same time permit the operation of complex geomorphic processes directly determined by the agents of the weather. Such is the mechanism of *sebkha* (salt flat) formation. Agricultural development, which seasonally destroys the vegetation and leaves the soil entirely bare during more or less long intervals, has the same effect but fortunately to a lesser degree. Cultivated plants indeed modify the regolith during their period of growth and maintain, as well as they can, a soil whose continued existence is facilitated by cultivation, especially ploughing and the use of fertilisers. The weather elements, it is true, act freely when the soil is bare, not on the rocks themselves, however, but on a horizon of pedologic alteration. The geomorphic conditions are therefore transitional because in one instance the weather elements act directly on the rocky surface, as in deserts; in the other they attack a soil where the presence of organic matter and a structure modify their action. Soil erosion, therefore, is marked by special characteristics: the atmospheric conditions of its development are of the same nature as those of desert regions, but their intensity is not the same because it rains more. Furthermore, it does not act on rock but on soil because the mineral environment is not the same. We may add that the techniques of cultivation, which are related to the economic system, have an influence on the physical and chemical properties of the soil and that soil erosion therefore ought to be studied separately in a course of geomorphology. Direct climatic influences only affect soil erosion partially, even when they explain landforms which converge with desert relief, such as dunes, gullies, and badlands. Moreover such landforms are particularly common when erosion is far enough advanced to have almost completely removed the regolith so that the bedrock is attacked directly.

It is worth while to compare this knowledge of the direct influence of climate on relief with the geomorphological concepts of the Davisian school. Davisian geomorphology, which is especially oriented towards theoretical speculation often from erroneous foundations, is not interested in soil erosion. The fact does not fail to be paradoxical when given the tragic importance this problem acquired in the United States. A lack of collaboration between soil engineers who have undertaken the fight against erosion and theoretical geomorphologists has been noteworthy. The considerable mass of information which has been assembled by soil scientists has hardly been put to use by academic geography. Inversely, the latter has hardly contributed to applied research. And yet only a close association between practice and theoretical research will prevent a shortsighted empiricism, which may be costly.

There are further problems. In so far as Davisian geomorphologists have noted man-induced accelerated erosion, they have misinterpreted it. They have taken it as proof of the action of runoff in the domain of so-called normal erosion. The gullying of the southern Alps, generated by man, appears in de Martonne's *Traité de géographie physique* as an example of the stage of youth which will progressively pass to gentler, more mature forms covered by vegetation. Typical in this regard is his Plate XVII A, which in the foreground shows a wild gullying in the clays of the area of Lake Bolsena (Italy) under the title 'youthful forms in clayey terrain', whereas bush covered hillsides of identical profile but completely devoid of gullies are seen in the background. This is a good example of how a too strong adherence to theory can hamper observation.

In the two morphoclimatic realms where climate has a direct impact on relief (deserts and glacial regions), the morphogenic system has a high mechanical predominance. The solid state of water in the one case and its scarcity in the other hinder the chemical processes. The near absence of vegetation reduces the biochemical actions to insignificance. Such special conditions reduce the number of morphogenic factors and make these morphoclimatic environments particularly simple; they are characterised by a restricted number of dialectic oppositions. Outside the domain of tectonic deformations the evolution of relief is explained by the sole opposition of lithologic and atmospheric factors.

We can now understand why these two morphoclimatic environments are the only ones which have been taken into consideration as such by Davisian geomorphologists. They constitute Cotton's 'climatic accidents', and to them are restricted the 'exceptions' to 'normal erosion', with which we are resolved to agree. The processes linked to frost, which characterise the periglacial realm, have not been ignored but reduced to a minor role. Solifluction lobes and patterned ground have not been replaced in a new morphogenic system but considered as isolated facts; they have therefore been reduced to an episodic role, and their real geomorphic significance is not clear. Because of insufficient analysis, the periglacial realm thus has been included within the realm of 'normal erosion', and the microforms produced by intense cold considered as mere curiosities rather than as elementary or complex processes of a specific morphogenic system.

The concessions made by Davisian geomorphologists to climatic geomorphology result from an oversimplified view of natural phenomena. Davis's followers have practically ignored the biosphere, the existence of soils produced with the help of living organisms, or the plant screen interposed between the weather and the ground. It is this position which has led them to develop the monster of normal erosion in which certain geomorphic mechanisms observed in semiarid regions or in areas subject to intensive man-induced erosion are put end to end with other phenomena of more humid lands and a dense plant cover—the man-induced gullies of badlands (considered to be in the stage of youth) with the regularly meandering streams fed by springs (stage of maturity). The origin of such confusion is to be sought in Davis's way of thinking: the flight of the excessive imagination, which is incompatible with sane scientific research. Nature is too complex to be imagined: our incomplete knowledge as much as our limited mind leads to successive simplifications which end in completely separating abstract reasoning from reality.

Even in hydrodynamics, a field actively pursued and supported by considerable monetary means, mathematical reasoning only produces approximate formulas of restricted validity, usable only in certain cases and always for short durations. It is wrong to extrapolate them to geologic time. Moreover, they deal with only one geomorphic mechanism among others: the evolution of a specific element of the relief. For the same reason it is an even greater wrong to extrapolate them to the evolution of the whole relief.

The indirect influence of climate

The direct influence of climate on relief is only felt in extreme deserts. Everywhere else there is interposed between the atmosphere and the lithosphere an intermediate layer that is inhabited by living matter, mainly plants, and is modified and transformed by them into soils, often very different from the parent rock. Outside the biological deserts the influence of climate on relief is mainly indirect. Plants modify the weather agents close to the ground, breaking the fall of raindrops and the velocity of the wind and reducing temperature differences. But the plants, in turn, are under the influence of climate.

The plant formation-types are distributed over the face of the earth according to a zonal principle: (in descending order in Eurafrica from the poles) tundras of the high latitudes, coniferous and deciduous forests of the midlatitudes, woodlands of the Mediterranean regions, steppes on the fringe of the deserts, tropical steppes and savannas, and lastly the tropical rainforests. The screen interposed by a dense forest, almost impenetrable to the rays of the sun, differs considerably from a sparse steppe too meagre to cover the soil adequately. Under natural conditions the climate determines the type of vegetation which, in turn, filters the agents of the weather that strike it. But the indirect influences of climate are not limited to the plant cover. The vegetation is indeed implanted into the soil from which it obtains its sustenance and to which it later returns its organic matter. There is an uninterrupted exchange: the plants draw solutions from the soil and restore to it organic matter in part elaborated from the atmosphere.

To these biochemical actions must be added the purely mechanical ones: the penetration of roots and the work of burrowing animals; these modify the superficial layer of the lithosphere, transforming it into soils which, rather than the parent rocks, bear the brunt of erosion. The vegetation not only modifies the agents of the weather but also the lithologic environment, which is the scene of the morphogenic processes. The world distribution of soils, which is, like vegetation, related to climate, is largely zonal, given analogous parent materials. But there is interaction between soils and vegetation: vegetation influences the soil-forming processes and vice versa. The plant cover changes more rapidly than the soil: residual plant associations, linked to palaeoclimates different from the present climate, occupy a more restricted area on the earth's surface than palaeosols. The Tertiary tropical climates of France have left behind thick granitic sands and flinty clays, but no savannas, only a few relict plant species (ivy, lianas, etc.). Whereas the influence of climate as recorded in the vegetation is felt without appreciable lag on the geologic time scale, the same is not true of the relationship between climate and soils: survivals, in the form of palaeosols, have a wide distribution.

The main features of the indirect influence of climate on geomorphology having been stated, it is now possible to deepen our analysis without the risk of distorting the interaction of the factors by isolating them. We will thus successively study the relationships between the morphogenic processes and vegetation, and between the morphogenic and the pedogenic processes.

Interactions between vegetation and the morphogenic processes

There is interaction between the plant cover and the morphogenic processes. In a given climatic environment the plant cover modifies the morphogenic processes, but, in turn, the latter influence the ecologic conditions (the habitat) and thus have repercussions on the vegetation. On the unstable reliefs of badlands only herbaceous plants can gain a foothold, and most of the vegetation becomes concentrated in the more stable gully bottoms. In some periglacial environments with intense solifluction, as in the high Andes (Puna de Atacama), plants take refuge on sands and gravels because they cannot gain a foothold on clays and silts which are subject to constant movements. Certain plants are adapted to a definite environment and to a specific intensity of the morphogenic processes. An example is the cork oak, which prospers in areas of light sheet erosion. If the erosion is accelerated and the roots are bared, the trees die and are replaced by another association, usually that of evergreen oak. Many plants are in this way indicators of certain dynamics. Studies carried out in teamwork by plant geographers, ecologists, and geomorphologists should be fruitful in this regard.

The question of the interaction between the plant cover and the morphogenic processes naturally leads us to insist on only one aspect of the interaction: the geomorphic one. We will thus successively examine the influence of vegetation on the agents of the weather and on the agents of transportation, and then study the relationships between the evolution of vegetation and the evolution of landforms.

THE EFFECT OF VEGETATION ON THE AGENTS OF THE WEATHER

The screening effect of the plant cover modifies the intensity of the agents of the weather and the processes which they set in motion.

Let us take the example of rain. The falling drops reach a maximum velocity that is all the higher as they are large: their kinetic energy increases as the square of the velocity. They literally pelt the bare soil. They scatter sand grains and aggregates a distance and a height of several decimetres. It is only necessary to observe the base of a wall, a board, or a stick stuck into the bare soil after a shower to see an example: their surface is completely covered with small earth fragments. Normally the latter again fall to the ground and repeatedly participate in the pelting; the largest in turn contribute to the disintegration of the soil surface and to the liberation of new particles. Even if the intensity of the shower remains constant its devastation increases with time, at least as long as there is material available. The freed particles are ready for transportation by overland flow or by the wind, which may later blow on the drained soil. The scattering of the particles on a slope automatically brings about a slow measurable downward migration, the trajectory of the downslope directed particles being larger than that of the ones going upslope, granted all other factors remain the same. This important mechanism is called *splash erosion*.

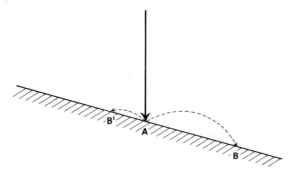

FIG. 2.1. Sketch illustrating splash erosion
On a slope the earth particles scattered by the effect of the impact of a raindrop travel much farther downslope than upslope. The result is a generalised displacement of material downslope.

Let us now examine what happens if, on the contrary, the soil is covered by vegetation. Two extreme cases must be considered: the soil is bare between the trees, or it is covered by grass or an uninterrupted layer of litter. In the first case only a small part of the rainwater reaches the ground directly: a large number of drops fall on leaves and branches and are scattered. When they finally reach the ground, often through fragmentation, each droplet has a smaller mass, a lesser velocity and, as a result, a much reduced kinetic energy. The soil particles which rebound are fewer and smaller with less intense pelting. Often particles which could have been displaced are agglomerated into aggregates too heavy to be ejected and too solid to be dissociated by slower falling drops. Furthermore, an important fraction of the rainwater is retained by leaves and branches and remains in the asperities of the bark, in moss, in lichens and in leaf veins, never reaching

FIG. 2.2. Dense forest forming an effective screen. Montane forest with bamboo. Bocaina, State of Rio de Janeiro, Brazil

Montane forest often shrouded in fog and loaded with epiphytes. The vegetation is sufficiently dense to intercept the rain. Raindrops never fall directly on the ground, which decreases splash erosion. Litter is abundant and impedes overland flow.

FIG. 2.3. Dense xerophytic brush, upper Jacuipe Valley, Bahia, Brazil

The plant cover is dense enough to serve as an effective screen although foliage is poorly developed. But the supply of plant debris on the ground is small, which impedes soil development.

68

the ground because of dispersion by evaporation after the end of the shower. The pluviosity at the ground (i.e. the quantity of rainwater that reaches the surface of the actual soil) is reduced; it is inferior to the precipitation recorded by weather stations. The deficit may be considerable. There are, unfortunately, only a very few measurements which even permit the existence of the phenomenon to be proven. At the experimental station of Clarinda, Iowa, from 10 August to 7 October 1935, the average precipitation recorded at the ground in a corn field was 30 per cent less than that recorded in an equal number of rain gauges located in the free air.

FIG. 2.4. Open subhumid forest offering a fairly good protection. Darling Ranges, Western Australia

The essential protective role is provided by the grasses of the underwood, in contrast to Fig. 2.2. Rainfall is about 600 mm (24 in).

Comparative measurements have been made under forest and in the free air in Berlin and in Italy. They reveal the seasonal variations of the interception. They are a function of the state of the foliage (less interception in winter under deciduous trees) as well as of the intensity of the showers. Heavy showers breach the screen more easily, and fine showers may be totally intercepted. For example, under beech trees in the winter interception was 19 per cent as against 40 per cent in the summer. From one type of climate to another interception varies as a function of the regime of precipitation. But within the same region it also varies from one type of plant cover to

FIG. 2.5. Degraded subhumid forest near Manaring, Western Australia

Rainfall is about 800 mm (32 in), but protection is less effective than in Fig. 2.4 because grazing and fire have reduced the grass cover. There are patches of bare soil which are subject to splash erosion and rillwash.

another. For example, in California it is 38 per cent in one type of chaparral and only 19 per cent in another. An important part of the intercepted water flows down the branches and trunks, especially when they are smooth, which is often the case in subhumid regions. In California, in the chaparrals mentioned above, 30 per cent and 8 per cent of the total rainfall reach the ground in this manner, and all this water generally infiltrates into the soil. Under higher plant covers, however, such as the rainforests, it often happens that aerial streamlets develop during heavy showers, concentrating part of the intercepted water at certain points on the ground, which are intensely pelted and sometimes subjected to splash erosion. The locations of such aerial streamlets does, of course, frequently vary.

At the present we have at our disposal enough data only for the temperate zone, in particular for the forests of the United States (studied as early as 1919 by Horton) and Germany. Furthermore, the problem is extremely complex; three elements must be carefully distinguished:

1. The total fraction intercepted, which not only depends on the plant cover but even more on the seasonal variations in foliage and shower intensity.
2. The proportion of intercepted water that falls on the ground in the form of aerial streamlets and which causes a certain amount of splash erosion if conditions favour it. This seems to occur especially under very large trees, less under small trees and shrubs.
3. The proportion of plant cover at the various forest tiers and the thickness of the litter. Part of the rain may fall directly on the ground in an open forest, as in the eucalyptus afforestations of Western Australia, or in savannas. Streamlets dropping from large trees strike the litter only in certain dense forests, such as those of western Europe.

Due to this complexity and insufficient data, it is therefore hazardous to generalise; but direct observation, for want of measurements, makes it possible to state the following qualitative characteristics:

(*a*) Interception is high in the rainforests, and there are no seasonal variations because the trees shed their leaves individually. Aerial streamlets play a certain role owing to the very poor development of an herbaceous ground cover and the rapid decomposition of the litter. Overland flow may be relatively important, as Rougerie (1960) has observed.
(*b*) In the deciduous seasonal tropical forests, showers encounter a poor screen at the end of the dry season; the effect of the storms is thus heightened, particularly as the rains are often violent. It is even greater in savannas with a lesser plant density.
(*c*) In dry regions the role of the plant screen diminishes with the spacing of the plants, and the lack of litter increases the possibilities of splash erosion.
(*d*) In the temperate zone the importance of the herbaceous ground cover plays a decisive role from the dense steppes onward, and overland flow is normally less than under certain rainforests. The effect is more one of protection than of interception.
(*e*) Open shrub formations and especially succulent plants offer the least interception and the weakest protection. This is the case, for example, of the xerophytic brush formations, even when they are quite dense, of the *sertão* of northeastern Brazil or the semiarid regions of Andean America and Mexico.

71

The second extreme case which we must consider is that of a well protected ground surface. A dense herbaceous cover is, of course, a very effective screen. Grass leaves play a protective role and probably cause a non-negligible loss of rainwater; some raindrops slide down gutter shaped leaves and are completely decelerated as they reach the ground. Such a cover is practically continuous, and few raindrops fall directly on the ground. The rebounding of fine soil particles is reduced to those places where, by exception, the ground is bare, as in a clearing caused by a burrowed hole or an uprooted tree. Very simple observations, which could easily be transformed into measurements, allow anyone to see the mechanism in action.

Some investigators have gone even further and tried to analyse the influence of vegetation on rainfall within the framework of mesoclimates. For a given region they have tried to determine the influence of deforestation and of afforestation on the annual rainfall. The results have been divergent and are difficult to interpret. Even on the same site the average annual rainfall varies from one decade to another, and it is very difficult to account for the part played by changes in vegetation in the differences observed. More demonstrative, it seems to us, is the method employed by Doignon in the Fontainebleau Forest. The method consists in comparing the climate of the forest with that of neighbouring stations whose conditions are as nearly analogous as possible. In this example the average precipitation, at 2 m (6.6 ft) above the ground, would be 17 per cent higher in the forest (measurement made in a small clearing), whereas thunderstorms would be fewer. Doignon's observations are confirmed by similar results obtained in other parts of the world. Thus in India afforestation of 16 000 sq. km (6 200 square miles) has resulted in the following differences in rainfall:

| | AVERAGE ANNUAL TOTALS | |
PERIOD	In the afforested region	In the general area
1869–75, before afforestation	1 215 mm (47·8 in)	1 072 mm (42·2 in)
1875–83, after afforestation	1 369 mm (53·9 in)	1 074 mm (42·3 in)

The increase would only have been 12·6 per cent here, which is less than in Fontainebleau, but this can be explained by the existence of a dry season during which the influence of the forest is naturally most reduced. It has often been observed that cloudiness is increased by forests in the temperate zone. A dense plant cover seems to increase the rainfall (at 2 m above the ground) and the regularity of the showers but to decrease their intensity; furthermore, it considerably diminishes their direct mechanical action on the ground.

The effect of the forest is also reflected on the snow cover. It is more evenly

spread because snowdrifts do not form so readily. Part of the snow is intercepted by the branches and foliage (on conifers) where it melts more rapidly,
thereby decreasing the amount of snow received at the ground. In dense
forests with a lot of shade it may melt later. Whatever the case may be the
meltwater runoff is less violent. In the sunny climate of California's Sierra
Nevada, Anderson and Gleason (1960) have found that the annual snowfall on open plots is 178 to 305 mm of water (7 to 12 in) more than on neighbouring plots under forest. On the average an additional 203 mm (8 in)
of water are drained from open plots. At Wagon Wheel Gap, Colorado,
deforestation has reduced the lag of the meltwater flood by three days and
increased its peak flow by 50%, according to F.A.O. (1962); melting is
faster, and the time necessary to concentrate runoff is reduced.

Vegetation also has a great influence on variations of soil temperature.
It diminishes the amount of solar radiation received and the movement of
the air, which are the two main factors affecting soil temperature. Foliage
and branches brake the air flow to the point that there is practically no
wind in the forest. Furthermore, thanks to chlorophyll, plants absorb a
considerable part of the solar radiation (approximately one third) and use
it in chemical syntheses and in transpiration; this further reduces the
temperature at the ground, and it is cool in the shade of the trees. Night
temperatures, on the contrary, are higher than those outside of the forest.
On the whole, temperature oscillations are reduced. Soil temperature
variations are minor (1° to 2°C) in the tropical rainforest, whereas they are
considerable in the arid tropics (up to 40° or 50°: 72° or 90°F). The temperature range on exposed bare rock, however, is considerable in the wet
tropics: the sun may heat the rock surface to 50° or 55°C (122 to 131°F).
The same influences are again found in a completely different domain,
that of the ground penetration of frost, in which forests register a lag in
comparison to neighbouring areas. The agents of the weather are slowed
down in the forest: frost is retarded owing, among other reasons, to the
absence of wind on the ground; it penetrates less deeply into the ground,
and thaw sets in later principally because of the shade of the trees. Snow,
therefore, generally persists longer in the forest during spring. In midlatitude climates the insulating role of the litter or the grass mat also
diminishes the depth of frost penetration. Thus, in Alsace during the
winter of 1953–54 the maximum depth reached by a frost of − 20° or
− 22° (− 4° or − 8°F) was 20% less under a grass mat than on the bare
soil. In very cold climates the mechanism operates in reverse order: it is
the propagation of thaw that is retarded, so that the upper surface of the
perennially frozen subsoil under a cover of shrubs or muskeg is often less
deep than it is under bare ground.

Just as the plant cover reduces the variations of soil temperature, it also
reduces variations in soil moisture. As it receives less heat, the soil takes
more time to dry. While it is already hardened and cracked in clearings,
it still preserves an appreciable amount of moisture in the forest.

73

At the experimental station of Trangbom, South Vietnam, comparative evaporation measurements have been made at intervals of 170m (560ft) in both forest and clearings. During the period of 1933 to 1937 the average annual evaporation was found to be about 2·5 times higher in the clearings than in the forest:

	In forest	In clearing
Diurnal evaporation	183·7mm (7·2in)	540·6mm (21·3in)
Nocturnal evaporation	113·8mm (4·5in)	161·6mm (6·4in)
Total	297·5mm (11·7in)	702·2mm (27·7in)

The mechanical stresses produced by desiccation are therefore less important in a soil under a dense plant cover. The accumulation of dissolved salts as a result of evaporation is also reduced, and later we will see what an important role it plays in the formation of concretions and cuirasses. But there are great differences in the amounts of water evaporated from different plant covers. FAO studies (1962) have indicated that certain Swiss forests have summer relative humidities that are considerably higher than those recorded in the free air: 9·35 per cent higher in a beech forest, and 7·85 per cent in a larch forest, but only 3·87 per cent in a pine forest. The hygrometric effects of the rainforests and the savannas should differ even more, and in the latter according to the seasons. Unfortunately we lack data with which a complete picture could be obtained.

Forests have a tendency to decrease the temperature of the air. In the Fontainebleau Forest the average annual temperature is 1·5°C (2·7°F) lower than that of the surrounding cultivated area, and the number of days with frost is 40 per cent higher.

Although precise studies are still wanting, it is nevertheless possible to conclude that on the whole the plant cover reduces the intensity of the agents of the weather. It therefore plays a role antagonistic to the mechanical factors in the morphogenic processes.

THE EFFECT OF VEGETATION ON THE AGENTS OF TRANSPORTATION

Relationships of the same order as those existing between vegetation and the agents of the weather exist between vegetation and the agents of transportation, which are running water and the wind.

Vegetation behaves in a complex manner in relation to running water. We have already seen that it causes the direct dissipation of part of the rainwater into the air even before it reaches the ground, thus diminishing the amount of water capable of causing overland flow during a shower. By preventing the drops from packing the soil and transforming its surface into an impermeable crust under the effect of evaporation, the plant cover facilitates the infiltration of the water into the soil. The numerous drops sliding down the grass leaves disappear into the tiny bare space of ground in which they are implanted. The activities of burrowing animals are added to the effects of the plants: worm earth-twirls and burrows are softened materials which favour infiltration. The plant cover not only decreases the morphologically active fraction of the atmospheric precipitation but also favours the infiltration of that part of the water which reaches the ground. Moreover, when the vegetative cover is sufficiently dense it modifies the rate at which the rain reaches the ground. On a bare or partly bare surface (steppe) the rain reaches the ground without breaking through any obstacles; the force of the shower is the same on the ground as in the free air. When the rainfall reaches a certain intensity it exceeds the rate of ground absorption, and overland flow occurs while the subsoil remains dry. It is common to find dry soil at a depth of 5 to 10 cm (2 to 4 in) in an arid region after a shower which produced an appreciable amount of runoff. As a result of the intensity of the shower and the condition of the ground surface, the water absorption and infiltration capacities of the soil have not been used by the subsoil. This mechanism seldom occurs when the plant cover is dense, especially under forest. Indeed in this case the plant screen modifies the density of the shower. At the ground its intensity is much lower than in the free air. The effect of the precipitation is reduced: the immediate evaporation of the water collected on branches and leaves is added to the lag imposed on the fall of the drops by the successive obstacles of leaves, twigs, and branches. It is a common experience that a tree may provide complete protection for several minutes from the raindrops of a cloudburst. Unfortunately, geomorphologists too fascinated by the tune of the Davisian siren have often forgotten these commonsense facts and neglected to specify them by measurements. It would be of the highest interest to make comparative studies of the amount and density of rain showers under different plant covers and in the open air in the various morphoclimatic zones. Its applications in the domain of soil erosion, hydrology and plant ecology would be important.

Methods for obtaining such measurements have already been perfected. Barat (1957) has elaborated on one of them: it consists in exposing a special roll of paper (treated with methylene blue) which dyes the raindrops upon impact. A system of coils permits the automatic winding and unwinding of the paper and, therefore, the registration of the shower intensity. The size of the impact is proportional to the diameter of the raindrop, which is spherical in the atmosphere. Unfortunately, only its author has applied the method, and the results are limited to the Montagne d'Ambre in the north of Madagascar. Comparative data would be most useful for different climates and types of vegetation.

In general vegetation is antagonistic to runoff. French civil engineers of the administration of 'Eaux et Forêts' had already clearly noted this more than a century ago, long before this fundamental concept was obscured in the minds of geomorphologists by the Davisian theories which unceasingly speak of the sculpturing of hillsides by runoff in regions of 'normal erosion' (i.e. in regions the majority of which were forested in their natural state).

Cailleux (1954) has shown how insignificant runoff is under forest in a temperate oceanic environment. Most of it occurs during thaw and the melting of snow, whereas summer thunderstorms do not have any influence. An objection has been made that the area investigated was sandy, therefore permeable, and that the example was therefore not very demonstrative. In fact,

FIG. 2.6. Open scrub permitting rillwash, Muntadgin, Western Australia. Rainfall 275 mm (11 in)

Pediments sculptured by rillwash. The rills form easily, but the vegetation forces them to divide, which prevents gullying. Such conditions are optimal in the formation of pediments.

runoff can very well gully sands that are much more permeable than those of the Dourdan Forest southwest of Paris. Tricart has observed truly abrupt gullies, sometimes one metre deep, in fixed Pleistocene dunes on slopes steeper than 12°; but this was in southwest Mauritania between Nouakchott and the Senegal delta, a poor steppe region with an annual rainfall of only 200 mm (8 in). Comparison with the Dourdan Forest indicates the role of the plant cover in the development of runoff, a role as important as that of the nature of the underlying rocks. Moreover Cailleux's observations are confirmed by Magomedov's measurements in western Ukraine (1950). In small basins of some 30 sq. km (12 square miles) 99·6 per cent of the water was absorbed by the grassy loams during a summer rainfall of 80 to 100 mm (2·4 to 4 in). The stream runoff was only 10 per cent on a podzolic argillaceous soil in the Priirpensk basin and 17 per cent on powdery argillaceous grey-brown forest soil in the Pridesniansk basin, both areas in which crops occupy a large acreage. These figures were recorded during a shower which dropped 53·5 mm (2·1 in) of water at the rate of 2·2 mm per minute on 27 July 1938. Here, too, most of the runoff occurs during thaw. However, to be demonstrative such measurements should only be made in very small basins of 10 to 30 sq. km (4 to 12 square miles), otherwise springwaters, whose contribution is difficult to evaluate, interfere too much.

Fig. 2.7. Open woodland, rillwash important, Green-hills area, Western Australia

The grass cover is insufficient. There are patches of bare soil subject to splash erosion. During showers discontinuous rillwash moves fine particles which form small silt flats.

Often, where the plant cover and the soil are sufficiently developed, an important fraction of the rainwater infiltrates into a rather permeable soil and finds itself blocked at its base at the contact of the bedrock or a less permeable weathered horizon. In hilly country the water in such a temporary waterlogged layer, well known to pedologists, migrates toward the base of the hills following the irregular plane of discontinuity which stops it. This phenomenon is known as *lateral drainage* in pedology and as *interflow, sub-surface (storm) flow*, or *underflow* in hydrology: it sustains the stream flow after the peak of the flood. This kind of drainage plays an important role from the geomorphological point of view: after a heavy rainfall the water content at the base of the soil or subsoil may exceed considerably the saturation capacity. Mud pockets then burst and form small slumps, as in Queyras (valley of the Guil River, French Alps) during the flood of June 1957.[1]

FIG. 2.8. Bushy woodland, very poorly protected soil, Greenhills area, Western Australia

In contrast to Figs 2.2 and 2.7 the substratum here is not formed by granites or gneisses but by a ferruginous cuirass which covers them. To the influence of drought is added that of the unfavourable nature of the soil (edaphic factor), which accounts for the poor plant cover. Absence of a grass tier under the *Casuarina* permits the unprotected soil to be subject to the effects of temperature and humidity variations and splash erosion. The cuirass disintegrates into rubble.

[1] See note, p. 57 above.

But the influence of the plant cover does not only affect the nature of run-off, it also modifies its geomorphic effect. In this matter the grass mat plays a more important role than trees. Under a dense forest, such as the primary rainforests or a secondary tropical forest at least a hundred years old, the light which reaches the ground is too weak to allow the growth of a grass cover; instead, the soil is almost completely bare because the forest litter is rapidly decomposed and the rain produces a rillwash capable of transporting loamy materials. This overland flow, which only represents a small proportion of the precipitation, is quite turbid when it reaches the stream courses. On a grass mat, soil protection is otherwise effective. The most violent downpours can wash the densest prairies without affecting the mineral soil, thus leaving the water perfectly clear as it reaches the base of the slopes. The effect of the sward is double in this case: it assures a mechanical protection to the soil and a reduction in the erosive capacity of the water. The mat may even resist powerful currents, as may be observed on a midlatitude bottomland during a flood. If the water does not carry pebbles capable of mauling the sward, the latter may resist currents of up to 10 km per hour (6 m.p.h.) (a case observed on the Bruche River near Strasbourg in the winter of 1953–54). In order to overcome the resistance of the grass mat the water flow must become concentrated, as happens on certain very steep alpine meadows when the concentration of water becomes such as to impair the grass cover and to cut a channel which undermines the sod. The erosive capacity of water flowing on a grass mat is also decreased by the irregular surface on which it flows; the grass leaves although bending under the effect of the current form innumerable asperities which break the velocity of the water and therefore reduce its erosive effectiveness. Experimental studies carried out by Ree and Palmer (1949) are particularly demonstrative in this respect.

Overland flow is more important on grasslands than in forests because of the lesser thickness of the plant screen. In an American experiment station a torrential rain producing about 50 mm (2 in) of water in one hour caused an overland flow of 2 per cent in a wood, 5 per cent on non-cultivated grassland, 25 per cent on wheat and oat fields, and 50 per cent on cotton and corn fields. But the denser grass mat nevertheless protects the soil more effectively; the experimental stations of the United States have shown that there is, normally, four to six times as much sheet erosion in a forest as on grass. On a 10 per cent slope in Texas the removal of a soil layer 20 cm (8 in) thick would take about 170 000 years under a grass mat, 25 000 years if it were forested, but only twenty to fifty years if it were ploughed.

Vegetation, therefore, strongly affects overland flow and fluvial morphology; the resistance of the vegetation to overland flow largely destroys the geomorphic effectiveness of running water in the case of dense grass or litter-covered surfaces. Plants also play a role on stream banks. The often impenetrable thickets which fringe the gallery forests of tropical rivers interpose themselves effectively between the water and the river banks. In some cases they are capable of completely preventing a lateral undermining by the current. Lacking abrasives the mechanical erosion of the streams of the Fouta Djallon is almost negligible no matter how strong the stream flow during the rainy season; overland flow does not provide any slope-waste, nor

FIG. 2.9. Bottomland in a forest near Kontoura (east of Kindia), Guinea (*drawing by D. Tricart after a photograph by J. Tricart*)

Typical example of a bottomland in its natural state occupied by evergreen forest extending as galleries into the savannas. Mechanical erosion is practically negligible; the waters are turbid in the midst of the rainy season; sedimentation of silts and clays is very slow.

can debris be torn from riverbanks that are effectively protected by the gallery forest. The waterfalls thus perpetuate themselves almost indefinitely. However, as soon as man has sufficiently despoiled the plant cover, the waters gully the slopes, incise the bottomlands, dig meanders, and abrade rockbars with sand and pit them with potholes. The riverbank vegetation also plays a role in temperate climates, but before defining what it could have become in the natural state one should take account of man's interference, which has profoundly altered the natural conditions. Most riverbanks have been terraced, lined with dikes, and often deforested; the banks of small streams are frequently artificially raised by dredgings imposed by law; in small valleys artificial channels have been dug where, earlier, the water possibly spread into a marsh. Unfortunately, what is artificial and what is natural has not been determined; until it is, one is limited to qualitative observations. For instance, weeds which grow as soon as stream bottoms

FIG. 2.10. The influence of man on important rivers in the Fouta Djallon, Guinea, Garabé Falls. Because of the shortening of the fallow interval in shifting cultivation, the stream carries sands which enable it to abrade sandstone ledges. Rocksteps become eroded, boulders polished and rounded, and potholes start to form. Contrast with Souma Falls (Fig. 1.4).

become stabilised break the current and counteract the renewed entrainment of sediment during floods; or groves of weeping willows which have established themselves on gravel banks will in the next flood favour the formation of obstructions as long as the current is not strong enough to sweep them away. Large recent floods in France have indicated that the role of riverbank vegetation may vary considerably. Trees are often undermined, toppling into the riverbed. They then increase the turbulence, and important semicircular holes are excavated by sapping; transported along the river, they damage the banks by ramming them and then heap up against obstacles. On the Guil, in the French Alps (June 1957), they completely obstructed certain bridges, which were later swept away, causing destructive spates. On the Adour, in southwest France, trees often pile up on gravel banks and, obstructing the stream flow, increase the jam and sometimes cause overflows, which may play a role in the cutting off of meanders. Flexible shrubs, on the contrary, hardly impede the current and offer a very effective protection to riverbanks.

Vegetation is not only an obstacle to running water but also to the wind, and in both cases the mechanisms are similar. Plants also break the wind, protecting the soil which they hold with their roots. The strength of the wind is reduced by 30 per cent behind a curtain of trees for a distance equal to ten times the height of the trees. The breaking of the wind is even more effective when there is a continuous plant cover, such as a forest, on the ground of which wind is practically non-existent. An Indian girl of a Tupi-Guarani forest tribe whom Dr Veillard had encountered appeared stunned when she came out of the forest: it was the first time in her life she had heard the wind! The weakness of the wind, particularly in its geomorphic action, is extreme in forests. The same is also true of dense and tall grasses. At their contact the wind is curbed, its velocity distribution curve, varying as a function of height, begins tangent to the ground where wind velocity is close to zero or slightly above it. At a higher level its velocity increases rapidly. The only geomorphic action the wind is able to exert in a forest is through the medium of the trees when they are uprooted and felled. In such cases a mass of soil is unearthed together with the roots, and the agents of the weather can then reach the surface of the parent material at the bottom of the hole and spread its substance.

Plants also hold the soil particles and impede their removal by the wind. To fix the coastal dunes of the *landes* of Gascony, the following principle, advocated in Brémontier, has been used: to curb the wind with wattle windbreaks in order to give lyme-grass[1] time to grow; then take advantage of the grass to plant maritime pines. Once implanted the sand blown from the bare surfaces strikes the grass tufts, which because of their curbing action on the wind cause the sand to accumulate. Clumps of vegetation in areas of intense wind action may even cause small accumulations or *nebkas* several metres high to form. This, of course, can only happen where the plants have adjusted themselves to this kind of environment and can resist the choking action of the sand by continuously lengthening their stems, as is the case of the lyme-grass of western Europe. Where such conditions are not fulfilled the wind will completely overcome the plants with its abrasives of sand, silt or even snow. In cold climates the wind may cause the destruction of the vegetation by locally removing the snow cover. The intense cold will then kill the plants as they become exposed to the rigours of the winter. The wind then has free sway on the soil, which lies at its mercy.

We will now study the interactions that exist between the plant formations and the morphogenic processes.

---

[1] Also called beach-grass, marram, or sea-bent (French: *oyats*, Dutch: *helm*). A creeping grass widely used in western Europe to fix aeolian sands. Usually the *arenaria* species of *Arundo*, *Psamma* or *Elymus* (KdeJ).

CONSIDERATIONS OF SCALE

Several scales corresponding to different types of interaction between vegetation and morphogenic processes should be distinguished:

1. On a *world scale* the morphogenic processes are greatly influenced by the distribution of the plant formation-types (such as the tropical rainforest). Each plant formation-type forms a specific geomorphic environment with its own characteristics. This is because the main plant formation-types roughly correspond to the major climatic zones and interpose between the agents of the weather and the lithosphere a screen of variable importance. The overall world pattern of plant formation-types is first of all controlled by climate. Soil differences (intrazonal and azonal) and the human impact make themselves felt at a lower taxonomic level (i.e. on a smaller scale) in the form of regional modifications in the plant formations (such as in the African tropical rainforest). The secondary tropical rainforest, of course, differs from the primary, but it is nevertheless a rainforest quite different from the savannas or even the midlatitude forests. The plant formations are climatic integrals. Their distribution depends on the interaction of various factors: seasonal distribution of temperatures, temperature ranges, frequency of certain critical temperatures and humidities, seasonal distribution of precipitation, length of droughts, wind, and evaporation. Quite often the available climatic data are not sufficient to explain the distribution of plant formations in detail. Furthermore, the means of measurement are sometimes deficient; for example, there still does not exist a satisfactory instrument for measuring evaporation, a measurement which is extremely important for understanding the regime of stream flow and even the alternations of drought and moisture in the soil.

In Israel Boyko (1949) has shown that a detailed statistical study of the plant formations makes possible the calculation of certain elements of climate, such as rainfall, with an excellent approximation. Amounts of rainfall calculated from the kind of plant cover have come to within 10 per cent of those independently recorded in the neighbourhood.

Each biochore, and within each biochore each plant formation-type is characterised by its own role *vis-à-vis* the weather and, as such, influences the morphogenic system in a definite way. The forests form a particularly effective screen and make possible a shady mesoclimate quite different from that of cleared or cultivated land. Because the mechanical actions are considerably reduced, an important fraction of the infiltrated water promotes intense pedogenic actions and chemical weathering. The savannas and steppes have less effective plant screens: the soil is not as well protected against the weather; the role of runoff is increased; and the wind acts as soon as the steppe is open. Savannas and steppes are further characterised by a seasonal desiccation of leaves and branches, which become the prey of fires in warm climates. Brush fires considerably influence the soil-forming process and, through it, the landforms. The deserts and their margins, lastly, have a sufficiently reduced plant life so that the morphogenic action of fire is insignificant; the same applies to polar fell fields and glacial regions.

2. On a *regional scale*, ranging from 10 to 1 000 km, the distribution of plant formations includes inconsistencies within a single morphoclimatic zone. For example, in Amazonia or in Ivory Coast the rainforest is broken by clearings occupied by savannas; climatic differences between the savannas and the surrounding forest are minimal and are approximately limited to the influence of the forest on the mesoclimate. The difference in vegetation, however, is enough to produce differences in the morphogenic processes. The agents of the weather are differently modified in a forest and in a savanna, and the same is true of the soil-forming processes which determine the nature of the solum upon which the morphogenic processes act. Sometimes it is the nature of the soil which determines the distribution of the plant formations in transitional areas. There is then an interplay of pedologic, palaeoclimatic, and biogeographic influences on the morphogenic processes. But the action of man is more often the cause of this distribution. The example of the Fouta Djallon is typical in this regard: the less peopled parts of the massif have preserved a typical tropical morphogenic system characterised by the near absence of mechanical erosion. Because there was no shortage of land the cropping system which included a prolonged period of fallow permitted the forest to re-establish itself. Cultivated plots never form large contiguous clearings and thus prevent the concentration of runoff and the development of man induced erosion, which is primarily mechanical. The historical agricultural period is only reflected by the higher rate of sedimentation of the bottomlands with an increased proportion of clastics in relation to organic products and a substitution of dark grey clays by light grey argillaceous sands. In the most populous districts, however, where the density of the population reaches and exceeds fifty inhabitants per square kilometre (125 per square mile), the more rapid rotation of cultivated land has eliminated the forest fallow interval. Between years of cropping the soil supports a poor grass cover; protection against the actions of the weather is insufficient, and a wave of pseudoclimatic mechanical erosion begins; gullies form, meanders appear, bottomlands rapidly fill with clastics, and the crests of waterfalls start to wear. Thus two neighbouring basins may have a quite different morphogenic system as a result of the modification of the plant cover by man. This fact should never be forgotten whenever a large area is being studied; measurements made without this understanding only reflect averages that may obscure quite different realities or realities that are differently influenced by man.

3. On the *local scale* of some tens or hundreds of metres, perhaps even one or two kilometres (sixth order of magnitude in our taxonomic classification), the vegetation often constitutes a useful indicator of the morphogenic processes; we must therefore study it in more detail. A knowledge of the distribution of plant formations is not enough; in general one should also take into account the distribution of the species (i.e. plant associations). When

the morphogenic processes are sufficiently active they impose on the environment a progressive or sudden modification to which, in turn, the plant cover adapts itself as well as it can. The vegetation may thus disclose the presently active morphogenic processes or permit the dating of a past sudden break in the natural equilibrium.

The use of vegetation as a means of dating is common in glacial geomorphology. The age of the plant association occupying a moraine makes it possible to know, within a few years, the age of the moraine itself. This method has frequently been used in the Alps, in Alaska, and in East Africa. Occupance begins with certain species such as lichens and proceeds with the appearance of pioneer associations still poor in species. Knowing their evolution it is possible to establish their approximate age. They may be followed by shrubs and trees, whose age can be determined from growth rings, at least in climates with sufficiently marked seasonal contrasts. One should of course make certain that other factors have not perturbed the normal sequence of plant life. The nature of the geological formations occupied by the plants should not include important differences, and the climate should be essentially the same, which implies comparisons at comparable elevations in a mountainous area. An excellent example of the application of this method is contained in de Heinzelin's (1953) study on the Ruwenzori Mountains. Up to now such researches have only dealt with moraines, but the method could also be used in the study of coastal deposits and the construction of alluvial fans and natural levees. The boulders transported by the disastrous flood of 1942 in the Catalan Pyrenees were still recognisable in 1954 by the light tone which differentiated them from the boulders that remained in place and were covered by lichens.

Recourse to the study of vegetation to determine the activity of the morphogenic processes is less common. The method may, however, be very useful. The *Laboratoire de l'Institut de Géographie* of the University of Strasbourg has successfully used it in the Senegal delta.[1] Here the distribution of plant species is largely a function of the length and frequency of the submersions caused by the annual flood of the river. A number of indicator plants may serve to disclose the intensity and the nature of the present depositional processes. Gonakies (*Acacia senegaliensis*) always occupy old natural levees which are no longer submerged except during the highest floods. They therefore mark the levees which have ceased to be functional. The same is true of *Indigofera sublongifolium. Spirobolus pyramidalis*, however, occupies basins that are regularly flooded during a period of several weeks (two to four months), thereby causing the settling of muds, a process that is moreover accelerated

[1] The Senegal delta has been the object of intensive study by Professor Tricart and his collaborators. Cf. Tricart, J.: 'Aspects géomorphologiques du delta du Sénégal', *Rev. Géom. Dyn.* (May–June 1956), pp. 65–86, and Tricart, J., 'Note explicative de la carte géomorphologique du delta du Sénégal,' *Mém. Bur. Rech. Geol. Min.*, no. 8 (1961), 137 p., 9 plates, 3 coloured maps, scale 1 : 100 000. The reader is also referred to the bibliographies of these works (KdeJ).

by a dense grass cover which curbs the current. The deepest parts of the basins where the waters remain longest are bare, and, before drying up, the salts reach a concentration that is too high to permit the growth of *Spirobolus pyramidalis*. These saline soils are often subject to aeolian deflation. Kuhnholtz-Lordat's (1952) studies of the vegetation of Provence contain similar examples; he cites the frequent occupation of the base of scree slopes by pines.

In some cases this biogeographic method might even help to predict catastrophes. For example, in Menton (Provence), which was devastated by landslides, geomorphologists have been able to demonstrate the influence of the stepped retaining walls that were once built to hold the soil but which have long since disappeared. The terraces were first occupied by a secondary vegetation typical of semiarid Mediterranean uplands and later by pines. During this period the interstices of the dry walls were not yet filled in, and soil drainage was satisfactory. Then the pines progressively disappeared and were finally reduced to a few puny survivors, while plants characteristic of humid soils, such as reeds, started to appear and spread. By this time the abandoned walls had ceased to be permeable and blocked the drainage. Exceptional rains were all that was necessary to cause them to give way to the pressure of the waterlogged argillaceous soil and for earthflows to form. A timely geomorphological study could have revealed the danger and by advocating improved drainage limited the catastrophe. Such a biogeographic method is capable of rendering great services to geomorphology but presupposes a good botanical knowledge of the area to be studied; the method would gain much by being carried out in teamwork.

Although the plant cover modifies the agents of the weather at the ground level, the role it plays is not limited to this effect. It is the most important agent of soil formation, and it also helps to modify the lithologic conditions of the work of erosion.

Interactions between morphogenic and pedogenic processes

The morphogenic processes do not act directly on the rocks except where the plant cover is sparse or absent; they operate, instead, on the soils which result from the weathering of the rocks and in most cases also from the activities of living organisms: plants, microbes, and burrowing animals. This transformation may be so thorough that the resulting soils display completely different mechanical and chemical properties from those of the parent material. On most of the earth, soils are in this way an unavoidable link between the lithosphere and the morphogenic processes. Misapprehension of this fact was a grave error in Davisian geomorphology.

In our description we will first recall the principles of soil formation, after which we will enlarge upon the relationships between soils and erosion, mechanical as well as chemical. Lastly, we will examine the value of pedology in the field of geomorphology.

THE FORMATION OF SOILS

In a vegetated area the uppermost surface of the lithosphere shows a more or less distinct, layered, pedologic alteration that is caused by weathering and biotic activity. A geomorphologist should be acquainted with the type profiles of these horizons which are called soils.

1. *Soil profiles* Let us examine a soil profile of a loess region free of anthropic erosion in a humid climate. We see, from top to bottom, the following horizons:

(a) An A_0, or blackish horizon, full of plant debris, whose thickness does not exceed 5 to 10 cm (2 to 4 in) in France. The high organic content is due to an important accumulation of slowly decaying leaves and branches forming a vegetable mould.
(b) An A_1 horizon, somewhat lighter in colour, moderately acid, and rich in humus.
(c) A leached A_2 horizon, ash-grey in colour, with little organic matter.
(d) A B horizon of accumulation; brown, hard, brittle when dry, sticky when moist; cracking into small vertical columns under the effect of desiccation, a consequence of its argillaceous composition (known as *lehm* in Europe).

The non-weathered subjacent loess contains calcareous concretions 1 to 5 cm (0·4 to 2 in) in diameter, sometimes more, in contorted shapes called *loess-dolls* or *puppets*. They result from the precipitation of lime dissolved higher up by percolating waters.

Such a succession of characteristic layers, called horizons by pedologists, is found in most typical soils. It is often blurred by ploughing, which either mixes the layers or produces a sufficiently intense erosion which removes them. Typical and well differentiated soil profiles are therefore mainly found in old forests and in uncultivated land. They are characterised by a similar succession to the one we have presented above and are the object of a special terminology. The upper horizons are generally characterised by an impoverishment: part of the soluble and colloidal matter having been removed by water. They are the *leached* or *eluviated* horizons. Conventionally they are designated by the letter A followed by a number. A_0 is reserved for the upper horizon made up of poorly decomposed organic debris (acid humus, leaf-mould), A_1 is the humic horizon, and A_2 a leached mineral horizon. Our *lehm* (B) and puppet (C) horizons are, on the contrary, characterised by an accumulation; here part of the minerals dissolved in the eluviated horizons is deposited at the contact of the parent material. They are *illuviated* horizons, particularly the B horizon. The C horizon is mainly composed of the more or less weathered subjacent rock. This rock is often, but not always, the parent material of the soil. Sometimes the soil is derived from a transported material, whether by wind or by fluvial action.

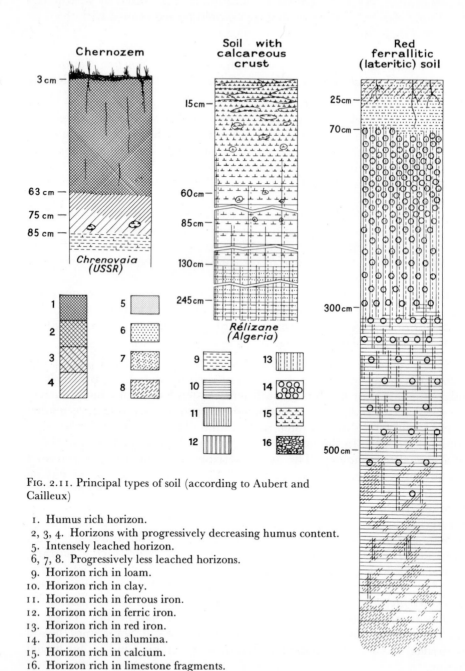

FIG. 2.11. Principal types of soil (according to Aubert and Cailleux)

1. Humus rich horizon.
2, 3, 4. Horizons with progressively decreasing humus content.
5. Intensely leached horizon.
6, 7, 8. Progressively less leached horizons.
9. Horizon rich in loam.
10. Horizon rich in clay.
11. Horizon rich in ferrous iron.
12. Horizon rich in ferric iron.
13. Horizon rich in red iron.
14. Horizon rich in alumina.
15. Horizon rich in calcium.
16. Horizon rich in limestone fragments.

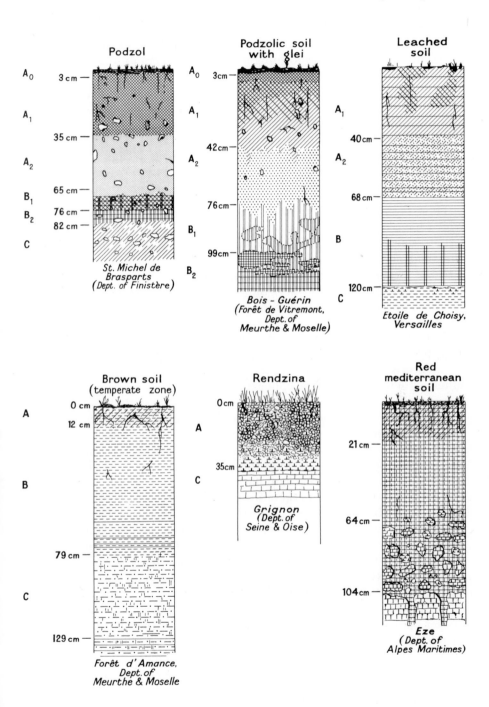

		Podzol
A_0	3 cm —	
A_1		
	35 cm —	
A_2		
B_1	65 cm —	
B_2	76 cm —	
	82 cm —	
C		

*St. Michel de
Brasparts
(Dept. of Finistère)*

Podzolic soil
with glei

A_0	3 cm —
A_1	
	42 cm —
A_2	
	76 cm —
B_1	99 cm —
B_2	

*Bois - Guérin
(Forêt de Vitremont,
Dept. of
Meurthe & Moselle)*

Leached
soil

A_1	
	40 cm —
A_2	
	68 cm —
B	
C	120 cm —

*Etoile de Choisy,
Versailles*

Brown soil
(temperate zone)

A	0 cm —
	12 cm —
B	
	79 cm —
C	129 cm —

*Forêt d'Amance,
Dept. of
Meurthe & Moselle*

Rendzina

A	0 cm —
C	35 cm —

*Grignon
(Dept. of
Seine & Oise)*

Red
mediterranean
soil

21 cm —
64 cm —
104 cm —

*Eze
(Dept. of
Alpes Maritimes)*

The profile is as follows in a podzol soil on sandstone:

(*a*) An A_0 horizon of black or brown plant debris; for example, half decomposed pine needles forming a layer 5 to 10 cm (2 to 4 in) thick.
(*b*) A humic A_1 horizon.
(*c*) An eluviated A_2 horizon of loamy, white ashy sand completely devoid of iron and which leaves the fingers covered with powder.
(*d*) An illuviated B horizon replete with small iron oxide concretions which sometimes form an almost continuous layer.
(*e*) A C horizon of fragmented sandstone with its original light colour, as in the case of the Grès Bigarré (Buntsandstein) of the Vosges Mountains, the fragmentation having been caused by Pleistocene frost.

These typical horizons are sometimes absent, not only in cultivated soils which have lost their original structure, but even in certain undisturbed natural soils such as rendzinas which develop on limestones in a temperate climate. Rendzinas often show a direct superposition of the A horizon (composed of a crumby soil mixed with clay loams, organic matter, and limestone fragments) and the C horizon (composed of limestone rubble due to the mechanical disintegration of the parent rock). In short, one speaks of an AC horizon.

The essential distinction however, which must serve as a guide, is that of an upper leached, humic horizon and a lower horizon of accumulation. This distinction is basic to soil science, and the two horizons determine what is the influence of the soil on the morphogenic processes.

2. *Pedogenesis* (*the soil-forming process*) is in most cases characterised by the following factors:
(*a*) The dead vegetal matter falls on the ground and persists until bacteria and insects dispose of it. The importance of this layer of debris depends on two variables: the rate of accumulation and the rate of destruction. The first depends on the type of vegetation: a coniferous forest sheds less organic matter than a beech forest or a prairie of tall annual grasses. The rate of destruction depends on the micro-organisms (bacteria, moulds, fungi), whose activity is a function of temperature and humidity. This activity is much reduced in cold climates; if there is any organic matter, it persists for a long time and forms a continuous and sometimes thick layer on the ground, even if the supply is limited. For example, in the podzols of central Sweden the A_1 horizon of pure black humus is 5 to 10 cm (2 to 4 in) thick, which is much thicker than in the more southerly temperate forests. In the steppes drought is added to the rigour of the winter to retard the decomposition of the organic matter; the humic horizons may reach a thickness of several decimetres in typical chernozems (etymologically 'black earths'). In the tropical rainforest heat and humidity are combined to accelerate the same process. A_0 horizons are very thin, often almost non-existent in the entire humid tropics in spite of the considerable mass of organic matter provided by the

often dense forests or tall grass savannas and whose weight per hectare by far exceeds that of the midlatitude forests or the chernozem steppes.

But the humus of soil profiles varies not only in quantity but also in quality. There are two kinds which determine the pH of the soil: sweet humus and acid humus. They depend on the vegetation, the microbial activity and the lithology of the parent material. Acid humus decomposes slowly because of a limited microbial activity. The humic acids which result (humolignin: little transformed lignin) do not affix themselves to colloidal minerals but remain in a dispersed state. They may combine with iron to form soluble iron humates, which migrate freely. Podzols, for instance, develop in this way; their ashy consistency is due to the fine grain, the incapacity of the humolignin to cement the particles into aggregates, and the intense leaching of the iron oxide. On the other hand, in a calcium rich environment (calcareous rocks) the humic acids are grey and solidly bound to the mineral colloids of the soil; together they produce a cement which joins the sand grains into aggregates. Brown humic acids appear in a moderately acid, intermediate environment and together with the mineral colloids form combinations of less stable aggregates. Bacteria gradually decompose the humus, producing ammonia (ammonification), which in a basic or slightly acid environment is in turn transformed into nitrous acid. Nitrous acid is further transformed, again by bacteria, into nitric acid. Each transformation is by a different bacterial species. Carbon dioxide, whose intensity varies considerably depending on the environment, is released simultaneously.

(*b*) The plant roots plunge into the A, B, and C soil horizons and also occasionally into the bedrock. They extract from them the mineral solutions which feed them. Through transpiration they again dissipate into the atmosphere an important fraction of the rainwater, which is lost to the geomorphic processes. This fraction increases with the density of the vegetation, with its biologic activity, and with certain physiologic characteristics. A midlatitude forest does not transpire nearly so much water as a tropical rainforest; transpiration ceases almost completely during several winter months when the trees are leafless; the density of the living matter per hectare is three to five times less, and the biotic activity is reduced because the trees grow slower and transpire less. Inversely, the plants of dry regions, as for example the acacias of Mauritania, transpire very little water; their foliage is most limited, their transpiration minor, and their activity slowed down during long periods between rains. All the plants of dry regions show adaptations to drought: increased absorption thanks to long roots and decreased transpiration. Some grow spines instead of leaves, while others develop a thick waxy cortical layer around leaves and branches. Current studies on papyrus thickets in the Nile Valley have indicated that the quantity of water dissipated by their transpiration is equal to several times the evaporation at the surface of a sheet of water of the same dimensions. The thickets of the Bahr-el-Ghazal thus considerably impoverish the White Nile. The same mechanism plays a role in the interior drainage of the Chad Basin.

But the role of water extraction by plant roots is not limited to these factors alone. The draw of the roots increases the quantity of rainwater that must first be infiltrated before overland flow can start. The transpiration of plants thus decreases the chances of overland flow as well as diminishing the quantity of infiltration and, therefore, the flow of springs. By accelerating the water circulation between the ground surface and the horizon tapped by plant roots, it also favours chemical migrations, especially leaching. The water percolating through the humus becomes loaded not only with organic matter, notably humic acids, but also with carbon dioxide and nitric ions which react with the mineral fraction of the soil and permit the formation of soluble compounds such as calcium bicarbonate from limestone or iron humates. Where the climate includes a marked dry season, the roots dry the soil at their contact. A supersaturation of dissolved products may result, causing their precipitation. It is in this way, for example, that calcareous jackets are formed around roots (as are presently forming in the coastal dunes of the Oran area in Algeria or were formed during the Pleistocene in some Alsatian *lehms*. They are also represented by *pseudo-mycorrhizas*).

(*c*) A third factor in the soil-forming process is the mechanical action of roots and the circulation of air in the soil. Roots exert a notable action on rocks; they penetrate into joints and often succeed in opening them even when the rock is well consolidated. After the death and rotting of the plant, they leave openings through which both air and rainwater penetrate. Burrowing animals which dig tunnels either for shelter or to find food are a powerful agent of aeration. Rodents dig hundreds of kilometres of galleries per hectare in grasslands, thus disturbing the soil and undermining it to the point that it becomes tiresome to walk on. Foxes, wild boars, wart-hogs, and ant-eaters dig galleries several decimetres in diameter and several metres long. Moles may damage cultivated prairies and, like voles, be a hindrance to agriculture. If earthworms, ants and termites are added to this list it can easily be understood that the soil is profoundly modified; all these animals bring to the surface, in a slow but continuous mixing, particles from various depths. Duché has collected some numerical data on these phenomena. Obtained from one locality, they only provide indications of a specific order of magnitude, but they are nevertheless useful. Thus in the prairie soils of the United States 7 million insects have been found per hectare (2·8 million per acre), and 100 to 500 kg (18 to 90 lb per acre) of earthworms. Prairie dogs (*Cynomys*) dig holes down to depths of 5 m (16 ft). In the Siberian steppes the pietschanka rodent accumulates 1 to 2·5 kg (2·2 to 5·5 lb) of hay in his burrow, and 500, even 1 000, of such burrows have been counted on one hectare (200, 400 per acre). *Nematodes*, microscopic earthworms, are counted by the hundreds of thousands per hectare in midlatitude forests (200 000 to 1 000 000 per hectare or 80 000 to 480 000 per acre). As to bacteria, their numbers have been estimated at 3 000 million per gram of superficial forest soil in the temperate zone. Density of all these animals increases with the humus content and with the quantity of organic matter available. On the average the root development of trees is two-thirds that of their aerial parts, which is less than that of herbaceous plants especially in semiarid regions where root systems are particularly well developed in order to obtain the indispensable water from a dry soil. The enormous mass of roots under a forest or even under a savanna or a dense prairie breaks up the mineral soil and after death increases the number of cavities and the soil porosity. The apparent density of a clod of earth varies between 1·2 and 1·5, which is more than a third less than that of sand or compact clay.

Roots penetrate like wedges into joints, widening them. They may even directly attack the solid rocks by chemical means. The roots of lyme-grass have even bored through thick pieces of glass. Micro-organisms are thought to be capable of fragmenting the grains of rocks and of transforming them into loams and colloids. *Bacillus extorquens*, which is very active, is known to coat particles with a corrosive substance. The action of most micro-organisms in a humid environment is optimum between 20° and 30°C (68° and 86°F), which are conditions that are realised in the wet tropics. Temperatures that are too high paralyse them: close to 60° or 65°C (140° or 149°F) fungi, protozoans and a number of bacteria cease to function. For this reason bare soils in the tropics and subtropics are characterised by a much reduced activity in comparison to those of the forests or even of the savannas. The same is true in arid environments and on cuirasses.

Some animals modify the texture and even the chemical composition of the soil: earthworms through their digestion and termites and ants through the impregnation of formic acid. Flocculation of the mineral colloids of the soil by humic acids also plays a primordial role by favouring the formation of aggregates. As aggregates are larger than colloidal micelles, the pores which separate them are large enough to circulate water or air. Owing to the presence of air, many aerobic microbes can live in the soil; they liberate important quantities of carbon dioxide and various acids which later contribute to the chemical decomposition of the mineral fraction. It is due to want of enough aeration and microbic life that the soils of cold climates are often boggy: the organic matter does not decompose and accumulates in place. Depending on the zonal ecologic conditions, the microbic fauna of soils varies considerably in species as well as in density, resulting in important differences in the biochemical processes and in the genesis of the soils. This domain is unfortunately still little explored. All that is known is that the number of microbes decreases rapidly with depth: to one half for animals with the first 10 cm (4 in), which indicates that the biotic activity is mainly concentrated close to the surface.

Due to the effect of the soil-forming processes, whose main characteristics we have just reviewed, rocks are subjected to increasing alterations with time. The more a soil is evolved, the more it differs from the parent rock. In western Europe the lithologic factor is important on the present soils: podzols and podzolic soils develop on sands and sandstones, rendzinas on limestones, and gleyed soils on clays and compact marls. Under the forests of tropical Africa where the soil-forming processes are more intense and more rapid than in Europe the lithologic influence seems to be least; granites, basalts, diorites, and decalcified limestones are all subjected to ferrallitisation. There, according to pedologists, the process of soil development tends to accentuate the characteristics of the soil which result from the vegetation and the climate and to leave only a secondary role to the lithologic factors, which only introduce varieties in the main zonal categories.

Born of the contact of the lithosphere and the biosphere, soils, like flora and fauna, adapt themselves to the climatic factors. They therefore play a double role *vis-à-vis* the morphogenic processes: that of a buffer between the rocks and the superficial processes of erosion, and that of agents of rock weathering, causing a deep chemical erosion (or weathering). We will first examine the relationships between soils and chemical erosion because the latter is closely related to soil development. Later we will examine the relationships between soils and mechanical erosion.

PEDOGENESIS AND CHEMICAL EROSION

The influence of soils on chemical erosion is twofold. First, the formation of aggregates, the mixing of the soil by animals, and the penetration of roots into the soil facilitate the infiltration of water, the principal medium necessary, or at least favourable, to chemical erosion. Second, the decomposition of the humus and the various micro-organisms which swarm in the soil liberate important quantities of carbon dioxide and various acids; these are often very active when in contact with the mineral matter of the soil and the subjacent rocks towards which they are transported by infiltrating water.

The material removed by water from the eluviated horizon has two different destinies. Part follows the water towards deeper layers, later reappearing on the surface in springs; the rest is abandoned at the base of the soil itself where it forms the illuvial horizon. The infiltration waters after having percolated through the illuvial horizon and abandoned in it part of the products carried in solution may pick up other products and even attack the underlying bedrock, helping to transform it into the C horizon. One should, from the point of view of geomorphology, therefore distinguish between leaching and accumulation or, in other words, between *eluviation* and *illuviation*.

1. *Eluviation (leaching)* Chemical erosion and pedogenesis generally go hand in hand on the earth's surface. Limestone, however, may present an exception: it may dissolve when in association with atmospheric water as happens, for example, at the contact of snow patches and in the formation of mountain lapiés. A slow permeation of meltwaters may be enough to cause appreciable chemical erosion. The same is also true in subterranean cavities. As we have already indicated, this process of pure mineral chemistry is favoured by low temperature. It is most often found in cold regions other-

wise unfavourable to the development of a plant cover on rocks. It is one of the main mechanisms of karst formation, but it is not the only one. Under a plant cover the chemical erosion of limestone is also facilitated by the liberation of carbon dioxide and the formation of acid humus. Whereas pure chemical erosion is favoured by low temperature, biochemical erosion increases with temperature and rainfall. This explains why the lime content of spring water, according to Schoeller's observations, increases from southern Sweden to Tunisia, and why the effects of dissolution are much more intense in tropical karsts than in karsts of temperate or cold regions. The rate of limestone dissolution on the equator is equal to several times that of the midlatitudes. In the wet tropics, as Renault has shown, limestone is only attacked close to the ground surface: its dissolution is a function especially of the supply of CO_2, which is closely linked to that of soil formation. Tropical karsts are superficial karsts with sink holes which through deepening and intersection evolve into needle or dome karsts but produce very few caverns.

The intensity of leaching is directly dependent on temperature and the quantity of infiltrated water. In temperate climates a moderate average temperature and a considerable slowing down of the biotic activities during winter produce a moderate amount of leaching. It mainly involves alkaline ions, lime and iron, as is shown by the two examples of *lehms* and podzols which we have already analysed. The attack of granitic rocks thus principally involves two minerals: the micas, which are dissociated through the formation of iron humates; and the feldspars (especially the soda-lime feldspars), which are dissociated through the departure of lime and alkalis. Quartz seems to be scarcely attacked, if at all. The principal product of rock weathering is a clay loam. The clay fraction has different characteristics depending on the environmental conditions: kaolinites in an acid environment but more often montmorillonites and illites. These clays have different mechanical properties which affect the morphogenic processes. For example, montmorillonite varies considerably in volume according to its moisture content. However, less evolved types of clays, mainly illites and hydromicas, predominate in young soils. Kaolinite is most often found under conditions of poor drainage and acidic soils, as in bogs. A large part of the kaolinite of the midlatitude soils is inherited from Tertiary epochs of subtropical climate. Montmorillonite forms in a basic environment especially in the presence of Mg ions; it mainly characterises the soils of semiarid regions. The lithology having been taken into account, it is the proportions of the several clays rather than the presence of certain particular clays in the whole that characterise the soils of a given climatic environment.

In very cold climates the tundra permits the formation of a soil, but the soil-forming process is not as active as in the temperate zone and results in a very slow decomposition of the humus and a high acidity. Soil types are the same as those of the midlatitudes, but they form very slowly and only reach a limited thickness. There are tundra podzolic soils, some grey soils, and especially many gleys because aeration is generally deficient and the structure compact; in an anaerobic environment iron occurs in the ferrous state combined, for example, in sulphides.

In the wet tropics high temperatures, abundance of plant debris, and water all favour leaching. It reaches an intensity unknown in the temperate zone; minerals stable in the midlatitudes are less stable here. Quartz grains are often corroded, saccharoidal, or covered with a plush of precipitated silica. Under certain conditions silica is dissolved, enriching many iron ores from rocks that were initially too siliceous to be exploited (the ores of Mt Nimba, Cerro Bolivar, Belo Horizonte, are examples). The migration of silica is facilitated in two cases: in an alkaline environment (the weathered mantle on basalts or diabases, for example, or the initial phase of the weathering of granite), or in combinations of silica and iron oxide. The latter

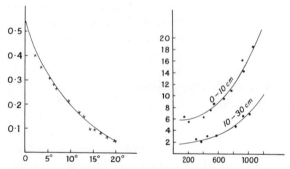

FIG. 2.12. Intensity of pedogenesis as a function of climatic factors (after Demolon)

Left. Variation in nitrogen content in constantly humid regions as a function of annual temperature. Ordinate: percentage of nitrogen in the soil. Abscissa: temperature in degrees Centigrade. *Right*. Variation of the content or organic matter in the soil with elevation; Indonesia. Ordinate: percentage of organic matter. Abscissa: altitude in metres.

penetrates into the quartz and is later leached out by humic acids. The wet tropics are therefore characterised by an intense argillaceous weathering of kaolinitic tendency. The wet–dry tropics are favourable to the formation of cuirasses, whether ferruginous or bauxitic. The latter are always found on basic or neutral rocks and only form as a result of prolonged leaching. Bauxitic cuirasses are thus, generally, high level cuirasses, capping the dissected remnants of former erosion surfaces.

The leaching of silica seems to be caused mainly by the weathering of feldspars rather than by a direct attack on the quartz. Quartz is not very soluble in water; it dissolves at the rate of 10 milligrams per litre at a temperature of $15°$ to $20°C$ ($59°$ to $68°F$). Opal and chalcedony are much more soluble (130 mg/l). Plants, however, extract silica from the soil and abandon it in the form of organic gels which become transformed into opal and chalcedony. Biotic activities, as Erhart suggests, are probably very important in the geochemical cycle of silica.

In the wet–dry tropics with a prolonged dry season the soil-forming process is less intense due to a water deficiency during half a year or more. Leaching is thereby reduced and less silica dissolved. It almost ceases in the semiarid zone. Soils are more like those of the warm temperate zone; brown soils form on limestones, for example. There is not as much chemical erosion.

The intensity of leaching in general increases with temperature and humidity. According to Demolon (1949), for a mean annual temperature of 11°C (52°F) the quantity of nitrogen which remains in the soil in spite of leaching varies as $0.00655(P - 0.023)$, in which P = annual rainfall in millimetres. Hydrolysis plays a major role in the leaching: the metallic cations unite with the ions (OH^-) to form basic hydroxides. The dissociation of H_2O into H^+ and (OH^-) ions, however, is simultaneously a function of the temperature and the content of CO_2 dissolved in the water; at 34°C (93°F) the dissociation is four times as rapid as at 16°C (61°F). At the same time the mobility of H^+ increases, which explains the greater thickness of tropical soils as well as their peculiar dynamics.

2. *Illuviation (accumulation by infiltrated water)*   We have distinguished two kinds of dissolved or removed products in the leaching process. One is redeposited somewhat later in the soil itself where it forms the illuviated horizon, the other is carried away in the ground water and perhaps plays a role in the lithifying process (e.g. the cementation of conglomerates). In chemical as well as in mechanical erosion the material is successively removed and left behind; sometimes it is even removed again. Dissolution and the formation of concretions alternate in the development of karsts. In soils, which are our present concern, there is intense leaching only in the higher horizons; lower down there is the zone of accumulation, the illuvial horizon. Some pedologists, however, think that a further dissolution takes place in the C horizon where the subjacent bedrock is weathered. This is perhaps the case of thick zones of rotten rock which typify the weathered profiles of granitic rocks in the wet tropics. Their aspect is the same as that of the unweathered rock, which denies a change in volume; the material is friable, however, which indicates a considerable amount of leaching. Millot and Bonifas (1955) call it *isovolumetric weathering*. As Ruxton and Berry (1957) have shown, in the humid tropics and subtropics there are true weathered profiles below the soil profiles. They are due to an intense chemical action which profoundly modifies the mechanical properties and permeability of the rocks, which in turn greatly influence the morphogenic processes.

Whereas the leached horizons are friable, the horizons of accumulation may be compact and even quite hard in some cases (calcareous crusts and tropical cuirasses) and play a major role in relation to erosion. It is therefore necessary briefly to explain their formation.

The most important illuvial horizons from the geomorphological point of view are those which are formed by iron and aluminium hydroxides. They are very hard, almost insoluble, and therefore difficult to destroy. They occupy a rather important part of the earth's surface. The problem of iron migration in the soil has recently been studied in an experimental way by Bétrémieux. We will follow his analysis in our discussion.

The simultaneous presence of organic matter and sufficient moisture causes a fermentation whose consequences on the chemical dynamics of the soil are important. The mobility of nearly all the mineral elements is thereby increased to variable degrees, depending on the element; it is particularly increased in elements which are normally almost insoluble but susceptible to being reduced in environmental conditions with a high biotic activity. The reducing action causes fermentation to be more than a simple producer of solvents of soil minerals, producing mineral acids and especially organic acids some of which as well as their salts are endowed with an unusual effectiveness. The fermenting action is much more intense and is capable of producing important physico-chemical reactions whose detailed mechanisms are not well understood.[1]

Processes of oxidation and reduction which are related to the fermentation of organic matter in the soil theoretically include the following type succession:

1. The iron of ferric compounds is reduced, especially in the oxides.
2. The reduced iron is dissolved in the form of complex electronegative ions which permit the migration of the metal in an environment otherwise unfavourable to the migration of the electropositive cation.
3. The ferrous iron in solution is rapidly oxidised into ferric hydroxide in an oxygenated environment, especially if the protective substances have disappeared.

Two climatic factors play an important role in these successive processes. *Temperature* is significant in that it directly determines the degree of microbial activity in a sufficiently humid intertropical environment. The bacteria consume large quantities of organic matter, producing an acidic soil which facilitates the migration of iron. The solubility of silica although remaining inferior to that of iron is much superior to that of alumina. Oxides and carbonates are the most easily dissociated. *Soil moisture*, the second factor, is a function of climate and of permeability. It often plays a determinant role in the precipitation of iron and other dissolved products. Soil structure, especially burrows; desiccation cracks; the holes left behind by former roots, by worms, and by insects; these all facilitate a gaseous exchange between the deep layers and the atmosphere. The result is an oxygen content that may be higher in deep layers than close to the surface and which facilitates

---

[1] *Annales agronomiques* (1951), p. 266. It should be noted that by 'element' Bétrémieux here means 'constituent'.

oxidation. Desiccation not only causes the disappearance of the solvents and a decrease in microbial activity, both of which also favour precipitation, but it also explains a certain amount of induration by iron oxides spreading outward from old cavities left by roots and burrows (e.g. termitaries), which are frequent in the formation of ferruginous cuirasses.

The mechanism of iron migration is not peculiar to the tropics, but it is much more intense there than elsewhere. Although common in podzols, it is most characteristic of tropical cuirasses. Concentrations of iron oxides are observed especially in the following cases:

(*a*) At a certain depth in desiccation cracks of the soil, occasionally ending in the formation of coatings on their walls. Such coatings create discontinuities and guide the localisation of new cracks the following season. This phenomenon has been observed in the Landes of Gascony and in Holland.
(*b*) Along decomposing roots, which, like cracks, facilitate the penetration of air. Iron jackets have been observed in Algeria; they sometimes look like an assemblage of pisolites. Under grasses the accumulation of iron is not as deep and occurs in the form of small concretions rather regularly distributed but most abundant in the layer richest in roots. Concretions seem to be larger and deeper under forests. It has been suggested that such criteria be used in palaeogeographic reconstructions, in tracing vanished forests, for example.
(*c*) In subhorizonal sheets of iron oxides which form at the level of the watertable and may locally greatly reduce permeability; the drainage of the rainwater then becomes increasingly difficult, causing the watertable to rise; this is unfavourable to roots, which then spread at a higher level where they cause the development of another horizon of concretions which gradually coalesce to form a new sheet of iron. In this way some authors explain the successive concretionary layers separated by friable horizons more or less regularly spaced which may be observed, for example, in the *alios* (iron pans) of the Landes of Gascony; but the discontinuity of such horizons remains to be explained.
(*d*) In the wet–dry tropics, where the migration of iron is active but seasonal. Drought seems to favour the precipitation of iron in the form of pisolites. Their formation has been observed in the natural levees of the Senegal delta and the Macina on the middle Niger. Such pisolites may later become concentrated at the surface if the plant cover is removed and the overlying friable soil eroded. Tricart has observed such a case in an extensive area in the neighbourhood of Lake R'Kiz in southern Mauritania. The pisolites may also be cemented into a cuirass at a certain depth. According to Bétrémieux, pisolites and cuirass would grow simultaneously through an influx from the upper soil horizons and also from the subjacent ground water, a growth wherein the dissolved iron would have arrived with the water from higher, more or less distant areas by subterranean flow. Such would be the origin of *ground water* cuirasses which occur only on valley bottoms close to rivers in

the wet–dry tropics. Bétrémieux has described one of these along the Logone in the Chad region. There the cuirass is limited to the vicinity of the river-bank and passes into a layer of innumerable pisolites towards the foot of the hills; it rests on a somewhat impermeable yellow sandy clay forming an aquifer rich in seepages during low water.

Illuviation would thus occur at least in the following three ways: by the progressive destruction of the organic compounds in their migration to the depths of the soil; by the progressive concentration of the soil solutions as a result of seasonal drought and root drainage; and by a swinging watertable producing alternate wetting and drying of the soil. Illuviation is not very intense in the midlatitudes; it results especially in the formation of soil concretions like those found in podzols, the puppets of loess, or local accumulations such as bog iron (*ortstein*) and iron pan (*alios*).

In semiarid regions, where the quantity of organic matter is smaller and humidity deficient during a very long dry season, it is mainly lime which forms the illuvial horizon. Chernozems include well developed calcareous concretions. In the Magreb there are thick crusts of Pleistocene age which were probably formed at the end of the pluvials under a climate more contrasted than the present one. In Algeria such crusts were probably caused by overland flow, according to Jacques Durand. They are to be found in Mediterranean regions, Australia, Mexico and the southwest of the United States. In the Magreb, principally in Morocco (Raynal, 1961) and Tunisia (Coque, 1962), it has been demonstrated that calcareous crusts are characteristic of a certain rather dry but not arid fringe. They are absent in humid regions and disappear towards the margins of the Sahara. They were formed during well defined intervals at the end of cold periods locally designated as pluvials. There is no terminal-Würm crust, however. The oldest crusts are the thickest and the most resistant to erosion. Often they are also the most complex because they have been subjected to considerable lithification from the moment they ceased to form. Lime was dissolved from them and recrystallised under the effect of the circulation of rainwater. All these crusts, except the occasional gypsiferous crusts of salt flats, were formed *per descensum* through the precipitation of carbonates supplied by the superficial waters percolating in friable materials. The old theory of pedologists (formation *per ascensum* through rising salts from the ground water subjected to high evaporation) has now been abandoned because it is in contradiction to several established facts. These genetic conditions explain the development of crusts in sites favourable to the infiltration of overland flow: bajadas, pediments, and colluvial slopes. The calcareous crusts mould a residual relief and tend to assure its preservation, but their resistance to mechanical and chemical erosion is most often inferior to that of the intertropical cuirasses. A colder climate subjects them to frost; a more humid climate to dissolution.

Cuirasses of relative accumulation result from the concentration of the indurating element through leaching of the other elements. They are therefore formed by relatively insoluble products, mostly alumina. They are bauxitic cuirasses whose formation requires specific geomorphic conditions (high remnants of old dissected planation surfaces long subjected to intense drainage). *Cuirasses of absolute accumulation* result from the induration of a relatively mobile element, usually iron, sometimes manganese. The silicified ground water cuirasses of semiarid regions may also be classified with this type.

Ferruginous cuirasses are by far the most widespread and are usually cuirasses of absolute accumulation. Some ridge crests, however, composed of ferruginous quartzite or silico-ferruginous metamorphic rocks (e.g. itabirite) are covered by ferruginous cuirasses of relative accumulation formed through the leaching of silica. Ferruginous cuirasses of absolute accumulation always characterise Sudano–Sahelian-like climates with a high seasonal evaporation. During the rainy season the iron migrates owing to the rapid decomposition of the organic matter which produces humic compounds; during the dry season these compounds are destroyed and the iron precipitated. Where the iron is very abundant, at the foot of a cuirassed escarpment, for example, or on a hillside rich in iron, cuirasses can form on a slope, especially through the cementation of colluvium. But they most often form on bottomlands flooded during the rainy season and drying during the dry season. Inland alluviation by rivers originating in forests, such as the middle Niger, which has only emptied into the Gulf of Guinea since the Riss (Saalian), particularly favour the formation of cuirasses.

Cuirasses of absolute accumulation are thus a facies of chemical accumulation peculiar to the wet–dry tropics. Their genesis has nothing to do with the lithology of the substratum but is influenced by that of the regolith in which they form. Clays are not favourable to the cuirassing process; iron coats the crystals and deflocculates the clay without producing a cuirass. Sands, boulders and materials of mixed calibre are favourable.

Generally, cuirasses as well as the calcareous crusts of semiarid regions were formed during a particular period of geomorphic evolution. In West Africa Vogt, Michel, Rougerie, and Tricart (see Michel, 1959) have demonstrated the existence of several cuirassing periods during the Pleistocene. On the middle Niger they correspond to the humid phases which increased the chemical influx from the Guinea Arch, whereas the dry phases were characterised by mechanical morphogenic processes. In the basins of the Senegal and the Gambia, the cuirassing phases followed the deposition of the four recognised Pleistocene detrital formations (Michel, 1959): three cuirassed pediplane levels correspond to the high, medium, and low terraces, whereas local consolidations take place in the rubble of the 'underbank gravels' of the present river channels.

Cuirasses constitute one of the main elements of tropical geomorphology and are as important as periglacial deposits are in midlatitude regions. Geomorphologists led astray by Davisian concepts have long ignored them, and only pedologists have studied them. A number of errors have resulted, as in the case of calcareous crusts: they were believed to be the result of accumulations *per ascensum*; if this were so they would be of local origin and therefore related to the substratum. Decisive progress has been made with the introduction of the geomorphological point of view.

Distinction between bauxitic cuirasses (essentially formed of alumina and in which iron only plays a subordinate role) and ferruginous cuirasses (in which iron is the main element) is of the utmost importance. Unfortunately, this distinction has not always been made, and there is still talk of 'laterite' to designate not only both types of cuirasses but the non-consolidated alumina of 'lateritic' clays as well. In fact they are two formations which have developed in distinct geomorphic environments. Both types of cuirasses, however, play the role of resistant rocks in relation to erosion.[1]

THE REGOLITH AND MECHANICAL EROSION

The high degree of chemical erosion which accompanies the weathering and soil-forming processes produces important changes in the mechanical properties of the regolith. In one place the solid rocks are transformed into friable weathered products; in another, friable horizons are solidly cemented into cuirasses or into crusts. Wherever there is a residual or transported regolith, this, not the rocks, is subjected to the mechanical agents of the morphogenic system acting on the topographic surface. Except for river banks and sea cliffs, this regolith is the unavoidable intermediary between the bedrock and erosion. The properties of the regolith (including the soil) in relation to mechanical erosion are therefore as important as those of the rocks. Outside the arid and glacial regions, the bedrock has only a direct influence on erosion in stream beds, on abrupt rocky slopes, on shores, and at the foot of sea cliffs. Everywhere else (i.e. on almost all land surfaces) the work of mechanical erosion is determined by the properties of the soils and subsoils.

Within these properties of the regolith two cases must be considered: a friable regolith, and well cemented formations referred to as crusts and cuirasses.

1. *The friable regolith* The friable regolith is by far the more extensive of the two categories. Compact crusts and cuirasses are the result of a particular, limited type of evolution.

[1] Because of the imprecision of the term *latérite* (and *latéritique*) (whose English original is just as imprecise), Tricart has proposed to abandon the word altogether to avoid any further confusion (see *Treatise*, vol. 5). The term or its adjectival form should be replaced by latosols (a catchall) or by more precise terms such as ferrallitic or kaolinitic in connection with tropical soils, and by cuirass in connection with indurated materials found at or close to the ground surface and having a tropical origin. The French have long used the excellent term *cuirasse*. Cuirasses may furthermore be qualified as to their chemical composition, whether ferruginous, bauxitic, or ferrallitic. If the composition is unknown the qualifier is simply omitted. Not only the abandonment of the term 'laterite' but also the adoption, in English, of the term 'cuirass' is highly recommended (KdeJ).

The resistance of soil to mechanical erosion is much greater than that of purely mineral materials of the same texture even when the vegetation, whose existence determines the nature of the soil, is destroyed. Soils owe this property to their *structure*. In an unconsolidated rock the particles are inter-mixed; they may, however, show a preferred orientation and a disposition in beds. In a soil the disposition is much more complex. Colloidal humus coats the clay particles, cements them together, and then agglutinates them into aggregates and lumps which have a certain cohesion. Lime plays an im-portant role in the formation of stable aggregates; it combines with humus and retards its decomposition. Humic calcareous soils generally have good cohesion and are resistant to mechanical erosion (e.g. the chernozems, rend-zinas, and rankers). This is easily demonstrated in picking up a lump of soil: it does not disperse like a handful of sand. This structure also has a greater permeability. In an unconsolidated rock the only voids are those of the interstices between the grains packed together like balls in a sack. The soil, however, is traversed by innumerable fissures, cracks, and tubes; these result either from a slight contraction due to drought, from frost, from the action of roots and animals, or from variations in volume due to the re-moval of certain substances by leaching and the swelling of other substances as through hydration. Incessant changes of volume resulting from chemical reactions and the growth and rotting of roots prevent the soil from being as compact as a rock. Permeability and aggregation combine to make soil resis-tant to the mechanical agents of erosion. Permeability facilitates infiltration and reduces overland flow, and aggregation increases the resistance of the particles to entrainment by running water or by wind.

The dimension of the aggregates is the main determining factor on the in-fluence of the wind. The most favourable texture for aeolian deflation is be-tween 0·5 and 0·02 mm. But aggregates seldom have such a small dimension, and in general the presence of humus reinforces the resistance to aeolian erosion. Thus in northern Germany the soils of cultivated sandy loams are practically immune to deflation as soon as the proportion of humus reaches 6 to 10 per cent. But sometimes the proportion of humus is less, and the aggregates being smaller may become prey to the wind. This is the case of the Canadian prairie where the deflation of large areas is facilitated because the humus here produces small aggregates. All that is then necessary for de-flation to take place is an extensive enough interruption of the vegetation. On overgrown dunes the minimal dimensions of an open space which would allow a measurable amount of erosion are of the order of 2·5 by 9 m (8 by 30 ft); on more solidly aggregated soils they may attain 400 m (1 300 ft) and more. Such numerical data are significant; outside of complete denudation as caused by extensive cultivation, aeolian deflation can only act in deserts and in the surrounding steppes with an open and discontinuous plant cover. This is confirmed by the infrequent occurrence of round and frosted aeolian sands in the Jurassic, Cretaceous, and Tertiary deposits of Europe and North America. Not only are plants a direct obstacle to the wind, but the presence of a certain amount of humus in the soil impedes deflation.

The properties of soils differ even more in relation to the influence of overland flow. They are dependent upon the relationships between permeability and aggregation. Aggregates play an important role in splash erosion. When they are fragile, the pelting of the raindrops breaks them and the debris is thrown into the air as if there had been an explosion. Moved by overland flow they eventually fill the interstices between the particles. The soil therefore becomes rapidly impermeable and presents a unified surface on which overland flow is facilitated. When dry it is 'puddled' and has a shiny smooth surface. Its aggregates, however, are as unstable as those of any soil. The presence of humus and lime prevents puddling, whereas the presence of chlorides, dispersing the clay, facilitates it. Saline soils are naturally puddled; water hardly penetrates them when it rains. Clay has a certain cohesion when it is deflocculated. Silts and fine sands with little humus are much more sensitive to splash erosion and overland flow than clays.

Compact gley soils are not as permeable as rendzinas or as grey-brown podzolic soils which have a much better structure. Podzols, characterised by a highly leached A_2 horizon poor in humus, clay, and iron oxides, sometimes become compact due to moisture; their erosion is facilitated by an insufficient development of aggregates. In the west of the Ukraine, according to Magomedov (1950), a grass covered soil subjected to intense showers dropping at least 40 mm (1·6 in) of water, with an overland flow equal to '1' on argillaceous soils, registered '3' on chernozems, and '24' on grey powdery argillaceous soils originally formed under forest. The overland flow which was 0·4 per cent on the grass covered argillaceous soils thus reached 9·6 per cent on the grey argillaceous soils poor in structure. The most erodable soils were those which had either compact or powdery upper horizons: on the compact ones erodibility was caused by restricted infiltration; on the powdery ones by the mobility of the particles. Fletcher and Beutner's studies (1941) in the upper basin of the Gila, on the border between New Mexico and Arizona, appropriately emphasise the double influence of soil texture and of type of soil of climatic origin on erosion. There are in that region two types of soils: relict brown soils formed during a humid phase of the late Pleistocene; and the present desert soils which are much poorer in humus. The erodibility coefficient varies according to the parent rock:

PARENT ROCK	DESERT SOILS	BROWN SOILS
Quartzite	6·8	3·7
Basalt	12·8	3·4
Rhyolite	10·0	4·8
Limestone	14·8	3·2
Granite	9·9	9·9

The main fact is the greater erodibility, on the average, of desert soils; they have less humus and a poorer structure than grey-brown podzolic soils. In semiarid climates the lesser intensity of the soil-forming process produces soils that are less resistant to erosion than soils formed in humid climates; soil erosion is therefore easily catastrophic. Another consequence of such limited soil development is that the differences from one parent rock to another are greater in desert soils than in grey-brown podzolic soils. Granite seems to be the only unexplained exception.

2. *Crusts and cuirasses* The development of a thick and compact illuvial horizon has quite different consequences. Such a development impedes erosion in the same way as resistant rocks and becomes a truly *stabilising relief factor*. As soon as the horizon is sufficiently developed, it interferes with the infiltration of rainwater and consequently increases the runoff and the variations in the water content of the deeper layers of the regolith. The result is an acceleration in the formation, or at least in the denudation, of the crusts and cuirasses because more runoff increases mechanical erosion on the surface. Some authors think that due to this increased erosion the illuvial horizon is found at lesser and lesser depths and thereby becomes increasingly subjected to superficial variations of temperature and humidity, conditions which probably favour its formation. Owing to this effect, the illuvial horizon would become increasingly consolidated and impermeable. It is certain that the cuirasses of Guinea are at present being denuded by erosion. The French refer to such processes by the term *bowalisation*.[1] It is accelerated by brush fires which destroy the humus and thereby facilitate the erosion of the superficial soil.

It is necessary to distinguish between calcareous and saline crusts as well as between bauxitic and ferruginous cuirasses. Calcareous and saline crusts are formed by soluble materials and are therefore less resistant than cuirasses. In the Magreb calcareous crusts are not forming at the present time; they date back to the end of the successive cold Pleistocene periods (pluvials). Today they are often covered by vegetation which is a spontaneous open Mediterranean woodland at least in the north. The roots of the trees successfully penetrate the crusts and begin to break them up as if they were limestone. The degradation of the vegetation by man accelerates splash and aeolian erosion which sometimes bares the crusts, which then break up slowly under the influence of the weather, particularly frost, if they occur in regions of cold winters. As the crusts are relatively thin (0·5 to 3m thick: 1·5 to 10ft), their influence on the relief is not as marked and durable as that of cuirasses. Nevertheless, they do contribute significantly to the preservation of palaeoforms.

The bauxitic and ferruginous cuirasses of the humid tropics play a much more important geomorphic role. It is quite possible that they are still

---

[1] After the vernacular *bowal* (plural: *bowé*) (KdeJ).

forming in the wet–dry tropics with the help of brush fires. The geomorphic role of cuirasses is due to three specific properties:

1. Their thickness, which is always several metres and may reach and even exceed 10m (33ft). This thickness is due to the intensity of the chemical processes in a humid tropical environment.

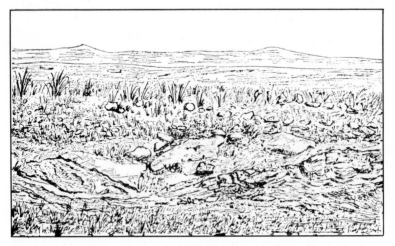

FIG. 2.13. A calcareous crust in the Magreb, near Oujda, eastern Morocco (*drawing by D. Tricart after a photograph by J. Tricart*)
Middle Pleistocene crust. Braided layers enveloping less indurated lenticular masses. Superficial disintegration. In the background, beyond the piedmont slope stabilised by the crust, the Beni-Snassen Range with inselbergs.

2. Their chemical nature; because they are not composed of soluble products but of residues that are not easily dissolved and removed: oxides of iron, manganese, and aluminium. They may undergo some change, it is true, under certain favourable circumstances. Near Dagana, in Senegal, the early Pleistocene ferruginous cuirass underwent some weathering into loam before the appearance of the middle Pleistocene dunes (Tricart). But such weathering is slow and difficult, and cuirasses may persist as outcrops during entire geological epochs, as is the case of the highland bowals of Guinea.

3. Their very compact structure. Cuirasses are particularly hard when affected by periodic brush fires. Their mechanical resistance is much superior to that of calcareous crusts. Fissures are few and fragmentation is all the more difficult as there is no frost. There is, however, a certain amount of degradation on outcropping cuirasses; it produces a rubble of a few centimetres which seems to be due to thermoclastism[1] (temperature ranges are particularly large on dark coloured outcrops). Roots and termites also help their destruction. Lastly, on the bowal margins removal of the subjacent friable

[1] Physical weathering due to temperature variations above 0°C (32°F) (KdeJ).

materials occasions an underground drainage and piping[1] which undermine the cuirass.

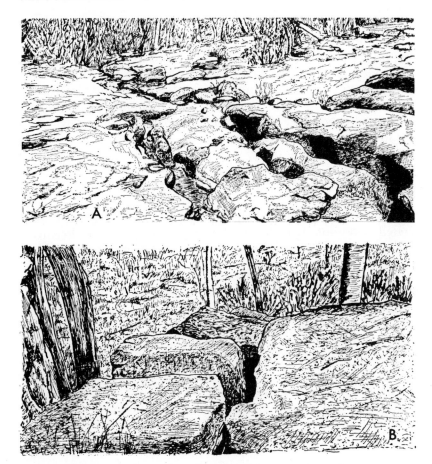

Fig. 2.14. Surface of a cuirass, after de Chetelat (*drawn from a photograph by J. Tricart*)
A. Gullies and potholes dug by runoff on the surface of a denuded cuirass slab south of Négaré, Guinea.
B. Dismantling of the cuirass on a bowal margin. The destruction is caused by the removal and consequent undermining effect of the friable subjacent materials. At Foumbaya, Guinea.

The cuirasses of the humid tropics are therefore of the utmost geographical importance. They effectively protect the relief, permitting an almost perfect preservation of old erosion surfaces over a very long period of time. They influence the distribution of plant formations and are a barrier to forests: cuirasses are generally covered by savannas whose less effective pro-

[1] French *suffosion*; concerning the term 'piping', see Parker (1963) (KdeJ).

tection favours the continuation of the processes of illuviation. Cuirasses produce a whole series of special reliefs that are characterised by an important amount of runoff which may reach 70 to 80 per cent of the rainfall on certain bowals of Guinea. Nevertheless, mechanical erosion remains limited due to the exceptional toughness of the material; the running water remains clear. Weathering proceeds mainly on the margins, inward from incised canyons through an undermining of the friable materials below the cuirass. The settling which results from undermining breaks the cuirass into blocks, and the margins retreat little by little. Ferrallitic or ferruginous cuirasses introduce into the landscape of the humid tropics, tablelands and surfaces of a remarkable uniformity which often dominate concave slopes. Cuirasses are resistant par excellence. Most of the distinctive 'structural' reliefs such as tablelands and cuestas are produced by cuirasses. After sandstones and quartzites, cuirasses are the most important formations producing this type of relief in the tropics.

Thus the soil-forming processes help to create the environmental conditions exploited by erosion. Knowledge of the particular laws of pedogenesis may be a precious help to the geomorphologist and assist him in making progress in this domain.

THE USE OF PEDOLOGIC DATA IN GEOMORPHOLOGY

Soils reflect a fragile equilibrium between relief, climate, and vegetation. If man destroys the vegetation, the soils, which took thousands of years to form, may be squandered in a few years through accelerated erosion. If the relief evolves very rapidly, soils may not have the time to form and remain embryonic. If the climate changes new soils are formed, and their differences with the old soils may give some measure of the amplitude of the oscillation. Soils may therefore help the geomorphologist to unravel the modifications which have taken place in the evolution of the relief as well as to reconstruct the nature of the past morphogenic processes.

1. *Analysis of the morphogenic processes in relation to soils*   The most important law in pedologic geomorphology is: 'Chemical erosion is approximately proportional to the intensity of the soil-forming process, and the normal evolution of soils is all the more advanced as mechanical erosion is restricted.' This law can still only be enunciated in a qualitative manner, but the increasing number of numerical data will soon permit it to be stated quantitatively.

The existence in a given place of a typical, complete soil with its well defined horizons implies that mechanical erosion is slower than the soil-forming process. Let us take the example of a midlatitude forested slope: overland flow is minor although not entirely absent; from place to place falling raindrops hit bare spots between dead leaves, such as on soil brought to the surface by burrowing animals, worms, or a fallen tree (a frequent case in virgin forests). The drops scatter the particles or even entrain them downslope. But such grains of sand, silt, and colloidal aggregates mix with the plant debris and, as long as they are present in small enough quantities, are incorporated in the superficial humus. If the process is rather slow nothing is changed in the soil profile of the area of accumulation at the foot of the slope. Neither is anything changed in the soil profile of the upper slope; new leaves will fall and reconstitute the humus which, together with the superficial mineral matter, covers the soil.

It may happen that slope erosion is accelerated. This can occur, for example, following deforestation, or along a forest path, or in fields. On the upper slope the water no longer carries only the scattered friable mineral particles over the ground surface; the most abundant runoff becomes channelled, and the rills incise themselves into the humic horizon attacking the subjacent soil horizons and, eventually, even the bedrock if it is unconsolidated. At the foot of the slope, where the gradient is less, the running water can no longer transport the load it carries and deposition takes place. The old soil, if it has not also been eroded, is then buried under the debris; this debris arrives too rapidly for a new soil to form in the interval between two accumulations and *colluvium* is formed. The slope is now composed of two soil profiles: in its eroded part there is a *truncated profile* exposing what otherwise would have been deeply buried layers of the subsoil; and in the area of accumulation there is a *buried profile* where the normally superficial soil horizons are now covered by colluvium (materials of recent transport, as explained above, or those whose pedologic evolution is not far advanced—Fig. 2.16A). In the ploughed incipient vales of Beauce (central Paris Basin) colluvium attains a thickness of 100 to 150 cm (3 to 5 ft).

Normal profiles, truncated profiles, and buried profiles are indices of the intensity of mechanical erosion to the geomorphologist. They all have their precise significance:

Normal soil profiles indicate either an absence of mechanical erosion or an equality in the rates of soil formation and of biotic or mechanical erosion. The soil gains at the expense of the bedrock that which it loses through down wastage; if there is no soil, none will develop. A perfectly rigorous equality is exceptional; usually there are small anomalies which indicate the direction of the geomorphic evolution. When mechanical erosion is nil (e.g. on a near plane surface covered with dense vegetation) the regolith progressively thickens at the expense of the bedrock and the soil horizons may reach an even fuller development (planosols). At the foot of a slope where colluvium accumulates, the soil thickens, but the disposition of the horizons changes: the A horizon alimented from particles coming from above profits much more from the general thickening than the eventual B and C horizons which only result from *in situ* evolution. Finally, when the soil undergoes some erosion but not enough to impede its development, it is thinner; this is the case of steep slopes. It is especially the A horizon which is affected by erosion and which supplies the major part of the removed particles. If all other factors remain equal, this provides an explanation for the thinner soils of mountain slopes. Chemical differences in the soil profiles of a given uniform area, depending on the site, must be added to these differences in thickness: on the upper part of the hillsides leaching is more intense than on the margins of the bottomlands. Thus in the Göttingen area of Germany where erosion is moderate, the soils developed on limestones are more evolved on the higher part of the hillsides because of better drainage. But such a location implies a near absence of superficial erosion. If this were not so sheet erosion would impede the development of the soil. Pedologists, for this reason, more and more frequently compare the soil profiles of topographic cross-sections perpendicular to the relief. They call such soil series *soil catenas*. Geomorphologists have a vital interest in following this work. But they should also show

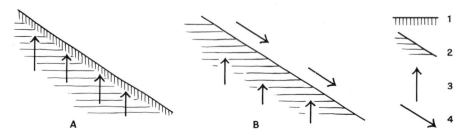

FIG. 2.15. Origin of the absence of soil on a slope

1. Soil. 2. Bedrock. 3. Intensity of the soil-forming process. 4. Rate of superficial mechanical erosion. The length of the arrows is proportional to the intensity of the processes.
A. The slope is soil covered because the soil-forming process is faster than mechanical erosion.
B. The soil on the slope has been destroyed and is incapable of forming again because the soil-forming process is slower than mechanical erosion.

pedologists the shortcomings of such a purely topographic interpretation. It is, in fact, only strictly applicable to an environment that is lithologically and geomorphologically uniform. If the relief is composed of units formed at different epochs, the soil-forming process on each one of them has its own particular history. Thus a terrace cut by valleys often has more or less re-adapted palaeosols older than the soils of the vale slopes.

The simple difference in topographic location is therefore not enough for a correct understanding of the catena. By giving it an exclusively topographic meaning a great error could be committed: one might end up attributing the differences between the soils of the terrace and of the valleys only to differences in leaching due to relief and not, as is the case, to differences in age. The catena concept must therefore be redefined in order to utilise it properly and extensively. *Topographic catenas*, composed of contemporaneous soils which are the consequence of dynamic differences only (differences affected by morphogenic factors), should be distinguished from *morphogenic catenas*, which are related to the complex evolution of the relief wherein several generations of soils were more or less completely developed.

Buried soil profiles may in certain cases reveal irregular rates of accumulation. Excellent examples in the form of palaeosols (*lehms* and chernozems) are found imbedded in loess; others in the form of coal developed from subaqueous organic soils are found interbedded with sandstones and shales; still others in the form of red palaeosols, representing short interruptions in a subsiding sedimentary basin, are found imbedded in the silts of the fluvial silt and gravel formations of the Valensole plateau of Provence. They may have a tectonic significance as do those of the Valensole, but such an origin should not be imputed to all buried soil profiles. One should not forget that on fluviomarine plains or in continental basins, sedimentation is often irregular due to the unceasing shifting of stream channels and breaching of

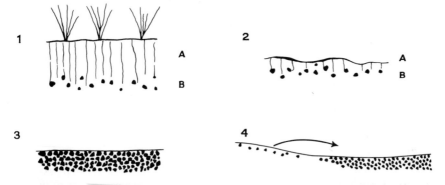

FIG. 2.16. Sketch of the formation of the ferruginous cuirass of northern Senegal
1. Development of a soil with ferruginous concretions under a savanna vegetation.
2. Disappearance of the vegetation due to climatic desiccation, destruction of the A horizon, and hardening of the concretions.
3. Concentration of the concretions by erosion and cementation into a cuirass.
4. Removal of the concretions by runoff in upland areas and their concentration in topographic depressions.

natural levees. But a repetition of palaeosols in a thick sedimentary series either indicates a progressive drowning or a general subsidence. Climatic oscillations may also be the cause of them, as is the case of the palaeosols imbedded in loess, which have developed during periods of non-deposition occasioned by climatic ameliorations (interglacials or simply warmer episodes during the same cold period). Buried soil profiles may be truncated just before a renewed accumulation when the upper soil horizons are reworked. In this way solifluction on loess has often beheaded the profiles of interglacial soils prior to the deposition of more loess. In the Valensole deposits, running water produced some erosion at the top of the palaeosols before burying them with gravel. Such palaeosols are often of great value. In the loess of central Europe they have provided the basis of a stratigraphy, each buried soil having regional climatic characteristics. Often they also contain pollen, which makes possible the reconstruction of the contemporaneous vegetation and, indirectly, of the climate. Finally, their extension reveals the area on which only very little erosion has occurred before the deposition of the fossilising sediments.

Truncated soil profiles indicate resumed erosion; during the preceding period mechanical erosion was not important enough to impede the soil-forming process, but then its action increased to the point that it overtook the latter in rate of speed. The soil horizon that usually resists best is a B horizon with concretions. Often it alone persists and sometimes is reworked. Some authors think that concretions become indurated at the contact of air, which would facilitate their preservation. This may be the case of the ferruginous pisolites (old loose pedologic concentrations of a B horizon) which occasionally appear on the ground surface in the Sahelian zone of West Africa (in the area of Lake R'Kiz in Mauritania). The ferruginised products which occur on the Oligo-Miocene surface and on certain Pliocene levels in the Paris Basin are quite similar. Sometimes only the C horizon, composed of weathered rock, is all that is left. Such are a number of granitic corestone horizons of the Harz, the Vosges and the Massif Central, which erosion has nearly everywhere cleared of the mantle of regolith which must have covered them during the Tertiary. The amount of soil erosion varies a great deal from place to place. It is reflected in the denudation of the regolith, which varies from one area to another and, as each horizon has its own colour, in the colour differences of the ground surface. On whitish marls, for example, the most affected areas appear quite light coloured, betraying the proximity of the parent rock, whereas less eroded surfaces preserve remnants of brown or red soil. Aerial photos clearly indicate such differences in colour; for example, in chalky regions such as Picardy. On a more resistant bedrock there are bare outcrops, tors or blockpiles. Such sudden waves of erosion are nearly always climatic or anthropic (destruction of the vegetation by man). When soil horizons are less distinct, methods of pedologic reconnaissance can be used to discover erosional episodes. In this way Lemée (1950) was able to map the extent of anthropic erosion in the loess regions of Alsace by studying the variations in the pH, which are higher in loess bared by erosion than in the *lehms* which normally cover them.

This pedologic method may become quantitative: knowing the rate of soil formation, it is possible to determine an order of magnitude for the rate of erosion or deposition. On sand or sandstone a podzolic profile requires a minimum of 1 000 to 1 500 years to develop. In Sweden podzolic profiles are perfectly typical in regions cleared by the ice only 6 000 years ago. If a slope is evenly blanketed by a podzol some 50 cm (20 in) thick, one may say that it does not suffer more erosion than a few millimetres per century at the most; if more, the soil profile could not have formed. With such information it has been possible to locate eighteenth-century clearings in the Vosges forests; on such clearings podzols have not yet had the time to reconstitute themselves,

while they continue to exist in the surrounding country that remained forested. In northern France Demolon's studies indicate that on the average one
hectare of loess lost 250 to 500 kg of calcium bicarbonate per year (215 to
430 lb per acre). With a calcium content of 20 per cent this represents the de-
calcification of a 7 to 25 mm (0·3 to 1 in) layer per century. Non-eroded
lehms, which have an average thickness of 0·75 to 1 m (30 to 40 in), would
according to Demolon be 6 000 to 18 000 years old; lehmification would
have begun between 16 000 and 4 000 B.C., on the average 10 000 B.C. This
evaluation is quite satisfactory because the alteration of loess into *lehm* must
have begun right after the end of loess deposition which probably occurred
near the end of the late Würm (Weichsel), or about 8000 B.C.

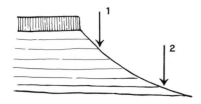

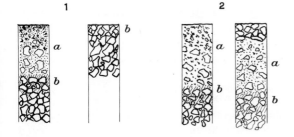

FIG. 2.17A. Formation of truncated and buried
soil profiles.

At 1, upper part of the slope, a rendzina soil has
developed on calcareous scree (left hand profile):
   a horizon: few rocks persisting in a clay-loam
   matrix.
   b horizon: slide-rock coated with calcite.
   Later a renewal of erosion attacks the slope; the
result is a truncated soil profile, as shown in the
right hand profile: the friable *a* horizon has completely disappeared.
   At 2, at the foot of the slope, a rendzina soil has
also developed up to the time of the renewal of
erosion. Later the upper part of the horizon has
been truncated and the debris of the upper slope
has buried the profile. On the remains of the loamy
a horizon there is a layer of unweathered debris
(right hand profile).

113

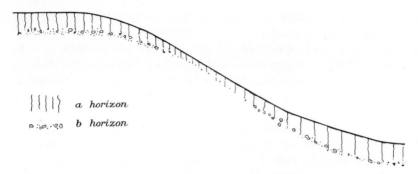

FIG. 2.17B. Variations in the thickness of a soil on a hillside (soil catena)

On top of the ridge there is active leaching and no mechanical erosion: the *a* and *b* horizons are equally well developed. On the upper slope, with the steepest gradient, leaching is active but the *a* horizon is thinned by wastage (creep, splash, or anthropic erosion). Slope waste accumulates at the foot of the slope, enriching the *a* horizon and making it proportionally thicker than the *b* horizon.

Pedogenic data can therefore, with prudence, be used to make quantitative evaluations; but with our present knowledge one should not attempt to go beyond the finding of orders of magnitude. Nevertheless the method is valid and will become more and more accurate with the further development of soil science. To pedology such geomorphological observations as well as absolute datings (varves, radiocarbon) should be of inestimable value and should provide a check on the data obtained in the laboratory or with a lysimeter box.[1]

2. *Geomorphic evolution* We have seen above that soils and palaeosols constitute a means of relative dating and of palaeoclimatic investigation. Both are useful in geomorphological reconstructions. Palaeosols may be preserved for a very long period of time if fossilised by overlying strata. Many geologic anomalies may be cleared up if pedology is called upon to explain them. The preservation of non-fossilised palaeosols is of course more difficult. If highly consolidated illuvial horizons are possibly several tens of millions of years old (the case of some bauxitic terraces), friable deposits are normally of Quaternary or Neogene age. In France, for example, quite a number of well preserved remnants of Pleistocene soils as well as important residual soils from the upper Tertiary have been recognised. They include certain sandy soils of the ancient massifs, Pliocene clays, Neogene formations with ferruginous concretions, certain siliceous limestones (the Pliocene *meulière* of

[1] A lysimeter box is a device to measure evapotranspiration. A tank is implanted into the ground and the displaced soil and plants are replaced in such a way that they are level with the surface. At the bottom of the box the rainwater that has percolated through the soil and the mat on which it rests is collected in an open space. The difference between a series of inputs (measured by a rain gauge) represents the quantity of water lost through evapotranspiration (KdeJ).

Tardenois), and the red clays of weathered molasse[1] and basalt. An essential factor in the preservation of Tertiary soils in temperate lands is their thickness; some which have developed under warm climates during a long period of time are thicker than Pleistocene soils which have developed under chemically less active climates and during a shorter period of time. The great thickness of Tertiary soils may have contributed to their preservation.

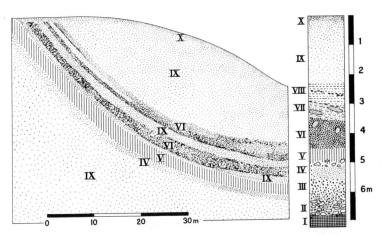

FIG. 2.18. Profiles of palaeosols in the loess of Lower Austria, after Brandtner (1954)
Vertical scale highly exaggerated.
    X Recent soil
   IX Typical ochre aeolian loess with calcareous veinlets
VIII Grey sandy alluvial loess with charcoal remains
  VII Light brown sandy soliflucted loess with lenses of reworked chernozem
   VI Very dark chernozem with loess structure
    V Weathered soil of number IV sand, red rusty brown with angular calcareous concretions
   IV Fine aeolian sand, slightly calcareous
  III Rust coloured Tertiary sand merging with IV
   II Tertiary rubble
    I Blue-grey Tertiary marls
    The left hand cross-section and the right hand profile have not been made in the same place and show some differences. On the cross-section *lehm* and chernozem horizons alternate with loess.

While the upper horizons have been more or less deeply eroded or modified by later soil-forming processes, deep horizons have sometimes remained identifiable. Frequently the whole soil can no longer be observed but only the C horizon or parent material.

    The preservation of palaeosols is of course first a function of the degree of erosion and of the total importance of its action during a given period. The

[1] Late geosynclinal (*Aubouin*) marine or continental often calcareous sandstones and conglomerates (KdeJ).

chances of finding old non-fossilised palaeosols on slopes are therefore rather remote. The best circumstances for their preservation are found on planation surfaces which have suffered a minimum of mechanical erosion. The Tertiary soils of the Massif Central only persist on very level, little dissected erosion surfaces. Detailed study is one of the best means of correlation between the remnants of these surfaces which have become isolated either by erosion or by differential uplift due to crustal deformations. Baulig has used this method with profit, and current progress in pedology makes it increasingly accurate. It should be emphasised, however, that when using it one should make use of positive arguments: if a slope does not have the same soil as the neighbouring tableland it is either because the dissection came after the formation of the upland soil, or that the thinner slope soil was removed as fast as it was formed by increased erosion. The formation of inselbergs during the not very humid climates of the end of the Tertiary makes this last case particularly frequent.

Pleistocene soils, which have suffered a less prolonged erosion, are usually better preserved than Tertiary soils. A detailed geomorphology of recent geologic history may be based upon them, because they are found not only on level surfaces but even on slopes. Wider correlations are therefore possible not only between planation surfaces but also between varied landscapes. For example, the erg formed by the wind in the south of Mauritania and in the north of Senegal during the pre-Eemian low sea-level had its surface weathered into a leached red soil 1·5 to 2 m thick (5 to 7 ft) during the Eemian (Ipswichian).[1] During the pre-Flandrian (Weichselian) low sea-level, a new dry recurrence permitted a moderate deflation of the sands, but no new red soil has formed since the Eemian.[2] The red soil can therefore be used to date the two generations of dunes and to map their extension. F. Bourdier, following A. Penck, has also shown what use can be made of the red soils (*ferretto*) of the early Pleistocene in order to distinguish the moraines of the earliest Alpine glaciations.

It seems that the possible use of palaeosols as a means of dating may be particularly useful in the temperate, arid, and semiarid zones. In the wet tropics excessive activity of the soil-forming process is unfavourable to the preservation of palaeosols with the exception of illuvial horizons. Tricart's observations, however, seem to indicate that even under evergreen rainforests the time necessary for the development of a typical kaolinitic weathering profile is considerable. Such profiles exist on terraces not younger than early Pleistocene near Salvador, Brazil and south of Lake Maracaibo. Detailed studies may reveal as yet hardly suspected possibilities in this morphoclimatic environment.

[1] The *Eemian* is the highest sea-level stage of the last interglacial (in Germany and the Netherlands), corresponding to the *Tyrrhenian II* in the Mediterranean region and the *Oujlian* of M. Gigout (1957) in Morocco. It was 5 to 8 m above the present sea-level. Refer to Tricart, J., 'Les variations Quaternaires du niveau marin', *Inf. Géogr.*, **22** (1958), 100–4 (KdeJ).

[2] See footnote page 173.

In temperate and cold climates the humic horizons of palaeosols are some-times preserved. Such residual horizons may hold pollen whose study will yield an idea of the flora and may occasionally provide a means of dating. They are therefore extremely valuable, and samples should be collected from them and studied by pollen specialists. It is due to the intensive develop-ment of palynology that the Pleistocene of Germany and the USSR, for example, is better known than that of France.

Pollen furthermore provides palaeoclimatic indications which help the interpretation of landforms. In the absence of pollen, soils play the same role to a certain extent. Taking into account their later eventual alteration under different climates, they provide a clue to the original climate and so make possible a morphoclimatic interpretation of the relief. Thus the leached red soils of the lower Senegal (mentioned above) imply a more humid climate during the Eemian than exists at the present; they are, however, also charac-terised by a marked dry season during which the desiccation of the soil pro-duced the dehydration of the iron oxides. Instead of belonging to the Sahe-lian zone, as at present, the area was then subject to an alternating wet–dry Sudanese type of climate.[1] In a similar way the essential data for reconstruct-ing the succession of the Pleistocene climates of central and western Europe have been obtained from the study of palaeosols buried in loess. The remark-able studies of Schönhals (1951) may be cited as an example.

Conclusion

The influence of climate on relief is fundamental. It is manifest in the most diverse and interrelated ways. It determines in a direct way the distribution of certain processes such as frost-weathering and the intensity of a number of geomorphic agents (wind, splash erosion, stream flow . . .). But its action is mainly indirect through its influence on vegetation and on the activities of animals and micro-organisms. The actions of organisms are all important in the sculpturing of the earth's surface: they disintegrate rocks, soften the earth, add to it a mass of organic matter, and liberate carbon dioxide and acids which react on the mineral matter. They are basic to the superficial transformation of rocks into soils, which have special geomorphic properties.

We are now far removed from the concept of normal erosion. The variety of the conditions existing on the earth's surface contrasts with the uniformity postulated by oversimplified and excessively theoretical schemes. Davis's view of the world was not that of a naturalist; he had excluded from it the most essential factor: life. Vegetation and the development of soils do not have a place in 'normal erosion' as he conceived it, and the only 'anoma-lies' which he admitted were those of the desolate regions of the world: the glacial and sandy deserts. He had failed to see the essential characteristic of our earth: the overpowering force of organisms in full expansion since the most remote known geologic past.

[1] See footnote page 173.

It is therefore important to reintegrate geography into the natural sciences by abandoning the overly theoretical and unnatural concept of 'normal erosion'. It must be replaced by two basic principles derived from the facts:

1. The land-forming processes, which reflect the geologically changing interaction of forces which operate on the earth's surface, are dependent on tectonic forces, climate, and the development of organisms.

2. The plant cover plays a fundamental geomorphic role because it determines the superficial modification of the lithosphere by the soil-forming processes (pedogenesis). Through these processes the plant cover influences chemical erosion. Inversely, the plant screen, together with the soils, is opposed to the development of mechanical erosion.

Bibliographic orientation

The matter treated in this chapter has not yet been the object of any systematic treatment, and it cannot, therefore, be found in any textbook. The documentation is, moreover, very dispersed and is scattered in the reports of agricultural experiment stations, textbooks of pedology and certain ecologic studies. As is the case for the other chapters of this volume, there are in the original French volumes of the Tricart *Traité* a greater number of references devoted to the various morphoclimatic zones. Here we present a selection which may be useful to the student for the completion of his own research.

General works

The following general works are listed especially from the point of view of their historical interest:

BARRELL, J. (1908) 'Relation between climate and terrestrial deposits', *J. Geol.* **16**, 159–90, 225–95, and 363–84.
TRICART, J. (1953) 'Climat et géomorphologie', *Cah. Inf. Géogr.*, no. 2, 39–51.
VISHER, STEPHEN S. (1941) 'Climate and geomorphology', *J. Geomorph.* **4**, 54–64.

Concerning the concept of the system of erosion, refer to Cholley, 1950.

Climatic data useful to geomorphology

Examples of works which attempt to define the ecologic climate

BERG, H. (1947) *Einführung in die Bioklimatologie*, Bonn, Bouvier, 131 p.
  Most complete exposition of the subject. Useful data.
BERG, L. S. (1947) *Climate and Life*, 2nd ed, Moscow, Geografguiz, 356 p.
BOYKO, H. (1949) 'On climatic extremes as decisive factors for plant distribution', *Palest. J. Bot. Rehovot* **7**, 41–52.
DOROGANEVSKA, E. A. (1954) 'The question of the hydrothermal coefficient of the period of vegetation of cultivated plants', *Izv. Akad. Nauk. SSSR*, Geogr. ser., no. 6, pp. 51–61.
GAUSSEN, H. (1949) 'Projets pour diverses cartes du monde au 1/1 000 000. La carte écologique du tapis végétal', *Ann. Agron.*, pp. 78–104.
GEIGER, R. (1930) 'Mikroklima und Pflanzenklima', *Handbuch der Klimatologie*, Berlin, Bornträger.
  Interesting summation although somewhat out of date. Numerous data.
GODARD, M. (1949) 'Microclimats et mésoclimats du point de vue agronomique', *Ann. Agron.* **19**, no. 4.

PHILLIPS, J. (1959) *Agriculture and Ecology in Africa*, London, Faber, 424 p., one separate map.

REMPP, G. (1937) 'Sur les frontières et les relations entre le macroclimat, le mésoclimat et le microclimat et entre le climat physique et le bioclimat', *Météorologie*, pp. 265–74 and 379–91.
Excellent general view of the problem.

SHAW, ROBERT H. ed. (1967) *Ground Level Climatology*, Amer. Assoc. Adv. Sc. 408 p.

WARMING, E. (1909) *Oecology of Plants : an introduction to the study of plant communities*, English ed by Percy Groom and I. B. Balfour, Oxford, Clarendon Press, 422 p.

On relationships of the atmospheric climate and the climate at the ground

BRAZIER, C. and EBLÉ, L. (1934) 'Introduction à l'étude des températures de l'air et du sol au voisinage de la surface terrestre', *Météorologie*, pp. 97 *et seq.*

BRAZIER, C. and EBLÉ, L. (1935) 'Sur une particularité de la transmission de la chaleur dans le sol', *C. R. Acad. Agr.*

CRITCHFIELD, H. J. (1960) *General Climatology*, New York, Prentice-Hall, 465 p., 180 fig.

DEFANT, A. (1916) 'Über die nächtliche Abkühlung der unteren Luftschichten und der Erdoberfläche in Abhängigkeit von Wasserdampfgehalt der Atmosphäre', *Sitzungber. Wien. Akad.*, Sect. 11a, pp. 1537 and following.

EBLÉ, L. (1946) 'Etudes expérimentales sur la propagation de la chaleur dans le sol', *Météorologie*, pp. 269–91.

GEIGER, R. (1959) *The Climate Near the Ground*, trans. Steward, Harvard University Press, xxii, 494 p.

GESLIN, H. (1935) *La température du sol*, Rapport à la Commission de l'Association Internationale de la Science du Sol, Versailles, 1934, *Météorologie*.

HAUDE, W. (1924) 'Temperatur und Austausch der bodennahen Luft über einer Wüste', *Beiträge Phys. frei. Atmos.* **21**, 129 et seq.

HEYER, E. (1938) 'Über Frostwechselzahlen in Luft und Boden', *Gerlands Beitr. Geophys.* **52**, 68–112.

MILOSAVLJEVIC, M. (1950) 'Relation entre la température minimum à 2m et à 5cm au-dessus du sol à Belgrade', *Bull. Soc. Serbe Géogr.* **30**, no. 1, pp. 11–25.

On methods to use to determine the ecologic climate

BARAT, C. (1956) 'Les données de la pluviologie dans la zone intertropicale', *Bull. Ass. Géogr. Franc.* **261–2**, 175–84.

BARAT, C. (1957) *Pluviologie et aquimétrie dans la zone intertropicale*, Mem. IFAN, Dakar, no. 49, 80 p.

BILANCINI, R. (1948) 'Su un tipo di analisi statistica della variabili meteorologiche', *Riv. Meteorol. Aeron.*, Rome, **8**, no. 4.

BELEHRADEK, J. (1935) 'Temperature and living matter', *Protoplasma Monogr.* **5**, no. 8, Berlin.

KUCERA, C. L. (1954) 'Some relationships of evaporation rate to vapor pressure deficit and low wind velocity', *Ecology* **35**, 71–5.
Basic work in the study of humidity variations.

LEOPOLD, L. B. (1951) 'Rainfall frequency: an aspect of climatic variation', *Trans. Amer. Geophys. Union* **32**, 347–57.

SERRA, L. (1951) 'Interprétation des mesures pluviométriques, lois de la pluviosité', *Int. Union Geod. Geophys.*, Brussels meeting.

Examples of regional monographs about ecologic climate

BERNARD, E. (1945) *Le Climat écologique de la cuvette congolaise*, Publ. INEAC (Yamgambi).

GENTILLI, J. (1949) 'Foundations of the Australian bird geography', *Emu*, Oct., pp. 85–130 (cf. pp. 106–13).

GENTILLI, J. (1952) 'Southwestern forests of Western Australia, a study in geographical environment', *Queensland Geogr. J.* **54**, 12 p.

SELTZER, P. (1935) 'Etudes micrométéorologiques en Alsace', doct. diss. Fac. Sc., Univ. Strasbourg, 58 p.

Examples of rainshower intensity

DAMMAN, W. (1948) 'Zur Physiognomie der Niederschläge in Nordwestdeutschland', *Göttinger Geogr. Abhandl.* **1**, 58–69.

INST. NAT. ETUDE AGRON. CONGO (Yangambi) (1951) 'Chutes de pluie au Congo Belge et au Ruanda-Urundi pendant la décade 1940–1949', Brussels, *Bur. climatologique, communication* no. 3, 248 p.

HARTKE, W. and RUPPERT, K. (1959) 'Die ergiebigen Stark- und Dauerregen in Süddeutschland nördlich der Alpen', *Forsch z. Dt. Landeskunde*, **115**, 39 p., 32 maps, Bad Godesberg [cf. *Erdkunde*, **15**, (1961) p. 246]. Also consult other publications by Hartke.

KOCH, P. (1950) 'La violence des orages dans ses relations avec le débit des égouts urbains', *Houille Blanche*, **5**, spec. no. B, 679–82.

REYA, O. (1948–49) 'The intensity of precipitation on the Slovenian coast', *Geogr. Vestnik*, **20–21**.

SEEYLE, C. *The Frequency of Heavy Rainfalls in New Zealand*, Wellington, N.Z. Inst. Eng.

VIDAL, J. M. and POTAU, M. (1951) *Intensidad de las lluvias en Barcelona*, Serv. Meteorol. Nac., 14 p.

Direct influence of climate on the morphogenic processes

One should first of all refer to the three following general works:

FOURNIER, F. (1949) 'Les facteurs climatiques de l'érosion du sol', *Bull. Ass. Géogr. Franç.*, pp. 97–103.
Excellent study with numerical data, and also an example in method.

FOURNIER, F. (1960) *Climat et érosion; la relation entre l'érosion du sol par l'eau et les précipitations atmosphériques*, Paris, PUF, 201 p, 15 graphs.

POUQUET, J. (1951) *L'Erosion*, Paris, PUF, Coll. Que sais-je?, 128 p.
Very interesting general views, applying especially to anthropic erosion.

The following works, more limited and dealing especially with anthropic erosion, include useful examples:

BARNETT, A. P. (1958) 'How intensive rainfall affects run-off and erosion', *Agr. Eng.* (St. Joseph), **39**, 703–7 and 711.

BASU, J. K. (1952) 'Soil and moisture conservation in the dry regions of the Bombay State', *Empire J. Exp. Agr.*, Oct., pp. 326–33.

EKERN, P. C. (1954) 'Rainfall intensity as a measure of storm erosivity', *Proc. Soil Sc. Soc. Amer.* **18**, 212–16.

FREE, G. (1960) 'Erosion characteristics of rainfall', *Agr. Eng.* (St Joseph), **41**, 447–9.

GALEVSKI, M. (1955) 'La corrélation entre les pluies torrentielles et l'intensité de l'érosion', *Ann. Ec. Nat. Eaux Forêts*, **14**, 384–427.

GEGENWARTH, W. (1952) 'Die ergiebigen Stark- und Dauerregen im Rhein-Maingebiet und die Gefährung der landwirtschaftlichen Nutzflächen durch die Bodenzerstörung', *Rhein-Main Forsch.*, no. 36, 52 p.

HARTKE, W. (1954) 'Kartierung von Starkregenzügen auf Grund ihrer bodenzerstörenden Wirkung', *Erdkunde*, **8**, 202–6.

OSBORN, B. (1954) 'Soil splash by raindrop impact on bare soils', *J. Soil Water Conserv.*, US, **9**, 33–8, 43 and 49.

SCHULTZE, J. H. (1953) 'Neuere theoretische und praktische Ergebnisse der Boden-Erosionsforschung in Deutschland', *Forsch. Fortschr.* **27**, no. 1, 1–7.

SCHMID, J. (1955) *Der Bodenfrost als morphologischer Faktor*, Heidelberg, Hüthig, 144 p., 27 fig.

Vegetation and the morphogenic processes

Influence of vegetation on climate

BEIRNAERT, A. (1941) *La Technique culturale sous l'équateur*, Brussels, INEAC.
Data applicable to the Congo.

BENCHETRIT, M. (1954) 'L'érosion anthropogène: couverture végétale et conséquences du mode d'exploitation du sol', *Inf. Géogr.*, pp. 100–8.

DOIGNON, P. (1946–51) *Le Mésoclimat forestier de Fontainebleau*, Fontainebleau, 3 vols, 142, 128 and 132 p.
The most detailed monograph on the climatic influence of a forest in a temperate environment.

FELT, E. J. (1953) 'Influence of vegetation on soil moisture contents and resulting soil volume changes', *Proc. 3rd Int. Soil Conf.*, vol. 1, pp. 24–7.

FOOD AND AGRICULTURE ORGANISATION (1962) *Influences exercées par la forêt sur son milieu*, Etud. Forêts Prod. Forest., no. 15, 341 p., 56 fig.

GEIGER, R. and AMANN, H. (1931) 'Forstmeteorologische Messungen in einem Eichenbestand', *Forstwiss. Zentralbl.*

LEPLAE, E. (1937) *Comptes rendus et rapports au VIIe Congr. d'Agric. Trop. et Subtrop.*, Paris.

MUSGRAVE, G. and NORTON, R. (1937) 'Soil and water conservation investigations at the soil experiment station, Missouri loess valley region, Clarinda, Iowa', *US Dept. Agr. Tech. Bull.*, no. 558, 181 p.

OVINGTON, J. D. (1954) 'A comparison of rainfall in different woodlands', *Forestry*, **27**, 41–53.

TAMHANE, R. V., BISWAS, T. D., DAS, B. and NASKAR, G. C. (1959) 'Effect of intensity of rainfall on the soil loss and run off', *J. Indian Soc. Soil Sc.*, **7**, no. 4, 231–8.

SCAETTA, H., SCHOEP, A., and MEURICE, R. (1937) *La Genèse climatique des sols montagnards d'Afrique Centrale*, Brussels, Inst. Roy. Congo Belge, 351 p.

Influence of vegetation on the mechanical agents:

ANDERSON, H. and GLEASON, C. (1960) 'Effects of logging and brush removal on snow water runoff', *Int. Ass. Sci. Hydrol.*, Helsinki, Surf. Waters Comm., pp. 478–89.

BLÖCHLINGER, G. (1932–33) 'Kleinlebenwesen und Gesteinverwitterung', *Z. Geomorph.*, **7**.

CAILLEUX, A. (1954) 'Le ruissellement en pays tempéré non montagneux', *Ann. Géogr.*, **57**, 21–39.

DABRALL, B., PREMNATH, and RAMSWAROP. (1963) 'Some preliminary investigations on the rainfall interception by leaf litter', *Indian Forester*, **89**, 112–16.

GOSSELIN, M. (1942) 'La défense des sols cultivés contre l'érosion', *Tunisie agr.*, March.

GUPTA, R., KHYBRI, M., and SINGH, B. (1963) 'Run-off plot studies with different grasses with special reference to conditions prevailing in the Himalayas and Siwalik region', *Indian Forester*, **89**, pp. 128–33.

HENIN, S. (1952) 'Quelques remarques concernant l'infiltration de l'eau dans les sols', *Rev. Gén. Hydrol.*, March–April, pp. 77–82.

HESMER, H. and FELDMANN, A. (1953) 'Der Oberflächenabfluss auf bewaldeten und unbewaldeten Hangflächen des südlichen Sauerlandes', *Forstarchiv.*, no. 11–12, pp. 245–56.

HORTON, R. E. (1919) 'Rainfall interception', U.S. Weather Bur., *Mon. Weather Rev.* **47**, 603–23.

HURSCH, C. R. (1951) 'Recherches sur les relations entre les forêts et les cours d'eau', *Unasylva*, pp. 3–10.

KENNEDY, A. P. (1952) 'Some factors influencing infiltration into soils under natural and artificial rainfall', *J. Soil Conserv. Serv.*, New South Wales, July.

KURON, H. and STEINMETZ, H. J. (1957) 'Die Plantschwirkung von regentropfen als ein Faktor der Bodenerosion', *Int. Union Geod. Geophys.*, Toronto meeting, vol. i, pp. 115–21.

MAGOMEDOV, A. D. (1950) 'Infiltration des eaux de pluie et de fusion dans les divers types de sols', *Pochvovedenie*, pp. 361–466 (French SIG trans. of Bur. Rech. Geol. Min.).

OSBORN, B. (1954) 'Effectiveness of cover in reducing soil splash by raindrop impact', *J. Soil Water Conserv.*, US, **9**, 70–6.

REE, W. and PALMER, V. (1949) 'Flow of water in channels protected by vegetative linings', *U.S. Dept. Agr., Tech. Bull.* no. 967, 115 p.
A great number of numerical data.

SOPPER, W. E. and LULL, H. W. (1967) *International Symposium on Forest Hydrology*, New York, Pergamon, 813 p.
Contains a wealth of information on the effects of the forest on precipitation, on soil water, on evapotranspiration, on runoff, on soil stabilisation, and on techniques.

STALLINGS, J. H. (1953) 'Continuous plant cover, the key to soil and water conservation', *J. Soil Water Conserv.*, US, **8**, 37–43.

WANDEL, G. and MUCKENHAUSEN. (1951) 'Neue vergleichende Untersuchungen über den Bodenabtrag an bewaldeten und unbewaldeten Hängflächen in Nordrheinland', *Geol. Jahrbuch*, **65**.

Vegetation in relation to morphogenic activity

HEINZELIN, J. DE (1953) 'Les stades de récession du glacier Stanley occidental', *Explor. Parc Nat. Albert*, 2nd ser., no. 1, 25 p.
Has a bibliography of the most important works using the method of the use of plant associations as a means of geomorphic dating.

KUHNHOLZ-LORDAT, G. (1952) 'Le tapis végétal dans ses rapports avec les phénomènes actuels en Basse-Provence', *Encycl. biogéogr. écol.*, vol. 9, Paris, Le Chevalier, 208p.

MARRES, P. (1954) 'Phénomènes actuels de surface et l'équilibre du tapis végétal dans la région méditerranéenne', *Colloq. Int. CNRS*, Régions écologiques du globe, Paris, pp. 117–23.

SCHNELL, R. (1948) 'Observations sur l'instabilité de certaines forêts de la Haute-Guinée française en rapport avec le modelé et la nature du sol', *Bull. Agron. Congo Belge*, **40**, 671–6.

Combined influences in splash erosion

Splash erosion, which has been much studied according to the methods we advocate, is an excellent example to demonstrate the complex interaction of climate, vegetation, and the mechanical qualities of soils in the intensity of erosion. Below is a choice of essential references about this subject. A bibliography on the measurements of splash erosion, methods employed and results obtained may be found in Bur. Interafricain Sols Econ. Rurale, *Bull. Bibliogr. Mensuel*, **5**, no. 9, 1955, and in a complement to this bulletin published in June 1958 (3 p.).

ANDRÉ, J. E. and ANDERSON, H. W. (1961) 'Variation of soil erodibility with geology, geographic zone elevation and vegetation type in northern California wildlands', *J. Geophys. Res.* **66**, 3351–8.

BALLAL, D. K. and DESHPANDE, R. P. (1960) 'Erodibility studies by a rainfall simulator. Effect on slope, moisture condition and properties of soil on erosion', *J. Soil Water Conserv.*, India, **8**, 12–25.

ELLISON, W. D. (1945) 'Some effects of rain drops and surface flow on soil erosion and infiltration', *Trans. Amer. Geophys. Union*, **26**, 415–29.

HENIN, S. (1953) 'Mécanisme de caractère de l'érosion par l'eau', *Technique-Agr.* **24**, 1 Jy, pp. 2–11.
Mechanism of splash erosion.

HENIN, S. and MONNIER, G. (1956) 'Evaluation de la stabilité structurale du sol', *C.R. VI Congr. Sc. Sol*, Paris.

KURON, H. and JUNG, L. (1957) 'Über die Erodierbarkeit einiger Böden', *Int. Union Geod. Geophys.*, Toronto meeting, vol. 1, 161–5.

MCINTYRE, D. S. (1958) 'Soil splash and the formation of surface crusts by raindrop impact', *Soil Sc.* **85**, 261–6.

ROSE, C. W. (1960) 'Soil detachment caused by rainfall', *Soil Sc.* **89**, 28–35.

SELIVANOV, A. P. (1960) 'Water-stability of the structure of different soil groups and its dependence on agricultural practices', *Pochvovedenie*, pp. 280–6.

Vegetation and pedogenesis

Basic works on soil science

AUBERT, G. and CAILLEUX, A. (1950) 'Esquisse d'une étude des sols', *Rev. Gén. Sc.*, pp. 28–39.

BLANCK, E. (1930) *Handbuch der Bodenlehre*, Berlin, Springer, 10 vols.
Very detailed collective work, basic reference.

BLANCK, E. (1949) *Einführung in die genetische Bodenlehre als selbständige Naturwissenschaft und ihre Grundlagen*, Göttingen, 420 p.
Mainly a secondhand work containing generalisations that are sometimes somewhat hasty. Refer to its bibliography for German works.

CLARKE, G. R. (1941) *The Study of the Soil in the Field*, London University Press, 228 p.

DEMOLON, A. (1949) *La Génétique des sols*, Paris, PUF, Coll. Que sais-je?, 126 p.

DUCHAUFOUR, PH. (1965) *Précis de pédologie*, 2nd ed, Paris, Masson, 481 p.

DUCHÉ. J. (1949) *La Biologie des sols*, Paris, PUF, Coll. Que sais-je?, 128 p.

ERHART, H. (1935) *Traité de pédologie*, Strasbourg, Inst. pédologique, 2 vols., 260 p. (vol. ii, 1938).

ERHART, H. (1955) '"Biostasie" et "rhéxistasie", esquisse d'une théorie sur le rôle de la pédogenèse en tant que phénomène géologique', *C.R. Acad. Sc.* **241**, 1218–20.

ERHART, H. (1956) *La Genèse des sols en tant que phénomène géologique*, Paris, Masson, Coll. Evolution des Sciences, 90 p.

GARKUSCHA, I. F. (1953) *Bodenkunde* (trans. from Russian), Berlin, Deutscher Bauernverlag, 360 p.

GÈZE, B. (1959) 'La notion d'âge du sol, son application à quelques exemples régionaux', *Ann. Agron.*, pp. 237–55.

GUERASSIMOV, L. P. (1944) 'La carte mondiale des sols et les lois générales de la géographie du sol', *Ann. Agron.*, pp. 448–94.

JENNY, H. (1941) *Factors of Soil Formation*, New York, McGraw-Hill, 270 p.

KELLOGG, C. E. (1950) 'Principal soils', *Trans. 4th Int. Congr. Soil Sc.*, Amsterdam, vol. 1, 266–76.

KUBIENA, W. L. (1952) *Claves sistemáticos de suelos*, Madrid, 388 p., 26 pl. in colour.
Excellent repertory of soil types with clear and uncomplicated classification and remarkable illustrations. Indispensable work tool.

PLAISANCE, G. (1958) *Lexique pédologique trilingue*, Paris, CDU, 357 p.

POLYNOV, B. B. (1948, 1951) 'Modern ideas of soil formation and development' (trans. from the Russian in *Pochvovedenie*, 1948, pp. 3–13), *Soils Fertilizers*, **14**, 1951, no. 2, 7 p.

POUQUET, J. (1966) *Les Sols et la géographie; initiation géopédologique*, SEDES, Paris, 267 p.

PRASSOLOV, L. I. (1944) 'Géographie et aires occupées par les différents types de sols', *Ann. Agron.*, pp. 495–9.

ROBINSON, G. W. (1952) *Soils, their Origin, Constitution and Classification*, London, Murby, 573 p.

The following references add important details on various matters:

BÉTREMIEUX, R. (1951) 'Etude expérimentale de l'évolution du fer et du manganèse dans les sols', *Ann. Agron.*, pp. 193–5.

CAVAILLE, A. (1958) 'Sols et séquences de sols en côteaux de terrefort aquitain', *Ass. Franç. Etud. Sol*, no. 94, 4–22.

CHEVALIER, A. (1929) 'Sur la dégradation des sols tropicaux causée par les feux de brousse et sur les formations végétales régressives qui en résultent', *C.R. Acad. Sc.* **188**, 84–6.

D'HOORE, J. (1954) 'Essai de classification des zones d'accumulation des sesquioxydes libres sur les bases génétiques', *Sols Africains*, **3**, 66–81.

GAUCHER, G. (1948) 'Sur la notion d'optimum climatique d'une formation pédologique', *C.R. Acad. Sc.* **227**, 290–2.

JACKSON, M. L. and SHERMANN, G. (1953) 'Chemical weathering of minerals in soils', *Advances in Agron.* **5**, 219–318.
Excellent summation, basic for the study of chemical erosion. Complete and compare with Reiche's work (1947).

JANSE, A. R. P. and HULSBOS, W. C. (1956) 'Influence de quelques plantes de couverture sur certaines propriétés physiques du sol', *Bull. Bibliogr. Mensuel Bur. Interafricain Sols Econ. Rurale*, no. 11–12, 5–6.

LANGLE, P. (1954) 'Sur l'application des études morphologiques pour la prospection des sols', *Soc. Sc. Nat. Maroc*, Trav. Sect. Pédol. **8–9**, 131–4.

NYE, P. H. (1954) 'Some soil forming processes in the humid tropics, I: a field study of a catena in the West African forest', *J. Soil Sc.* **5**, 7–21.

PLAISANCE, G. (1953) 'Les chaînes de sol', *Rev. Forest. Franç.*, pp. 565–77.
Includes a bibliography.

PITOT, A. and MASSON, H. (1951) 'Quelques données sur la température au cours des feux de brousse aux environs de Dakar', *Bull. IFAN*, **13**, no. 3.

REICHE, P. (1947) 'A study of weathering processes and products', *Univ. New Mexico, Geol. Public.* no. 1.

RUHE, R. V. and DANIELS, R. B. (1958) 'Soils, paleosols and soil-horizon nomenclature', *Proc. Soil Sc. Soc. Amer.* **20**, 66–9.

STRZEMSKI, M. (1956) 'A schema of soil distribution over the surface of the globe according to climatic zones', *Przeglad Geogr.* **28**, 131–42.

WETZEL, W. (1952) 'Sediment und Boden einer Grenzbestimmung', *Z. Pflanzenernährung, Düngung, Bodenkunde.*

The influence of soils on erosion

CHETELAT, E. DE (1938) 'Le modelé latéritique de l'ouest de la Guinée française', *Rev. Géogr. Phys. Géol. Dyn.* **11**, 5–120.
Essential study with excellent descriptions and good photographs.

FLETCHER, J. and BEUTNER, E. (1941) 'Erodibility investigations on some soils of the upper Gila watershed', *U.S. Dept. Agr. Tech. Bull.* no. 794, 31 p.

GROSSE, B. (1953) 'Untersuchungen über die Winderosion in Niedersachsen', *Mitt. Inst. Raumforsch.*, Bonn, **20**, 137–45.

The use of palaeosols as a method of geomorphological investigation

BRUNACKER, K. (1953) 'Die bodenkundlichen Verhältnisse der Würmeiszeitlichen Schotterfluren im Illgebiet', *Geol. Bavarica*, **18**, 113–30.

BRYAN, K. and ALBRITTON, C. C. (1943) 'Soil phenomena as evidence of climatic changes', *Amer. J. Sc.* **241**, 469–90.

DUBOIS, J. and TRICART, J. (1954) 'Esquisse de stratigraphie du Quaternaire du Sénégal et de la Mauritanie du Sud', *C.R. Acad. Sc.* **238**, 2183–5.

LEMÉE, G. and WEY, R. (1950) 'Observations pédologiques sur les sols actuels de loess aux environs de Strasbourg', *Ann. Agron.*, pp. 1–12.

SCHÖNHALS, E. (1951) 'Fossile gleiartige Böden des Pleistozäns im Usinnger Becken und am Rand des Vogelsberges', *Notizbl. Hessischen Landesamt Bodenforsch.* **6**, 160–83.

schönhals, e. (1951) 'Über einige wichtige Lössprofile und begrabene Böden im Rhein-gau', *Notizbl. Hessischen Landesamt Bodenforsch.* **6**, 243–59.

schönhals, e. (1951) 'Uber fossile Böden im nicht-vereisten Gebiet', *Eiszeitalter Gegenwart*, **1**, 109–30.

tricart, j. (1949, 1952) *La Partie orientale du Bassin de Paris, étude morphologique*, Paris, SEDES, vols 1 and 2, 474 p.

tricart, j. (1965) *Principes et méthodes de la géomorphologie*, Paris, Masson, 496 p.

Refer to the bibliography.

3

Criteria for a morphoclimatic division of the earth

In Chapter 2 we described the links that exist between climate and geomorphology. We must now define the criteria by which the land areas of the earth can be classified into different morphoclimatic regions.

The principal difficulty lies in the combination of natural causes: precipitation, temperature, and wind intensity, all factors that are in mutual interaction. We have seen that the indirect effects of climate in the form of plant cover and soil-forming processes prevail over the direct influence in importance and in areal extent. The morphoclimatic division of the earth must therefore be based on the integration of these several factors. Nevertheless each morphoclimatic process, whether frost-weathering or contrasts in temperature or humidity, plays its specific role in the shaping of the earth's surface and has its own areal extent, even though it is integrated into a specific morphogenic system. Furthermore, depending on the taxonomic scale on which we place ourselves, the criteria for a morphoclimatic division of the earth cannot be the same. In a first section we make a comparative study of the areal distribution of morphogenic processes and morphogenic systems; in a later section we show how zonal, azonal, and extrazonal elements combine in these morphogenic systems.

Comparative areal distributions of morphogenic processes and morphogenic systems

The various morphogenic processes do not have an equal importance within one and the same morphogenic system. Some play both a very general and a very intensive role; for example, variations in temperature and humidity affecting the rocks of arid regions, or frost-weathering in periglacial environments. Such general and intensive action in large measure determines the characteristics of the morphogenic system and the physiognomy of the relief. These processes may therefore be qualified as *predominating*. But there are also other subordinate processes. They either affect certain rocks that are particularly liable to them, or their action is felt only occasionally. Frost-splitting (riving, wedging) is an example. In cold periglacial regions it is a common process which more or less affects all solid rocks. More than any other process it assures the rock's reduction to movable debris. In these regions it is therefore a predominant process. Frost-splitting, however, is not unknown in maritime temperate climates; it occurs occasionally, but its action is impeded by the plant cover and the soil. Special conditions are necessary for the action of frost-splitting to be felt with enough intensity to influence the sculpturing of the relief; for example, a rock knob emerging from the ground surface or a rock escarpment. Elsewhere frost does not reach the rock buried by the regolith. Frost-splitting under such conditions is a limited process and is less effective, for instance, than chemical weathering. It is therefore only an *accessory* process in the maritime temperate zone.

This distinction between predominant and accessory processes is very important in geomorphology; it will be justified by increasing numerical evaluations as our discipline progresses, and will help to establish the criteria for the morphoclimatic division of the earth. We therefore examine it in more detail, with the aid of examples, before indicating what its consequences are.

Predominant process and accessory process

Let us go back to the example of frost-splitting, or the action exerted on rocks by changes in the volume of water due to freezing and thawing. This action should not be confused with effects of sub-zero (below 32°F) temperature variations which by sufficiently reducing the temperature in large enough volumes produce a fragmentation by retraction.

Frost-splitting is the very type of process directly dependent on climate. It depends on a temperature swinging above and below 0°C (32°F), or on the number of annual freeze–thaw cycles. Secondarily, it also depends on the intensity and the length of frost and thaw, which determine the depth of penetration of thermal waves into the ground, thereby introducing areal differentiation into the phenomena and its geomorphic effects.

In 'ice-cap' climates and their immediate surroundings (Greenland, Canadian archipelago, Spitzbergen, Siberian islands, Antarctica) the number of freeze–thaw cycles is small. Uninterrupted frost is the rule most of the year. Summer thaw is short, a few weeks only. Frost often occurs during night hours, which are a period of intense frost-splitting. The number of freeze–thaw cycles, however, remains below forty. It is often even less on the Greenland ice sheet and on the Antarctic continent. In many places of Antarctica the air temperature, during periods of observation of one or two years, has not risen above freezing point. Only dark surfaces of objects warmed by the direct rays of the sun may pass the zero curtain (32°F). The nunataks of polar ice sheets are therefore not much affected by frost-splitting. The lack of above zero temperatures explains the extreme scarcity of supraglacial moraines, even close to nunataks. Frost-splitting, however, is important on the margins of the ice sheets, as on the coasts of Greenland and the Antarctic peninsula; the arctic archipelagoes undergo a seasonal thaw whose intensity increases toward the south. Frost-splitting, unimportant on Peary Land in the northeast of Greenland, becomes very important in the Eqe[1] and Upernivik areas. As it is the predominant weathering process in all these climates, the intensity of the morphogenic processes depends on the intensity of frost-splitting. In Peary Land the evolution of the relief is very slow, and the wind plays an essential role but is only effective on friable materials which are provided in small quantities by infrequent frost-splitting. In the Eqe area and further south the action of frost-splitting is, on the contrary, very important and has been the object of classic descriptions. In this area the relief evolves rapidly owing to the intense fragmentation of even the most solid rocks.

It is in the circumpolar fell-fields where the average annual temperature is around 0°C (32°F) that the number of freeze–thaw cycles, at sea-level, is largest. One should, however, distinguish between maritime and continental locations. In the maritime variety, such as that affecting Iceland or the Kerguelen Islands, the number of freeze–thaw cycles reaches world records of 150 and sometimes even more. Each one is very short. The regime is essentially a daily one; in Iceland it may thaw during the day with a temperature of a few degrees above freezing and freeze 5 or 7 degrees (9° to 12°F) during the night. Thermal waves, of brief period, only penetrate a few centimetres into the ground. Frost-splitting which is intense but superficial is a function, especially, of the porosity of the rocks and their microstructure; it is a *micro frost-splitting*. In the continental variety, on the contrary, as in Canada or in Siberia, temperature oscillations about the freezing point are less numerous but have a larger range; frost-splitting tends to follow to a seasonal rhythm. In winter temperatures drop to −30°C (−22°F), and even −50°C (−58°F), and remain below freezing for weeks; thawing weather is rare and only affects the uppermost part of the ground. In the summer temperatures are relatively high; diurnal maxima are between 10° and 20°C (50° and 68°F). There are a few nocturnal frosts, but they are infrequent. It is, therefore, during the transitional seasons of spring and autumn that the freeze–thaw cycles most often occur. They are then an almost daily phenomenon. But these seasons are very short because of the continentality of the climate; in three or four weeks the weather passes from winter into summer, from weather of almost uninterrupted frost to almost continuously 'warm' weather with an occasional frost, or vice versa. For this reason there are three or four times fewer freeze–thaw cycles than in the maritime variety of identical latitudes; in northern Siberia and in north-western Canada they are of the order of twenty to fifty per year. Furthermore, summer thaw penetrates deeply into the ground; the 'active layer' subjected to the fluctuations is 1 to 2 m deep (3 to 7 ft). Frost exploits weaknesses in the rock structure and loosens blocks delimited by joints. This is *macro frost-splitting*, strikingly revealed on photographs taken by Boyé (1950) in Greenland.

[1] The Eqe is on the northeast coast of Greenland; it was one of the operational sites of the French Polar Expeditions conducted by Paul-Emile Victor (KdeJ).

Beyond the fell-fields, in climates experiencing a certain summer warmth, a new factor intervenes: vegetation. The tundras are followed by the forest, which appears approximately at the 10·5°C isotherm (51°F) for the warmest month. Frost-splitting and vegetation are opposed to one another. If frost-splitting is intense enough to eliminate the vegetation, which frequently takes refuge on sands or gravels little affected by geliturbation, it does not succeed in colonising clays and silts too much affected by changes in volume caused by freezing and thawing. Vegetation, however, also delays the progression of thermal waves into the ground. This delay occurs not only because of the shielding action of the plants but also because of the peaty layer which it produces from its own debris and whose decomposition is extremely slow. Summer thaw, therefore, penetrates less deeply into the ground under tundra: 0·5 m (20 in) in Siberia instead of 1 to 1·5 m (40 to 60 in) in barren land. A certain number of short freeze–thaw cycles do not reach the bedrock and only affect the upper part of the soil cover. With equal intensity and frequency freeze–thaw alternations have a lesser effect than in the polar barrens of frost-splitting.

The opposition between the purely climatic factor and vegetation is, of course, also felt in the realm of the forest. The peaty taiga with a soft marshy floor constitutes an excellent thermal insulator; to assure its food supply the vegetation absorbs a good part of the solar energy; *permafrost* (perennially frozen subsoil) occurs at varying depths in a large part of Siberia where it does not correspond any more to the present climate. Formed during the last glacial period of the Pleistocene it perpetuates itself as a typical survival because of the absorption of the solar energy by the plant cover. In Alaska a partial clearing is all that is necessary to cause the development of a chaotic microrelief generated by the unequal melting of permafrost due to the rapid downward spread of the melting process.

In the temperate zone vegetation plays the predominant geomorphic role. It often helps the formation of soils thick enough to prevent frost waves from reaching the bedrock. Thus in Alsace, where there are on the average some thirty freeze–thaw cycles with temperatures falling to − 15° or − 20°C (+ 5° or − 4°F) every year, the action of frost-splitting is negligible almost everywhere. During the coldest winters frost does not penetrate more than 60 cm (24 in) into the ground and, as a result, only affects the bedrock in exceptional places such as rock escarpments and riverbanks. It is an apparent paradox that frost-splitting is more effective in the garrigues of Languedoc where minimal temperatures seldom fall below − 5° or − 7°C (23° or 19°F) and where the number of freeze-thaw cycles is much smaller. The reason for such a paradox is that the anthropic degradation of the vegetation has in many places left the ground bare and the rocks outcropping. The latter are therefore subject to intense disintegration due to microfrost-splitting. The same is true on the High Plains of the Magreb. Whereas the small annual temperature range of maritime climates is not conducive to frost-splitting of the bedrock, the same is not true in regions of continental steppes; in eastern Europe and in central Sweden ground frost reaches depths of 0·5 to 1 m (20 to 40 in), and even more. Drought and wind are favourable factors because they eliminate that other effective thermal insulator which is the snow cover. This is the case, for example, on the high plains of Chergui, in western Algeria, where present frost slowly breaks up Pleistocene calcareous crusts.

The example of the relative importance of frost-splitting illustrates the principle of morphoclimatic zonation:

1. In a part of the polar zone (the periglacial zone) frost-splitting is the dominant process which, together with geliturbation, determines the originality of the periglacial morphogenic system. Its intensity affects the morphology and evolution of all the landforms.
2. In another part of the polar zone (the glacial zone) where temperature is almost permanently below freezing frost-splitting is only an accessory process, the principal process being iceflow in the form of glaciers.
3. In part of the boreal forest (Siberia, Alaska) frost-splitting is still a predominant morphogenic process and is associated with geliturbation which is mainly related to the survival of a deep permafrost.
4. Lastly, in the maritime regions of the forested midlatitudes frost-splitting is only an accessory process with a minimal importance. Though still accessory, it is more important in the continental regions of the forested midlatitudes and the cold steppes with sparse plant cover.

Raising ourselves from the level of morphogenic process to the level of morphogenic system, we also note that if the latter has a physical predominance the biochemical processes are subordinate, or vice versa.

Morphogenic systems with physical or biochemical predominance

The example of frost-splitting, a process directly dependent on climate, shows that climatic geomorphology is not a simple copy of climate. Even by carrying the analysis beyond the climatic data normally furnished and substituting the rather coarse notion of 'average' for that of 'frequency' and 'intensity' combined (number of freeze–thaw cycles and temperature ranges), there is no complete coincidence between an elementary morphogenic process and its climatic factor. The intensity of frost-splitting is a simple function of climate only in barren lands. Furthermore, it is extremely dependent on microclimate or site. Elsewhere the vegetation and the soils, which are in part a product of the vegetation, interpose themselves and create more complex relationships which may even cause frost-splitting to be more effective in a region with infrequent frost than in a region where frost is a common occurrence (Languedoc as opposed to Alsace). This illustrates a specific case of the fundamental distinction in nature between the physical world and the biotic world. There are, therefore, two major types of morphogenic systems: one with a physical predominance and the other with a biotic predominance.

1. Morphogenic systems with a physical predominance correspond to regions with a sparse and discontinuous plant cover: cold or arid regions at low or high elevations. The influence of climate on the evolution of the relief is felt directly. Low temperature or scarcity of water both hamper mineral chemical actions and the development of life with all its biochemical manifestations. Soils are generally thin and form very slowly; their screening effect is very limited, and the originality of the distinctive environment which they create is crude in comparison to soils due to biochemical processes. Because of this direct influence of climate the morphoclimatic systems are characterised by the predominance of the mechanical processes over the chemical and, even more, over the biochemical processes. The limits of the regions over which they extend are of an ecological nature; they correspond to the appearance of a plant cover dense enough to intervene effectively in the morphogenic processes.

Such limits are of course seldom sharp. Most often there are more or less wide zones of transition. Thus the tundras are interposed between the arctic barrens and the boreal forest. In this transitional zone the opposition between mechanical and biotic processes is particularly clear; the vegetation is absent in certain places because of drifting snow and the mechanical actions of geliturbation. In other places the mechanical processes are modified, causing special forms of solifluction below tundra or grass and impeding the penetration of thermal waves into the ground. An analogous transitional zone is found in the steppes on the margins of the dry zone in the middle as well as in the low latitudes. In this belt the sparse vegetation does not impede the action of the wind or of runoff but only modifies it. The agents of the weather have an equal intensity (the wind), if not greater (showers, wet–dry alternations) than in the neighbouring waste lands with ineffective plant covers. The result is a specially precarious balance between the vegetation and the mechanical processes; this transitional zone is particularly vulnerable to variations in the morphogenic conditions. A climatic oscillation or an intervention by man has considerable consequences of a size that may appear disproportionate to the cause. The opening up to cultivation of the steppes east of the Rocky Mountains in the United States has had as a consequence a veritable encroachment of the desert with the development of dunes within a few years. Earlier we have noted the rapid changes which accompany the clearing of the tundra. The reason is that the plants of the tundra live at the limit of their possibilities and are only able to persist through an unceasing struggle against the environment; their elimination suddenly removes the only obstruction to powerful mechanical processes.

The limits of the morphoclimatic regions with a physical predominance are therefore set by an ecologic complex determined by several factors: the weather and climate, the nature of the soils, including palaeosols, and the plant cover.

It should be emphasised that a single factor never intervenes alone in nature; if low temperatures do play an essential role in the formation of glaciers, precipitation is also important, particularly through its distribution in time. The snowline, which is the climatic criterion for separating the periglacial from the glacial regions, is determined by both temperature and precipitation. Some very cold regions, such as Peary Land, are largely free of glaciers because of drought, whereas high midlatitude mountains which are less cold but receive much more snowfall have impressive ones. Similarly, Pleistocene ice sheets were well developed in Scandinavia and northeastern North America whereas they were limited to small isolated systems in Siberia where drought is more pronounced and where a summer precipitation maximum does not favour the preservation of snow.

In like fashion, contrary to what was admitted by Davis, the evolution of relief is not dependent on a single process in a zone that has a morphogenic system with a physical predominance. Desert relief depends as little on the sole action of the wind as glacial relief depends on the sole action of ice abrasion.

2. The morphoclimatic regions with a biotic and pedologic predominance are characterised by the existence of soils whose genesis results from a profound modification of the parent rock. This fact denotes a predominance of biotic and chemical actions, the only ones capable of producing and maintaining such soils against the mechanical actions which, on the contrary, tend to destroy them.

Erosion in humid regions proceeds mainly through biotic or chemical means, and the mass of the substance removed to the ocean in solution exceeds that which is entrained as solid load. The mechanical actions are slowed down by the plant cover although normally they operate at the expense of the top soil; sheet erosion is limited under forest and even more below dense grasses; creep is neither general nor very effective; burrowing and the uprooting of trees by the wind together with landslides are still often the most active processes sculpturing the hillsides. The mechanical action of running water only affects the stream bed itself. The wind rarely has an effect on the soil.

In the morphogenic systems with a biotic predominance one should distinguish several varieties determined by the degree of this predominance.

Biotic predominance is most marked in the dense tropical forest. Heat and humidity enable plant life to attain its maximum development on the earth's surface. The plant cover is both thick and dense and its interruptions are as few as they are small. Rock outcrops are the exception, even on riverbanks. Overland flow under trees often has only a limited morphogenic effect, and the filling of bottomlands is very slow. The scarcity of rock outcrops on riverbanks, the near absence of lateral influxes other than clay and the rapidity of corrosion all contribute to the meagre solid load carried by streams. The most important natural mechanical action is that of landslides and the settling effects along steep slopes caused by the massive departure of dissolved substances, by uprooted trees, and by the sapping of riverbanks.

Biotic and chemical actions are less active, as a whole, in the temperate forests. First of all the vegetation is affected by a dormant season, winter or dry season (mediterranean climates), which causes a slowing down of the soil-forming process. The lesser density of the plant cover and fewer constantly high temperatures have the same effect, whereas the persistence of humus has the opposite effect. The regolith as a whole is thinner on the average than in the humid tropics, and weathering is different and less protracted. Because the bedrock is at shallower depth it can be reached more easily by mechanical agents, as happens in the beds and on the banks of streams. Lower temperature slows down the chemical processes. Most mid-latitude streams carry a solid load which enables them to corrade their beds. This solid load constitutes the principal mechanical action of this climatic environment; it determines the mechanical processes of slope development: rockfalls, solifluction, landslides, earthflows, and creep. It is indeed the incision of streams which sets in motion the most dynamic processes of slope development, for overland flow is as unimportant here as in the tropical forest; if trees are a lesser protection to the soil than in the tropics, the underwood is matted with a thicker litter composed of dead leaves and other debris in decomposition such as mosses, lichens, and fungi in which the water infiltrates as in a sponge.

Savannas, even more, are characterised by a certain balance between the biochemical and the mechanical actions. The lesser height of the plant cover permits important variations in the moisture content of the soil. The dry season and a herbaceous vegetation favour fires, which are sometimes spontaneous. The silica contained in the grasses is then liberated and later entrained by mechanical means, whether overland flow, wind, or merely through dissolution; it is an amorphous hydrated silica (opal) ten to twenty times more soluble than quartz. Furthermore, the small water-holding capacity and impermeability of cuirassed soils favour runoff, and the latter plays a less negligible role than in the rainforest. Small streams sap the margins of cuirasses and contribute to their dismantling, which is helped by the removal of the friable subjacent materials in the vicinity of springs or seepages.

As one passes from the savannas towards the steppes, the mechanical processes of the morphogenic system become increasingly important; there is less corrosion by water and streams start to carve their beds mechanically, making them more like those of the humid temperate zone.

Limits of the morphogenic systems (morphoclimatic zones)

The limits of the morphogenic systems may correspond in their main outlines to certain climatic data. Thus the boreal limit of the forests corresponds approximately to the 10·5°C (51°F) isotherm for the warmest month. But in general it is not a single climatic factor that determines the limit but a combination of factors, which brings us closer to the concept of an ecologic climate. Two systems of climate classification, those of W. Köppen and of C. W. Thornthwaite, whose conceptions are synthetic, are helpful in determining the limits of the morphoclimatic systems.

The Köppen system is the older of the two (1923). It combines average values and critical limits. The world's climates are first divided into five major zonal climates on the basis of temperature criteria alone, except for the dry climates.

The A climates represent the humid tropics in which the coolest month has an average temperature above 18°C (64·4°F). The B climates represent the dry climates in which yearly evaporation exceeds precipitation. The C climates represent the warm temperate climates in which the average temperature of the coldest month falls between −3°C (26·6°F) and 18°C (64·4°F). The D climates represent the snowy temperate climates in which the average temperature of the coldest month is below −3°C (26·6°F). In both the C and the D climates at least one month must have an average temperature above 10°C (50°F). The D climates are characterised by a seasonally or permanently frozen subsoil and a snow cover which lasts several months. The E climates correspond to the polar climates in which no month has an average temperature above 10°C (50°F).

These major climatic zones are then subdivided on the basis of precipitation, and precipitation and temperature for the B climates: 'f' (*feucht*) stands for humid throughout the year, 'w' for winter dry, and 's' for summer dry. The following eleven basic climates result:

1. Af climate in which all months receive more than 60 mm (2·4 in) of rainfall. It corresponds to the wet tropics and the tropical rainforest.

2. Aw climate in which at least one month receives less than 60 mm (2·4 in) of rainfall, reflecting a seasonal distribution. It corresponds to the alternating wet–dry tropics and the deciduous seasonal forests or the savannas.

3. BW climate or desert (*Wüste*) climate
   in which rainfall (in cm) is below the average yearly temperature figure (in °C) for winter rain;
   in which rainfall is below $t + 7$ for even distribution;[1]
   in which rainfall is below $t + 14$ for summer rain.

4. BS or steppe climate
   in which rainfall is below $2t$ for winter rain;
   in which rainfall is below $2(t + 7)$ for even distribution;
   in which rainfall is below $2(t + 14)$ for summer rain.
   A third symbol may be added to the BW and BS climates to distinguish between tropical (h) and midlatitude (k) deserts and steppes:

[1] t = average annual temperature in degrees C.

134

h (*heiss*) indicates an average yearly temperature above 18°C (64·4°F);
k (*kalt*) indicates an average yearly temperature below 18°C (64·4°F).
BSh may be further qualified as winter dry (w) or summer dry (s); at
least 70 per cent of the rain must fall in either the summer or the winter
six months.

5. Cf climate in which the driest month receives at least 30 mm (1·2 in) of
   precipitation.
6. Cw climate in which the wettest summer month has at least ten times
   as much rainfall as the driest winter month (extra-tropical monsoon
   climate).
7. Cs climate in which the wettest winter month has at least three times as
   much precipitation as the driest summer month (which must have less
   than 30 mm). This type corresponds to 'mediterranean' climates.
8. Df climate (f is defined as in the C climates).
9. Dw climate (w is defined as in the C climates).
10. ET climate in which at least one month has an average temperature
    above freezing. It corresponds to the polar barrens and tundras of the
    periglacial zone.
11. EF climate in which no month has an average temperature above
    freezing. It corresponds to the glacial realm.

The C and D climates can be further subdivided according to monthly
temperature criteria represented by a third symbol: a, b, c, and d. The
addition of a third symbol (including the h and k symbols for B climates)
raises the total number of climates to twenty-one. One may furthermore
distinguish an Am climate for a monsoonal variety and an EM climate for a
maritime polar variety. Even further refinements are possible. The C, D,
and E climates may, of course, also be found in highlands.

Because it includes both temperature and precipitation, extreme monthly
and yearly averages, this system of classification is necessarily somewhat
complicated. Its author has tried to take into account the concept of ecologic
climate in the choice of the adopted limiting values: the humid tropical
climates are delimited by temperature; the dry climates are defined by
effectiveness of precipitation; the temperate (C) and the cold snowy
climates (D) of the midlatitudes are based on a combination of precipitation
and temperatures of each season; the polar climates are exclusively determined by the temperatures of the warmest month. The 10°C (50°F) isotherm of the warmest month has been used because it approximately coincides with the treeline on the margin of the polar zone. In the tropics the
18°C (64·4°F) isotherm of the coolest month approximately marks the
poleward limit of plants characteristic of the humid tropics.

Thornthwaite's system (1933, 1943, 1948) satisfies a similar principle; it
is ecologic in its preoccupations. Its author has first calculated the values of
the coefficients for the major vegetation zones and has later used these
values to delimit his climatic zones. The difference with Köppen's classifi-

cation is a greater effort at systematisation, using the same coefficients for all types of climate.

Thornthwaite's classification is founded upon two basic data and one accessory element. The basic data are the effectiveness of precipitation (P–E) and the thermal efficiency (T–E)[1] in which E represents the evaporation calculated from the precipitation and the average monthly temperature of the station. The accessory element is the seasonal distribution of the precipitation.

The values of the index of precipitation effectiveness have been classed into five groups which correspond to types of vegetation and whose limiting values (except the last) form a geometric progression:

SYMBOL	TYPE OF VEGETATION	P—E
A (Wet)	Rainforest	>128
B (Humid)	Forest	64–127
C (Subhumid)	Grassland	32–63
D (Semiarid)	Steppe	16–31
E (Arid)	Desert	0–15

The values of the index of thermal efficiency have been classed into six groups whose limits also correspond to transitional zones of vegetation:

SYMBOL AND NAME	TYPE OF VEGETATION	T—E
A' Tropical	Savanna or forest	>128
B' Mesothermal	Steppe or forest	64–127
C' Microthermal	Steppe or forest	32–63
D' Taiga	Taiga	16–31
E' Tundra	Tundra	1–15
F' Perpetual frost	Glacial desert	<1

In the warm temperate climates and the tropics it is the index of precipitation effectiveness which delimits the types of plant formations. In the cold climates, on the contrary, it is that of temperature.

The number of subdivisions is greater than in Köppen's classification: thirty-two as against twenty-three (as here listed), but the most important groups coincide. One should note, however, rather important differences within the tropics, notably in Amazonia and the Congo basin.

Thornthwaite's classification of climates, even more than Köppen's, can serve as a basis for dividing the globe into morphoclimatic zones. It has, however, been the object of important criticisms, and its value is rather variable from one part of the earth to another. For example, it proves to be

[1] P annual precipitation in inches; E annual evaporation (potential evapotranspiration) in inches; T average annual temperature in degrees Fahrenheit.

best in Australia but rather unrealistic in northeast Brazil, so that in the last few years new methods have been proposed. New indices have been worked out; some very complex or based on data that are rarely available, others valuable only in certain bioclimatic regions. Such, for instance, is Hiernaux's index (1955) or a number of those cited by Boudyko (1956).

The method adopted by Bagnouls and Gaussen (1957) is useful thanks to its simplicity. It consists of making *ombrothermal diagrams* or ordinary climographs for monthly precipitation and temperatures but on which a standard ratio of 2 mm of precipitation to every degree Centigrade has been adopted. These authors indeed define a dry month as a month which receives less than this ratio (for example, less than 50 mm if the average temperature is 25°), and inversely. However, they later modified their views.[1] As for us we will follow Birot who calls a dry month a month whose precipitation (in mm) is less than 4 times the mean monthly temperature in C° ($P < 4T$). Adoption of this formula is not any less artificial than adoption of other formulas that have been worked out in a more scholarly fashion, and it gives generally satisfying results. We will use it to determine the climatic variations of the tropics (in *Landforms of the Tropics, Forests and Savannas*).

From all these attempts, which are multiplied by ecologists and which will continue to grow, a certain number of conclusions are apparent:

1. The nature of the climatic factors which delimit the major types of vegetation is complex. Temperature and precipitation must be combined; for example, in order to appreciate the concept of drought.
2. Major factors exist and must be used to define the major bioclimatic zones: the continuously high temperatures of the tropical lowlands, the want of heat in the cold zone, and the lack of precipitation in the dry zone.
3. Variations and ranges in temperature as well as in humidity are important. For example, the irregularity of precipitation is one of the factors of the intensity of erosion by runoff, according to Fournier (1960).

The zonal concept in geomorphology

While geomorphology under Davis's influence was versed in theoretical abstraction, the zonal concept was developing in other branches of physical geography. It is mainly Dokuchaev and, following him, Berg and Grigoriev who have given it its form from research carried on in pedology, biogeography and, later, climatology. The fundamental concepts of modern pedology, which are due to Dokuchaev and the Russian school of the latter part of the nineteenth century, are based on the confrontation of soils and of climate; the world's soils are grouped in particular belts rather parallel to the present climatic and biogeographic zones.

The other branches of physical geography thus have a lead of more than

---

[1] Cf. Péguy (1961), *Précis de climatologie*, p. 250 (KdeJ).

half a century on geomorphology in their handling of the concept of zonality. It is therefore of the greatest benefit to utilise their experience and, particularly, to borrow part of their terminology, for the latter is apt to adapt itself well to climatic geomorphology and to render it a great service. Troll (1944), in particular, has done this in his general work on periglacial microforms. It is he who made the zonal vocabulary available to geomorphology.

Every phenomenon or process whose global extension is more or less conformable to latitude is qualified as *zonal*. The tropical rainforest, the coral reefs, and the polar ice sheets are all zonal phenomena. A relief which owes its particular forms to processes characteristic of a specific latitude is a zonal phenomenon, and its environment constitutes a *morphoclimatic zone*. By extension, zonal has become, for certain authors, approximately synonymous to climatic. Some authors call every process zonal which is linked to a determinate climate, even if the regions subjected to this climate do not form a latitudinal zone on the earth's surface. In this way pediments, which develop in a semiarid environment, whether in crystalline or sedimentary terrains, are sometimes considered as zonal. But some deserts are the result of their latitudinal position, such as the Sahara, whereas others result from their location in relation to seas and mountains, such as the deserts of western United States, central Asia, Anatolia, or the Andean highlands. It is preferable to restrict each of the two terms, *zonal* and *climatic*, to its exact definition.

The adjective *azonal*, or *worldwide*, helps to qualify phenomena or processes occurring in several or in all zones. For example, wave action on beaches is not necessarily the adjunct of a particular climatic zone; it exists on the equator as well as in the midlatitudes and, even, during the summer, in some parts of the polar regions. Vulcanism and the formation of fold mountains are, *a fortiori*, azonal manifestations, even if certain orogenic belts, such as the Tethys, have an east–west orientation.

A phenomenon is said to be *extrazonal* if, characteristic of a certain zone, it nevertheless occurs in another zone but in a more restricted, sporadic, infrequent, and less intense way. Thus glaciers on certain mountains in the equatorial zone, and the periglacial phenomena which accompany them, have been qualified as extrazonal by Troll. Another excellent example is that of savannas which form clearings in the midst of the tropical rainforest, as in Lower Ivory Coast or back of the coasts of the Guianas.

It is important to make a careful distinction between extrazonal and azonal. An azonal phenomenon is not bound to a specific zone but may occur, even frequently, in several other zones. An extrazonal phenomenon, on the other hand, is characteristic of a specific zone; it may occur in other zones but only where there are particular conditions which are exceptionally favourable. The same phenomenon, depending on where it has been observed, may therefore be zonal in one place and extrazonal in another, but it can never be zonal in one place and azonal in another.

138

One may speak of *polyzonal* or *plurizonal* in connection with phenomena or processes that occur in several specific zones without being worldwide. Running water is a good example as it mainly occurs in the temperate and intertropical zones and only plays a subordinate role in the glacial and arid zones. The characteristics of landform development under forest are typically polyzonal; they are found in the tropical rainforest as well as in the forests of the midlatitudes.

The terms having been defined, we will now examine the relationships that exist between the various zonal, azonal, and extrazonal morphogenic processes.

Relationships between zonal, azonal, and extrazonal factors

Definitions are generally based on simple examples and schematic cases. Nature is generally more complex and characterised by combinations in variable proportions of more or less zonal, azonal, or extrazonal elements. It is therefore necessary to go into some detail, with the help of examples, about the relationships between these various phenomena.

ZONAL, AZONAL, AND POLYZONAL

The relationships between zonal, azonal, and polyzonal phenomena determine the degree of zonality of a specific category of phenomena. At one end of the series the zonal characteristics are predominant, at the other end the azonal characteristics. Between the two there is a mixture in variable proportions.

The most typically azonal manifestations are found in the domain of tectonic, volcanic, and structural geomorphology; the determinant forces are internal forces, at least in their final stage. The best example is that of volcanoes whose distribution has nothing to do with zonality. The principal locus of emerged volcanoes in the world, that which surrounds the Pacific, is indeed located across the various morphoclimatic zones. The same is true of the volcanoes of the Mid-Atlantic Ridge and of the Indian Ocean. Zonality only becomes a part of vulcanology with the analysis of erosional forms; it is logical that these zonal elements, dependent on external forces, become more and more important with the slowing down of the constructional forces and the increased activity of the destructional forces of erosion. The relationship between zonal and azonal influences varies with time; during the active phase, that of the construction of the volcanic cone, the immediate internal forces are dominant and, consequently, the zonal influences reduced; during the passive or residual phase, that of the progressive destruction of the relief by erosion, they, on the contrary, take an increasingly important role tied to the predominance of the external forces. New volcanoes have a relief consequent upon the conditions of eruption, an azonal constructional relief. Old volcanoes attacked by erosion have, on the contrary, a dissected relief considerably influenced by zonal processes and mechanisms. This is the case, for example, in the glacial dissection of Bear Island, or the arid dissection of the volcanoes of Eritrea.

Tectonic actions, another category of the manifestations of the internal forces, seem to be less exclusively azonal. True, the forces which build mountains occur in the depths of the earth. As such, they would be azonal. Nevertheless, certain zonal influences on these manifestations are not excluded. The division in two parts of the most ancient parts of the continental plates, located each in one hemisphere and approximately aligned along the parallels of latitude, is a disturbing factor: in the northern hemisphere the Canadian, Scandinavian and Siberian shields; in the southern hemisphere, reaching beyond the Equator, the shields of Brazil and the Guianas, of Africa, of Australia, and of Borneo. This configuration, to which must be added the long Tethys fold system, has intrigued geologists and geophysicists and provoked elaborate explanatory hypotheses. An influence of the earth's rotation on the localisation of these major structural units is not excluded, but it is not demonstrated; we are still in the realm of speculation. Less conjectural is the influence of sedimentation, in part determined by zonal factors, on the formation of fold mountains and isostatic movements. The particularly high rate of landform evolution in semiarid continental climates perhaps accelerates the tectogenic evolution with the help of isostatic adjustments; the rapid denudation of raised blocks accelerates their uplift, and the equally swift filling of troughs increases their subsidence. In fact, it is under semiarid continental climates with a great deal of morphogenic activity that the clearest examples of regenerated residual geosynclinal mountains are found; for example, in western United States, central Asia, and the Andes of northwestern Argentina. There is nothing astonishing or mysterious about it; the tectogenesis of fold mountains is a complex mechanism which depends on the antagonism of internal and external forces, which implies a reaction of the one upon the other.

In the domain of the external forces the azonal influences are clear; they occur everywhere. Nevertheless, if we classify submarine relief, the littoral actions, the actions of the wind and of running water in the azonal category, some elucidation is needed. One should not consider this azonal character as exclusive, but only as predominant. For example, in the development of the recent submarine relief the essential fact is the liquid environment in which it is formed, which is azonal by nature. Of the present seas some are deep enough to have always existed in all latitudes from the poles to the Equator. The mechanisms of gravity fall which determine the depositional submarine relief are the same, as a first approximation, in all seas. In detail they undergo a modification due to temperature as influenced by climate, but the modification is limited to the upper layers of the sea water and is minimal in formulas; it is only reflected in parameters and does not necessitate the use of different mathematical formulas. On the other hand, the detrital material, to which formulas of sedimentation apply, varies with the influence of climate on the earth's surface. In the tropics and in the midlatitudes, continental sediments all come from rivers, and only their finest fractions are carried out to sea by marine currents. In the cold seas there is another form of transport: icebergs, which melt slowly during their displacements and abandon coarse morainic materials, including large erratics, far out at sea; whereas destruction of coral reefs is a great source of debris in the tropical seas. In short, the mechanism of detrital sedimentation is azonal in essence: the laws which regulate the fall of the particles are the same under all climates; only the rate of fall is affected by temperature; the material subject to these laws, however, is influenced by climate. The essential nature of the process is therefore azonal, but its forms include an accessory zonal component.

Wind action provides an analogous example. The laws of airflow on the earth's surface are the same whatever the climate. Wind eddies striking an obstacle are of the same nature at the poles as at the Equator; what changes is the dimension of the eddies and the velocity. It nevertheless remains true that the geomorphic actions of the wind differ with the climatic zones. The resistance put up by the plant cover is one factor; in forests the wind acts in an indirect way through the windfalls which it occasions; on glaciers and around them the wind blows the snow and modifies its distribution. The material transported also varies with the effect of the climate; it may be sand, silt, snow or silt aggregates cemented by salts (salt flats). The nature of the ground surface may also be subject to zonal modifications caused by purely physical phenomena; in sufficiently cold and humid climates the wind may act on icy surfaces and slide pebbles and gravel which it would not be able to move on a rougher surface. In very cold regions the wind produces mixed stratified accumulations of sand and snow, known as niveolian deposits. The characteristics which govern the aeolian mechanisms are therefore comparable to those found in marine sedimentation. Wind action is azonal in essence; its mechanism is quantitatively modified only by temperature; the modification is minor and may be recorded by parameters in formulas (viscosity). Zonal influences appear in the form of the environment on which the wind acts; this environment is not the standard one of the laboratory but a geographic milieu differentiated precisely as a function of climate, therefore in a zonal way. These zonal influences determine the nature of the ground surface; it may be more or less indurated snowfields, or more or less indurated sheets of sand or silt which supply the wind with its solid load, or an icy veneer or plant cover which determines the roughness of the surface and, through it, facilitates the transport of particles by the wind or on the contrary stop its velocity on contact with the ground.

Littoral actions provide another rather analogous example. The predominant process, the breaking of waves on the shore, is certainly azonal in essence. The zonal influences in this mechanism are perhaps even less pronounced than in the case of the wind. Differences in viscosity due to temperature are cancelled out by distant swells which affect shores far removed from their place of origin, depending on the configuration of the coastlines. Such remote swells are extrazonal, for they affect a different climatic zone from the one in which they originated. Here again it is through the immediate environment that the zonal factors intervene. Subaerial disintegration along the coast and local fluvial sediments provide the major part of the abrasive material used by the waves, and the nature of the material therefore varies with the morphoclimatic zone. Some Arctic shores are buried under alluvial deposits which are the products of frost-weathering brought by streams. The shores of the wet tropics, where biochemical weathering predominates, seldom have any pebbles, but mostly sands and muds. The shores of the midlatitudes, where mechanical factors of disintegration are important, are often strewn with pebbles, but they may be relicts of earlier actions. The geographic environment also nurtures biotic actions antagonistic to mechanical processes. In tropical seas building organisms are innumerable and very active; they make possible the construction of particular reliefs such as fringing and barrier reefs, which not only modify the littoral but influence the mechanical processes which fashion it. A littoral protected by a barrier reef is free from swells. Up to this point the parallel with wind action on the earth's surface does not produce any difficulties; disintegration determines the material available to wind and waves. The plant cover influences wind action just as building organisms influence littoral morphology. It is in the high latitudes that a new element intervenes; there, everywhere the wind blows from the land, whereas wave action is nonexistent or extremely rare because shores are perpetually frozen. On the shores of northern Greenland, on those of certain islands of the north Canadian and north Siberian archipelagoes, and on almost all the shores of Antarctica, the water remains permanently frozen; where it melts, it forms such a narrow channel that there are no waves. This time it is no longer a matter of a small quantitative modification in the action of an azonal process or only of the variation of a parameter, but it is a complete change in the nature of the mechanisms capable of producing original forms, such as the ice shelves of Antarctica, which are a true littoral nivation form because snow accumulates on the sea ice. A difference with the wind, present in all morphoclimatic zones, is that the glacial zone is not affected by littoral swell; this difference is formulated in the term *polyzonal*. The swell may be considered as polyzonal, for, without any change in its nature, it does not affect all of the morphoclimatic zones of the world but only some of them.

The zonal component of the predominantly azonal processes is even larger in the case of running water. In order to grasp it better, let us briefly recall what points running water has in common with the two morphologic agents already studied (wind and littoral processes): the laws of hydrodynamics also have a universal validity, for the climate only influences the parameters; the areal extent of the action of running water, however, is dependent on zonal factors because drainage characteristics result from the interaction of direct and indirect (type of soil, vegetation) climatic influences. But this areal extent is even smaller than that of the littoral swell. Two climatic factors contribute to it: cold temperatures (as in the case of swell) and drought. In the coldest zone of the globe drainage is in the solid form; glaciers advance and sculpture the relief in a quite different way from that of running water. Whereas in the most arid regions there is no drainage at all (areism), precipitation being insufficient to produce a drainage net. In some climatic zones running water participates in the evolution of the relief in an azonal form, whereas in others it is lacking. The term polyzonal (or plurizonal) well applies to this important intermediate stage between zonal and azonal manifestations.

We are now left with the analysis of the relationships between zonal and extrazonal phenomena.

ZONAL AND EXTRAZONAL

Extrazonal phenomena are, as we have seen, similar to zonal phenomena, but they occur outside of their own zone. They form, as it were, islands beyond the zone in which they normally occur. They are somewhat different from the corresponding zonal phenomena because of the influence of the different geographic environment in which they occur. Such differences may even be important enough to force a distinction between a zonal variety and an extrazonal variety of the same phenomenon. Typical in this regard is the case of glaciers in which two subdivisions must be made: local (or mountain) glaciers and ice sheets. Conditions of iceflow, ice temperature, and ways of nourishment and ablation differ in both cases.

There are good reasons for studying extrazonal phenomena as a function of the factors which make them possible. These factors may be grouped in two major categories: surviving phenomena and the characteristics of the environment.

1. *Extrazonal phenomena due to prolongations or survivals.* Surviving extrazonal phenomena are due to prolongations of certain physical geographic features either in space or in time. A first category of extrazonal phenomena is the result of a spatial prolongation. A good example is the existence of a fluvial morphology along the Nile in Egypt. The running waters which carved the Nile valley originate in precipitation falling not in the deserts which flank the river but in another much more humid climatic zone: the equatorial zone and the mountains of Ethiopia. The Nile valley, in the midst of the desert, is an extrazonal form of relief which results from the fact that the drainage of the high zonal valley is too abundant to be absorbed by the desert. But during its crossing, as the river scarcely receives any more water, it survives on its own reserves. These conditions also modify the fluvial mechanisms themselves. Even apart from the drain caused by irrigation, evaporation, and the absence of appreciable local water input, there is a progressive downstream decrease in the discharge of the Nile. This is the converse of what happens to a zonal river, karstic losses put aside. There are

many other exotic or extrazonal rivers besides the Nile: the Senegal, the Indus, and the Tigris and the Euphrates, to mention only the largest. Their morphogenic activity is highly influenced by a progressively decreasing stream flow. The Senegal delta, for example, reveals important peculiarities on account of the extrazonal character of the river: at low water, as the discharge no longer compensates evaporation, the sea water ascends the principal distributaries producing estuarine conditions in areas which are yearly flooded with several metres of fresh water; some bottomlands dry up at low stages and turn into salt flats, but during flood they function like decanting reservoirs. All of this of course affects the eventual agricultural development of the land.

Glaciers also offer typical examples of extrazonal phenomena due to prolongations; ice forms above the snowline and melts below it in much warmer periglacial or even temperate zones. Mountain glaciers often extend deep into forests. The Greenland ice sheet sends tongues all the way to the sea through barren frost deserts and tundras. The same was true of the Pleistocene ice sheets whose margins extended several hundreds of kilometres beyond the southern climatic limits of the glacial zone. As in the case of rivers, such migrations out of their zone of origin are accompanied by changes in the morphogenic processes: the melting ice of glacier tongues or ice sheet margins no longer has the same properties nor the same geomorphic effect as the sound ice of the glacial zone proper. Debris accumulates below the melting ice; running waters are abundant, at least during the warm season, digging channels and caverns, carving the ice, and removing its load. A completely distinct morphology different from that of the strictly glacial zones results from it and characterises the extrazonal parts of glaciers.

Such extrazonal manifestations are the rule with glaciers, whereas in the fluvial domain they are the attribute only of important rivers. The reason is that the slower flow of ice naturally favours the effects of inertia. But these effects in both cases assume the same form—a spatial one. There is another category of extrazonal phenomena in which the manifestations of inertia are temporal and are represented by survivals.

We have seen that some old ferruginous or bauxitic cuirasses formed under a climate which may have been markedly different from the present one perhaps continue to grow at their base if they are well drained; if this is indeed true then the cuirass is a typical survival, and the present cuirassing process is extrazonal. Another good example is that of the permafrost of the Siberian taiga and deciduous forests of the Soviet far east toward its southern margin; it was formed during the last cold phase of the Pleistocene and persists, or at least it melts only very slowly, due to the rigour of the present climate and the protection of the forest and often peaty soils. This extrazonal permafrost gives rise to peculiar periglacial forms called cryokarst by Soviet scholars. It is therefore still active morphologically.

The main point is that a still morphologically active surface makes it possible to distinguish between prolongations of extrazonal phenomena in time and in space, and another kind of survival whose effect is no longer felt and is, on the contrary, dead. This kind of survival manifests itself in *past forms* which persist in a *residual* way. They are *relict forms* inherited from previous periods characterised by conditions that do not exist today (inherited forms). Present processes do not develop them any more, but on the contrary contribute to their destruction. They result from a lag in the adjustment of the relief due either to a change in climate (*climatic residual forms*) or to the termination or marked slowing down of tectonic processes (*tectonic residual forms*). For example, the volcanic relief of Cantal, in the French Massif Central, is a relict form of a past relief which lava flows have ceased to construct and which is presently being slowly dismantled by erosion. The periglacial relief of the Paris Basin is also a relict form; geliturbation and solifluction no longer affect the slopes of chalky Champagne which is presently very slowly evolving by chemical erosion. An extrazonal form, on the other hand, is a live form or a *present form* which continues to evolve in the same way, even if it is slower, as cuirasses, or partially through the action of certain specific processes, such as the cryokarst of Siberia. Extrazonal landforms are sometimes associated with relict conditions. This is the case of the residual savannas of Lower Ivory Coast in which grasses continue to grow today but in which, nevertheless, the plant formation is in imbalance with the present climate. These residual savannas are remnants of once more widespread savannas which date back to a drier subrecent period; they are gradually being reduced at their margins through a spontaneous encroachment of the forest.

Spatial and temporal survivals do not exclude one another; they combine, as, for example, in the case of the Greenland ice sheet. Temperature measurements have indeed shown that the temperature is colder at a depth of 125 m (400 ft) than at the surface, which implies a lag in adjustment to the recent climatic amelioration. The Greenland ice sheet is, therefore, in part a relict, which does not prevent it from producing extrazonal tongues. Careful distinction of present and surviving phenomena, zonal and extrazonal, is therefore of great importance in connection with the action of man on nature.

2. *Extrazonal phenomena due to environment.* Some landforms with dimensions too small to be greatly affected by prolongations or survivals are occasionally extrazonal. They owe it to the geographic environment.

Some minor landforms, such as the badlands of semiarid regions, are due to the lithologic nature of the material from which they were formed. Badlands are zonal in dry regions but extrazonal in humid regions, whether temperate or tropical. In humid regions, however, they will only develop on cleared ground, not under forest or even dense grassland. Furthermore the terrain must be argillaceous or marly and unaffected by mass movements which is possible if there is sufficient drought. Violent showers will then readily produce gullies and badlands as occurred in the southern French

Alps and even as far north as Burgundy (near Châtillon-sur-Seine). Badlands, although not common, are also found in the periglacial zone. Some have been described by Malaurie (1949, pp. 2–8) in the Disko area of Greenland. Their presence also depends on the nature of the rocks: in this case sands rather than clays or marls. Gullying occurs on them not only because the texture of sand does not favour gelifluction but because the frozen subsurface is thoroughly impermeable. In northern Canada their topography stands in sharp contrast to that of the slopes fashioned by gelifluction which quickly destroys the incipient rills caused by rain. The solifluction of clays and marls serves as another extrazonal example in a temperate environment. For example, the pressure of a bed of limestone, sandstone, or basalt together with the waterlogging of its base by atmospheric water may cause a downslope mass movement of the underlying clays or marls. Such deep solifluction without frost and under plant cover of course differs in certain respects from gelifluction on permafrost, but its general character remains the same. The extrazonal character of the process is also underscored by terminology if we restrict the term *gelifluction* to soil flowage on permafrost and the term *solifluction* to all other kinds of very slow soil flowage whether in periglacial regions or in the tropics.[1]

Extrazonal phenomena may also be caused by topographic location. This is the case of certain groundwater cuirasses which develop in the wet tropics rather than in the wet–dry tropics. In the rainforest such extrazonal groundwater cuirasses occasionally occur in bottomlands where the watertable is not very deep and subject to fluctuations.

Tectodynamic factors are another cause of an extrazonal extension of a specific landform. Thus the sebkhas (salt flats) of the Oran area of western Algeria are located in a zone that is normally too humid to allow their development. Under the present climatic conditions there ought to be an intermittent exterior drainage rather than interior drainage. There are many indications, however, that the salt flats are unstable, yet they persist owing to the continuance of tectonic deformations that produce synclinal depressions which hinder the streams from reaching the sea. The sebkhas, normally found in a more arid zone, are here extrazonal and dependent on presentday tectonics. The same might have been true of certain salt deposits of the geological past.

Elevation, lastly, produces a final element of the geographic environment that causes extrazonal extensions. At a certain elevation periglacial mechanisms and glaciers exist in the mountains of the temperate zone and the tropics; they have sometimes been qualified as extrazonal. In reality the conditions are quite different and require a special designation.

[1] The literal meaning of solifluction is 'soil-flow' and Andersson, who introduced the term in 1906, did not restrict it to periglacial environments although he considered these to be optimum for the process. Many English and continental authors take a narrower view of the term than suggested here, some using it synonymously with gelifluction which, it should be noted, they would not confine to areas underlain by permafrost. (Editor's note).

Latitudinal and altitudinal zonation

Since Tournefort's accent of Mt. Ararat (1700) latitudinal zonation has often wrongfully been confused with altitudinal zonation. For a long time most authors have tried to show that as one proceeds to successively higher levels on mountains, one passes through similar types of climate as exist on the lowlands converging on the pole. In the tropics one would, therefore, successively experience temperate, periglacial and glacial climates with increasing elevation.

This is only a gross approximation. There are, it is true, similarities between the succession of the morphoclimatic environments encountered on mountains and in latitudinal morphoclimatic zones. But there are also important differences which led Flahault, the geobotanist, as early as 1900, to use the term *étage* (and *étagement*) rather than *zone*, previously in use, for mountains. This distinction is profitable to geomorphology and within this science the two concepts should no longer be confused as they are in other natural sciences.[1]

DISTINCT CHARACTERISTICS OF VERTICAL ZONATION

Vertical zonation is caused by the altitudinal variation of two basic elements of climate: temperature and precipitation. Temperature decreases almost regularly with elevation: mountains of sufficient height reach beyond the zero degree (32°F) isotherm. There is an important analogy between decrease in temperature and increase in latitude which, if observed unilaterally, as in the past, is at the origin of the confusion between latitudinal and vertical zones. The greater the elevation, however, the thinner the air; its heating capacity therefore decreases and its transparency to radiation in general increases. The range of daily variations of air temperature decreases and that of the ground surface increases. The solar radiation received is all the more intense as the air is thin. The result is that rocks are subject to much greater thermal effects in mountains than in lowlands, effects which increase with elevation. It often happens that close to 2 500 to 3 000 m (8 000 to 10 000 ft) in the midlatitudes, rock surfaces reach a temperature of some 20°C (68°F) in the early afternoon and −5°C (23°F) at night. Even larger ranges are frequent at higher elevations in the Mediterranean and intertropical zones. Such temperature variations are of course dependent on the degree of cloud cover. In cloudy climates it is on the contrary the decrease in the temperature range of the air that is important as rocks then maintain temperatures close to that of the air. This is so particularly on low mountains where the rarefaction of the air is insignificant, as in the Vosges, the Harz, and the Scottish Highlands, and on humid oceanic islands as, for instance, off the coast of Scotland.

[1] The English language has no equivalent term for *étage*; it is therefore best to refer to altitudinal or to vertical zonation. The word 'zone', alone, should be used with care, making certain no confusion is possible with latitudinal zones. The expression 'mountain zone' and the words 'level' and 'tier' may also be used (KdeJ).

Precipitation varies in a different manner in relation to elevation; starting at sea-level precipitation first increases with elevation and then passes through a maximum after which it progressively decreases. On sufficiently high mountains there is always a level characterised by more abundant precipitation and consequently greater cloudiness. This is because the dew point is reached, and therefore condensation occurs, as the rising air cools by expansion with elevation, though it holds a decreasing quantity of water vapour. Already at 2 000 m (6 500 ft) the level of condensation is reached for a water vapour content half that found at sea-level; at 4 000 m (13 000 ft) one quarter only. At the base of mountains the rising air of neighbouring plains reaches the dew point at progressively lower elevations as the water vapour content increases. There is, then, for each climate an optimum level of condensation which corresponds to the average elevation at which the rising air of the neighbouring lowlands condenses.

This level of *maximum* precipitation exists in all climates, even in the most arid regions. It causes an important difference between the montane morphogenic processes and those of the neighbouring lowlands. Even in arid regions there is at least one mountain level which has an organised drainage system, if not permanent streams. The Andes of northern Chile, the highlands of Ethiopia, and even the Ahaggar provide good examples. Watercourses are, of course, assured of a greater degree of perenniality if the mountains are high enough to carry glaciers which melt slowly and provide a reserve. Furthermore, as steepness of slopes favours runoff, mountains are the seat of a particularly intense stream flow which plays a most important geomorphic role. In arid regions mountain streams wither away at the foot of the ranges and become extrazonal. The Tafilelt, at the foot of the Moroccan High Atlas, is an example, as well as the fan oases at the foot of the ranges of Soviet central Asia.

An essential element of the geomorphological concept of vertical zonation is, of course, topography. Mountains high enough to have distinct climates are caused by intense uplift or important volcanic eruptions. The areas involved are restricted and have a scattered distribution. Because of the nature of the forces involved large elevation differences between crests and valley bottoms cause a predominance of steep slopes. There is action and reaction between this particularly rugged topography and its morphogenic system. Because of it there is a fundamental difference between latitudinal and altitudinal zonation; latitudinal zonation affects enormous regions and only a very few long rivers have sufficiently large basins to flow through several morphoclimatic zones. In mountain zonation, on the other hand, the rule is the coexistence of several different morphoclimatic tiers in the same basin of modest dimensions. It even frequently happens that a single slope spans several tiers, that the base of the slope is covered with forests and its crest reaches the alpine meadows or even the level of perpetual snow. Whereas lowland rivers differ from one morphoclimatic zone to the next, mountain streams, always more or less hybrid, resemble one another through the various tiers. Steepness of slope contributes to this affinity by favouring runoff; the *module* (discharge in relation to the surface of the basin) and the *drainage coefficient* (proportion of the precipitation drained by the streams) of mountain streams are always higher than those of the streams of neighbouring plains. A mountain stream system is all the more dynamic on account of it; it is reflected in a denser texture and a more violent stream flow. The morphogenic activity of the streams is thereby increased. Generalising its significance, the name *torrent* has been applied to such streams (in Latin *torrens* means a watercourse that dries up during the torrid heat of the summer). One commonly speaks of *torrential erosion* to emphasise the particular features of stream erosion in mountains. The importance of runoff, which is the consequence of steep slopes, increases the variations in discharge which play such an important role in the action of mountain streams; the result is an active dissection and the multiplication of steep slopes. Thus on the island of Réunion torrents carry pebbles that are nonexistent in lowland streams at the same latitude. Slope steepness also influences slope development. Being favourable to overland flow, it diminishes the intensity of soil formation already reduced by decrease of temperature with elevation. The soils of steep slopes, which are most frequent in mountains, are generally thin and poorly evolved. Areas of bare rock are also more frequent than in lowlands because the morphogenic processes more often, at least momentarily, impede the development of vegetation. The general result is that the mechanical processes are more active than in the lowlands. This, of course, applies especially to rockfalls and landslides which sculpture steep slopes.

These three fundamental factors: temperature, precipitation, and topography, functioning in all mountains worthy of the name, suffice to differentiate vertical and latitudinal zonation and to give the morphogenic systems a series of characteristics which differentiate them from those of neighbouring plains, creating for themselves a number of common features. But one should not be satisfied with this unilateral view; latitudinal zonation also influences vertical zonation.

INFLUENCE OF LATITUDINAL ZONATION ON VERTICAL ZONATION

One always encounters in all the mountains of the world, whatever their elevation, certain features of the zonal climate which might be called the 'basic climate'. These features are responsible for the preservation of a certain degree of zonal originality at the various mountain levels and are first of all reflected in the temperature regime. In the equatorial zone the predominance of daily oscillations over seasonal temperature oscillations that is typical of the lowlands occurs at all elevations and, for example, imparts a special regime to glaciers.

Two aspects must be considered in the influence of latitudinal zonation on vertical zonation. First, mountain zonation is not the same in all latitudes. Second, there are differences between vertical zones and corresponding latitudinal zones.

1. *The effect of the zonal climate on vertical zonation.* The main consideration in vertical zonation is the elevation of the level of maximum precipitation as a function of the zonal climate. Added to it is the influence of site: the precipitation maximum in the same mountain range is higher in the interior massifs than in the marginal ones (Prealps 1 300 m, Mont Blanc 2 500 m: 4 300 and 8 200 ft). The precipitation maximum in general is lowest in humid climates. Indeed all that is necessary to cause condensation is a slight rise of the air to make it reach the dew point. Thus in Hawaii where there is no marked dry season the level is only 700 m (2 300 ft); in Java only 800 m (2 600 ft); it is 1 300 m (4 300 ft) on the south flank of the Himalayas and in the Black Forest, 1 400 m (4 600 ft) in the Aigoual (Cévennes) and the Sierra Nevada of California. The precipitation maximum is therefore lower in the wet tropics than in the temperate zone and seems to be highest in the tropical deserts.

Of vital consequence to the morphogenic processes is the interval between the levels of the 0°C (32°F) isotherm and the precipitation maximum. This interval widens from the poles to the Equator. They almost coincide in mountains close to the Arctic Circle as well as in some of the high massifs of the Alps situated close to the axis of the mountain range as in the case of the Mont Blanc. Such conditions are, of course, optimum for the development of glaciers as the temperature is low enough for snow to fall and to be preserved at the level of maximum precipitation. That part of the mountains situated above the snowline receives enough precipitation to include important snowfalls, while the low temperature permits their preservation in a very high proportion. The large glaciers of Spitzbergen, Norwegian Lapland, Iceland, Alaska, and the large massifs of the inner Alps are thus easily explained.

On the Equator, on the other hand, there is an important interval between the 0°C (32°F) isotherm and the zone of maximum precipitation. This interval is more than 4 000 m (13 000 ft) in the Andes of Peru and Bolivia and in East Africa. Because the uppermost level does not receive much precipitation, conditions are unfavourable to glaciers, though temperatures are cold enough. Poorly alimented, glaciers do not extend far below the snowline. Furthermore there is a unique zone between it and the upper limit of the forest kept low by drought. In this broad zone rainfall is low, causing desert or subdesert conditions which impede the growth of vegetation. The extent of snowfields varies little because of the absence of a winter season due to latitude. Thus the soil is left unprotected to the weather, particularly to the large daily temperature range on the ground. This high tropical desert is favourable to periglacial processes: frost-needles (pipkrakes), frost-weathering, and solifluction. There are important differences with the subglacial zone of midlatitude mountains where there is sufficient humidity to permit the development of meadows which hamper solifluction, forcing it to form lobes and terracettes. A thick snow cover during winter stops the direct influence of the weather during a great part of the year. More precipitation and smaller variations in soil temperature produce a different balance between the various processes. The glaciers, for their part, better alimented, descend much farther below the snow line than in the tropics and a great part of their tongues is extrazonal from the climatic point of view.

The differences in vertical zonation caused by the zonal climate, however, are not limited to the tropics and the temperate zone. Friedland (1951), who made a systematic study of the vertical zonation of soils in the USSR, distinguishes a boreal type which, for instance, is well represented in the Sikhote Aline Mountains bordering the Sea of Japan. It is characterised by forest and podzolic soils followed by tundra. These mountain tundras correspond to the alpine meadows of more temperate mountains. These meadows are absent from this boreal type just as tundras are virtually non-existent in temperate mountains. The subtropical arid regions are characterised by a very reduced forest zone between the steppes on the lower mountain slopes and the meagre grasses of the highlands. The forest may even be completely absent as in Daghestan, to the west of the Caspian Sea, but this is also partly owing to the presence of limestone. The very great range of ground temperature, daily as well as seasonal, together with a light cloudiness causes intense mechanical weathering. Rockfalls and stone fields are therefore common, but they normally do not prevent a very open cedar forest and meagre grasslands. Frost-weathering is particularly important because the snow cover does not usually last the whole of the winter except near the snowline. In Corsica this zone of intense fragmentation, of rough stony mountains, begins at an elevation of 1 200 to 1 500 m (4 000 to 5 000 ft) although its lower limit, it is true, has been lowered by pastoral degradation. It is typical of Mediterranean mountains which are averse to grasslands because of summer drought.

2. *Differences between vertical zones and corresponding latitudinal zones.* The presence of glaciers on mountains and of a periglacial zone at a level immediately below them calls for comparisons with the glacial and periglacial zones of the high latitudes. In the past such comparisons have often gone too far: the mechanisms of mountain glaciers having been applied to ice sheets. The conclusions arrived at from the studies of one of the two types of glaciers have been applied without sufficient reservations to the other.

This lack of discrimination can hardly be justified for it does not correspond to the facts; it distorts them. The studies of glaciologists, notably Ahlmann, have shown that ice sheets and mountain glaciers belong to two very different physical types. The ice at the base of ice sheets is at a temperature well below the melting point, whereas the temperature of the base of mountain glaciers is closer to it. The regime is also very different.

Summer rain plays an important role in the ablation of midlatitude glaciers, whereas it is absent from the interior of Greenland and the entire Antarctic icesheet. The morphogenic effects are also very different; not only the ice temperature and, consequently, its mechanical properties differ, but the mass of the glaciers is not at all the same. A mountain glacier is strictly localised, forming a tongue of limited volume in a valley, whereas an ice sheet spreads over a vast surface, burying a varied relief below an ice cover 1 000 to 4 000 m (3 000 to 14 000 ft) thick. Neither are the rates of flow the same, they are usually more rapid in mountain glaciers than in ice sheets.

Gaps in our knowledge of glacial geomorphology are due in large part to confusions which have resulted in inaccurate extrapolations and have hampered the analysis of facts and their confrontation in a synthesis.

Similar differences are found between the latitudinal and altitudinal periglacial zones. They have been emphasised, in an occasionally adventurous manner, by Troll. There is, first of all, an essential difference between the two zones: in the circumpolar periglacial zone all of the relief is zonal. Even the watercourses which are altered in their features and in their morphogenic action have repercussions on erosion in general, especially on the conditions of slope development. In mountains, however, watercourses are altered little if at all; they have a torrential azonal character. The slopes alone evolve according to the laws of periglacial processes, and then, only partially, as these laws do not affect erosion as a whole. The base level of the slope is most often lowered by the incision of the torrent, whereas in the Arctic it may be raised under identical conditions by a progressive filling of the valley bottom; such depositional filling and raising of the base level is not so common in mountains.

These differences between mountain zone and corresponding latitudinal zone become even more complicated because the morphogenic conditions of a specific mountain zone are modified by the general characteristics of the latitudinal zone in which it is located, characteristics which differ from the latitudinal zone to which it corresponds. Thus land sculpture in a periglacial mountain zone is highly dependent on *exposure* whose influence may shift the zonal limits of two neighbouring slopes by as much as several hundred metres. In a temperate climate the shady side (*ubac*) preserves snow patches nearly the whole summer and is subject to frost-weathering; the sunny side (*adret*), at the same elevation is covered by a dense alpine meadow. The orientation of the mountain range also causes large contrasts in precipitation; the humid northwest facing limestone slopes of the Moroccan Middle Atlas are grass covered, whereas the southeast facing slopes overlooking the Moulouya valley are covered with rubble and frost-riven scree.

The influence of exposure varies considerably from one latitudinal zone to another. It has been said that it is limited to midlatitude mountains. This is not exact; Tricart has demonstrated its role in the Venezuelan Andes, which are equatorial mountains (see Tricart, Cailleux, and Raynal, 1962);

Dollfus (1959) in the tropical Peruvian Andes, and Raynal in the semiarid subtropical Atlas mountains. In the Venezuelan Andes, which have a high cloudiness, the southeast slopes stand in contrast to those facing the northwest. The slopes exposed to the east and southeast receive the first sun's rays before the ascent of the clouds; the nocturnal snow thus melts rapidly. Lower down a certain amount of desiccation takes place. The slopes facing northwest and west, on the contrary, rarely receive any sun; when they could be reached the sun is hidden behind the clouds which have formed in the forenoon. These slopes are very humid, with mossy forests and, higher up, glaciers which are in part alimented by direct condensation of the atmospheric moisture on the névés. In the Peruvian Andes analogous mechanisms function during the humid season. The eastern slopes do not preserve the night snowfall and are the locus of periglacial phenomena, whereas névés occupy the western slopes at identical elevations.

The contrasts in humidity observed on mountain slopes of opposite orientation throughout the tropics are emphasised by the particular conditions of the atmospheric circulation which moves in a much more constant direction than in the temperate zone: the *windward slopes* are humid, the *leeward slopes* are dry. In the Venezuelan Andes the contrast is observed on the crests between torrential basins only 500 m (1 700 ft) wide at the base and 200 m (700 ft) high. Their windward slopes are grown over with moss and algae covered forests, their leeward slopes with xerophytic shrubs and bushes. And whereas the leeward slopes experience intense runoff, creep is the predominant mass wasting process on the windward side.

In Morocco exposure simultaneously produces contrasts in temperature as in the midlatitudes, and contrasts in humidity as in the low latitudes. These contrasts, of course, decrease as one moves into more cloudy climates in higher latitudes: differences in direct heating by the sun's rays are then more or less eliminated. In the vicinity of the Arctic Circle Friedland (1951) has emphasised the influence of the wind; it sweeps the snow in the winter and allows a deep penetration of the ground by low temperatures, some of which are caused by atmospheric temperature inversions due to intense cooling. Valley bottoms are occupied by tundra, the valley sides by taiga. The hill crests have the least vegetation due to their exposure to the wind. Slope exposure plays some role although less than in the midlatitudes: peaty marshes develop on the shady side.

The latitudinal zonal element present on all levels of intertropical mountains is the absence of cold and warm seasons. The morphogenic mechanisms therefore have a daily rhythm and the snowline is very stable throughout the year. This stability is favourable to special ablation mechanisms on glaciers, such as the one which produces 'ice penitents'.

The differences observed between the periglacial morphogenic system of mountains and that of the Arctic and Antarctic periglacial zones, therefore, cannot be considered to be merely extrazonal. The mountain environments constitute a special class. To be sure, it has many features in common with the lowland periglacial zones, but above all it has many special characteristics and an extreme variety. The nature of the elementary processes and the limits of the various vertical zones vary from one slope to another due to the configuration of the mountains. A particular level highly developed in a certain massif is greatly reduced in a neighbouring massif. Added to configuration in the midlatitudes is the influence of exposure. Basic principles can serve as a guide in all these matters, but they do not permit a simple classification of all the observations. The want of systematic studies of the kind of Friedland on the characteristics of vertical zonation in the various mountains of the world is a serious gap in our knowledge.

Conclusion

The zonal concept, if used with discrimination, makes possible the analysis of the distribution of the morphogenic processes on the earth's surface. It enables the classification of the morphogenic systems, making allowance, in each concrete case, for what is particular and for what is general. As such it serves the regional as well as the systematic point of view in geography. The zonal concept does not exclude the play of other factors as we have shown by analysing its relationships with the manifestations of internal forces; if there is some zonality in certain tectonic phenomena, the latter produce modifications in the zonality through the construction of mountains whose morphogenic systems are different from those of the latitudinal zone in which they occur.

The principle of zonality also emphasises the ties that exist between geomorphology and the other sciences of nature. It contributes to the breakdown of the disastrous isolation of our discipline by replacing it in a wider and, therefore, better framework.

If the above is true it is because the principle of zonality is not like 'normal erosion' a schematisation which attempts to impose itself on the facts which cannot be enclosed, but a natural law which is one of the constituent elements of the intrinsic logic of nature.

Bibliographic orientation

Because geomorphology is a recent discipline, its fundamental concepts are still poorly defined and studies concerned with the principles of the morphoclimatic division of the earth are still few in number. Most of them have already been cited in chapters 1 and 2. We refer in particular to the publications of Büdel (1948), of Tricart (1952, 1953), of Birot (in Lehmann, 1954), of Aubert and Cailleux (1950), of Guerassimov (1944), and of Prassolov (1944). They should be completed with:

BÜDEL, J. (1961) 'Die Morphogenese des Festlandes in Abhängigkeit von Klimazonen', *Naturwiss*, no. 9, 313–18.
    Again uses his earlier conceptions, systematising them without adding anything new and disregarding publications which have appeared later.
KREBS, N. (1966) *Vergleichende Länderkunde* (3rd edn), Stuttgart, Köhler, 484 p. (pp. 65–77).
    Attempt at a climatic subdivision of the globe applied to geomorphology. Contains basic ideas and concepts. Bibliography to which we refer.

MACHATSCHEK, F. (1938–40) *Das Relief der Erde. Versuch einer regionalen Morphologie der Erdoberfläche*, 2 vols, Bornträger, Berlin (2nd edn, 1955).

PENCK, A. (1913) 'Versuch einer Klimaklassifikation auf physiographischer Grundlage', *Sitzungber. Preuss. Akad. Wiss.*, Berlin, 1st sem., pp. 77–97.
A study which ought to be better known. Many new ideas for its time.

TROLL, C. (1944) 'Diluvial-Geologie und Klima', Klimaheft der geologischen Rundschau Einführung, *Geol. Rundsch.* **34**, no. 7–8, 307–25.

TROLL, C. (1944) 'Strukturböden, Solifluktion und Frostklimate der Erde', *Geol. Rundsch.* **34**, 545–694.
Good example of the introduction of the zonal concept in the study of a geomorphic phenomenon.

On the distribution of various processes and on climatic indices

Here are a few examples listed for the sake of enabling the student to get some idea of the problem:

BAGNOULS, F. and GAUSSEN, H. (1953) 'Saison sèche et indice xérothermique', *Doc. pour les cartes des prod. végét., sér. généralités*, **3**, 193–239.

BEHR, F. M. (1918) 'Über geologisch wichtige Frosterscheinungen in gemässigten Klimaten', *Int. Mitt. Bodenkunde*, **8**, no. 3–4, 50–72.

BOUDYKO, M. I. (1956) 'Climatic indices of aridity', *Geogr. Essays, USSR Akad. Sc.*, pp. 142–9.
A study more general than is indicated by the title. The proposed index is, in fact, applied to all the major climatic zones and gives an account of their hydrologic balance.

EMBERGER, L. (1955) 'Une classification biogéographique des climats', *Recueil Trav. Lab. Géol. Zool. Univ. Montpellier*, sér. bot. no. 7, pp. 3–43.
Excellent compilation with numerous references.

GALON, R. (1954) 'Les principaux paysages morphologiques du monde du point de vue des profils synthétiques qui les caractérisent', *Czasopismo Geogr.* **25**, pp. 26–37.

HEMPEL, L. (1960) 'Bilanzen zur Reliefgestaltung der Erde', *Geogr. Ber.* **15**, 97–107.
Series of interesting diagrams showing the variations of the various morphoclimatic phenomena in the different parts of the world. Unfortunately savannas and intertropical forests are not differentiated. The intensities of the processes in the different zones, it seems to us, are often inexactly evaluated.

HIERNAUX, C. (1955) 'Sur un nouvel indice d'humidité proposé pour l'Afrique occidentale', *Bull. I. F. A. N.* **17**, pp. 1–6.

MANGENOT, G. (1951) 'Une formule simple permettant de caractériser les climats de l'Afrique intertropicale dans leurs rapports avec la végétation', *Rev. Gén. Bot.*, pp. 353–69.

SAPPER, K. and GEIGER, R. (1934) 'Die dauernd frostfreien Räume der Erde und ihre Begrenzung', *Meteorol. Z.*, pp. 465–8.

On the climatic and morphoclimatic zonation of the earth

The essential data are available in the following works:

UNESCO (1952) Consultative committee on arid zones. *Eastern Hemisphere Western Hemisphere, Distribution of arid homoclimates*, Paris, 2 maps.

CNRS, *Colloques Int.* **59** (1955) xii: *Les Divisions écologiques du monde*, Paris, 326 p.
International colloquium on the ecologic divisions of the world (means of expression, nomenclature, cartography).

BAGNOULS, F. and GAUSSEN, H. (1957) 'Les climates biologiques et leur classification', *Ann. Géogr.* **66**, 193–220.
Simple classification producing good results.

BERG, L. S. (1950) *The Natural Regions of the USSR*, trans. from Russian by Olga A. Titelbaum under the Russian transl. Project Amer. Council Learned Soc., edited by John A. Morrison and C. C. Nikiforoff. Macmillan, New York, 436 p.
This study provides a good understanding of the problems of the morphoclimatic division of the high midlatitudes.

BROCKMANN, H. (1913) 'Der Einfluss der Klimacharakters auf die Verbreitung der Pflanzen und Pflanzengesellschaften', *Englers Bot. Jahrb.*

CLAYTON, H. (1934) 'World weather records', *Smithonian Misc. Coll.* **79**, 1927, and **90**, 1934, Washington.
Collection of climatic data of the whole world constituting a useful reference work. Completes the tables given by Köppen and Geiger which remained incomplete.

CONRAD, V. (1944) *Methods in Climatology*, Harvard University Press, 21 p.

GAUSSEN, H. (1954) 'Expression des milieux par des formules écologiques. Leur représentation cartographique', *Colloques Int. CNRS, Régions écologiques du globe*, Paris, pp. 13–28.
The combination of simple indices and common data permits the characterisation of the ecologic factors of divers plants.

GAUSSEN, H. (1955) 'Les climats analogues à l'échelle du monde', *C.R. Acad. Agric.* (France), **41**, no. 5, March 9–16 meeting, pp. 234–8.

GENTILLI, J. (1953) 'Critique de la méthode de Thornthwaite pour la classification des climats', *Ann. Géogr.* **62**, 180–5.

GRIGORIEV, A. A. (1942) 'Attempt to characterize the basic types of physical-geographical environment, 3rd part: physical-geographical types of the Arctic zone', *Problems Phys. Geogr. USSR Akad. Sc.*

GRIGORIEV, A. A. (1948) 'Attempt to characterize the basic types of physical-geographical environment, 4th part: basic and general physical-geographical processes of the subarctic lands and the temperate zone, and justification for the division of the temperate latitudes into zones', *Problems Phys. Geogr., USSR Akad. Sc.*

GRIGORIEV, A. A. (1954) 'Geographical zonation and some of its applications', *Izv. Akad. Nauk. SSSR, Geogr. Ser.* no. 5, 17–39 and no. 6, 41–50.
Interesting views, unfortunately only applied to the USSR.

GRIGORIEV, A. A. (1956) 'Present state of the theory of zonality in nature', *Geogr. Essays, USSR Akad. Sc.*, pp. 365–71.

GRIGORIEV, A. A. (1962) 'Present state of the theory of geographic zonality', *Soviet Geogr. Amer. Geogr. Soc. Occas. Public.* no. 1, pp. 182–7.
The divisions are mainly based on temperature and humidity.

GUERASSIMOV, I. P. (1954) 'Analogies et différences dans la nature des déserts', *Priroda*, no. 2, 11–22 (French trans. by CEDP, no. 984).
Compares Sahara, Turfan, and Gobi deserts. Contrasts deserts with and deserts without rigorous winters.

HELIMANN, G. (1925) 'Grenzwerte der Klimaelemente auf Erde', *Naturwiss.* **13**, 845.

IVANOV, N. N. (1948) 'Zones of climatic landscapes of the world', *Mem. Geogr. Soc. USSR*, n.s., **1**.

KANONNIKOV, A. M. (1955) 'Concerning the question of natural zones', *Izv. Vsiess. geogr. Obchtch.* **87**, 529–34.

KENDREW, W. G. (1961) *Climate: a treatise on the principles of weather and climate*, 4th edn, Oxford, Clarendon Press, 329 p.

KÖPPEN, W. (1931) *Grundriss der Klimakunde*, Berlin, de Gruyter.
Presents a general view of the classification of climates.

MARKOV, K. K. (1948) *Fundamental Problems of Geomorphology*, USSR Akad. Sc.

MILLER, A. (1951) 'Three new climatic maps', *Inst. Brit. Geogr., Trans. Pap.*, no. 17, pp. 15–20.
Proposes simple indices and critical values. Interesting results.

MONOD, T. (1957) 'Les grandes divisions chronologiques de l'Afrique', *Publ. CSA*, London, 147 p., one map.

MORTENSEN, H. (1950) *Das Gesetz der Wüstenbildung*, Universitas, Stuttgart, vol. 5, no. 7, pp. 801–14.
This study analyses the problems of defining dry morphoclimatic regions.

MURZAIEV, E. M. (1953) 'Outline for the regional physical-geographical subdivision of Central Asia', *Izv. Akad. Nauk. SSSR, Geogr. Ser.* no. 6, 17–30.

PASSARGE, S. (1924) *Vergleichende Landschaftskunde*, Heft 4: *Der heisse Gürtel*, Berlin.

PASSARGE, S. (1926) 'Morphologie der Klimazonen oder Morphologie der Landschafts-gürtel?' *Peterm. Geogr. Mitt.*
Demonstrates that the regional divisions of climatic geomorphology should not be based on climates but on physical-geographical zones primarily determined by vegetation.

PERELMAN, A. I. (1954) 'The natural landscapes of the European part of the USSR and their geochemical properties', *Priroda*, no. 3, pp. 38–47, one map.

ROZOV, N. N. (1954) 'Development up to the present time of Dokuchaev's doctrines on zonal soils', *Izv. Akad. Nauk. SSSR, Geogr. Ser.* no. 4, pp. 3–17.

RUELLAN, F. (1953) 'O papel das anxurradas no modelo do relêvo brasileiro', *Bol. Paulista Geogr.* no. 13, pp. 5–18 and no. 14, pp. 3–25.

STRETTA, E. (1958) 'Délimitation des zones arides et semi-arides en Turquie', *Rev. Géomorph. Dyn.* **9**, 97–102.

TANNER, W. P. (1961) 'An alternate approach to morphogenetic climates', *Southeastern Geol.* **2**, pp. 251–7.

THORNTHWAITE, C. (1933) 'The climates of the earth', *Geogr. Rev.* **23**, 433–40.

THORNTHWAITE, C. (1943) 'Problems in the classification of climates', *Geogr. Rev.* **33**, 233–55.
Explanation of the author's method of classification and coloured map.

THORNTHWAITE, C. (1948) 'An approach toward a rational classification of climate', *Geogr. Rev.* **38**, 55–94.

TRICART, J. (1955) 'Types de fleuves et systèmes morphoclimatiques en Afrique occidentale', *Bull. Sect. Géogr., Com. Trav. Hist. Sc.* **68**, 303–45, 21 pl.

TRICART, J. (1957) 'Application du concept de zonalité à la géomorphologie', *Tijdschr. Kon. Ned. Aardrijksk. Gen.*, Festschrift Jacoba Hol, **74**, no. 3, 422–34, with map showing the morphoclimatic divisions of the earth.

TRICART, J. (1958) 'Division morphoclimatique du Brésil atlantique central', *Rev. Géomorph. Dyn.* **9**, 1–22.

TROLL, C. (1955) 'Der jahreszeitliche Ablauf des Naturgeschehens in den verschiedenen Klimagürteln der Erde', *Studium Generale*, **8**, no. 3, 713–33.
Classification of climates based on the utilisation of thermo-isopleths; climatic map of the Old World. Numerous figures.

VOLOBUIEF, V. P. (1953) *Soils and Climate*, Baku, Azerbaijan Acad. Sc., 319 p.
Unfortunately a not very accessible work. Excellent overall view of the division of the earth into major physical-geographical zones.

On vertical zonation

The above mentioned works should be completed with the following references:

FRIEDLAND, V. N. (1951) 'Essai de division géographique des sols des systèmes montagneux de l'URSS', *Pédologie*, **9**, pp. 521–53 (French trans. by CEDP).
Excellent analysis of an example and serving as a precious example of method.

HENRY, A. J. (1919) 'Increase of precipitation with altitude', *Mon. Weather Rev.* **44**, 33–41.

HERMES, K. (1955) 'Die Lage der oberen Waldgrenze in den Gebirgen der Erde und ihr Abstand zur Schneegrenze', *Kölner Geogr. Arb.* no. 5, 277 p. 4 maps.
The critical factors which determine the treeline are complex and vary according to latitudinal zones. Good analysis of the problem.

MILKOV, F. N. (1953) *Influence of Relief on Vegetation and Animal Life*, Moscow, Publ. Geogr. Lit., 164 p.
  Thorough analysis, unfortunately limited to Russian examples and a Russian bibliography, however important.
PEATTIE, R. (1936) *Mountain Geography: a critique and field study*, Harvard University Press, 257 p.
PEGUY, C. P. (1953) 'Hautes altitudes et hautes latitudes', *Cahiers Inf. Géogr.* **2**, 58–66.
TRICART, J., CAILLEUX, A., and RAYNAL, R. (1962) *Les particularités de la morphogenèse dans les régions de montagnes*, Paris, CDU, 136 p., mimeographed 'Cours' de l'Université de Strasbourg.
TROLL, C. (1955) 'Forschungen in Zentralmexiko 1954. Die Stellung des Landes in dreidimensionalen Landschaftaufbau der Erde', *Deutschen Geographentag*, Hamburg, pp. 191–213.
TROLL, C. (1959) 'Die tropischen Gebirge. Ihre dreidimensionale klimatische und pflanzengeographsiche Zonierung', *Bonner Geogr. Abh.*, Heft 25, 93 p.
  Abundant and accurate climatic data. Detailed discussion of the various vertical zones in tropical mountains.

On plant and animal geography

BIROT, P. (1965) *Formations végétales du globe*, Paris, SEDES, 508 p.
  Only profound study of its kind in French. Two-thirds of the work is concerned with a zonal study of the world's plant formations.
CAILLEUX, A. (1953) *Biogéographie mondiale*, Coll. Que sais-je?, 128 p.
  General introduction and abundant numerical data especially on the wealth and endemicity of floras and faunas.
CARLES, J. (1948) *Géographie botanique*, Coll. Que sais-je?, 120 p.
  Excellent introduction to the study of habitat and vertical zonation; methods; list of plant formations and associations of France.
CNRS (1952) 'Ecology 1950', *Colloques int.* Paris, 350 p.
  Numerous detailed studies and general views.
DELAMARRE, C. and DEBOUTTEVILLE, 'Microfaune du sol', *Actes Scientif. et Industr.*, 360 p.
  The conclusion (pp. 278–316) summarises the present status of a basic phenomenon concerning biogeography, distribution, palaeoclimates, and evolution.
GAUSSEN, H. (1954) *Géographie des plantes*, Paris, Coll. A. Colin, 223 p.
  Qualitative study of the world's plant formations.
PRENANT, M. *Géographie des animaux*, Coll. A. Colin.

4

Morphoclimatic equilibrium and interruptions in the morphoclimatic equilibrium

The morphoclimatic division of the earth, the criteria for which we have just set forth, constitutes a valid framework for the analysis of the present morphogenic processes and mechanisms. But as the morphogenic processes and mechanisms leave the time factor out of account, they do not provide a complete understanding of the relief of the various morphoclimatic regions. The Quaternary Epoch, as is well known, was characterised by exceptionally important climatic oscillations, the latest of which is extremely recent. Fifteen thousand years ago continental ice sheets covered northern Europe and northern North America. The tropics have been subject to equally important changes, but of a different nature; there were alternations of pluvials and arid phases in the Sahara and aeolian incursions well beyond the Tropics of Cancer and Capricorn in the direction of the Equator. A very large part of the present terrestrial relief has been sculptured during periods with climates different from the present one. Climatic geomorphology must therefore take into account the climates of the past; its progress depends in part on the methods of palaeogeographic reconstruction and dating.

Because of the last major climatic oscillation the relief of a great part of the earth is presently readjusting itself to the new morphoclimatic conditions. An investigation of the characteristics of this readjustment is therefore necessary. Being the most recent, the postglacial climate has left a profound imprint on the present landforms, but it is not the only one. Many different climates have preceded it: everybody agrees that there were several glacial periods in the temperate zones, pluvials in the semiarid zones, and that climates were generally warm in the midlatitudes during the Tertiary Period. These climatic oscillations have also been registered by the relief which presently survives as more or less well preserved relict forms depending upon the nature of the present climate, the lithology, and the geologic structure. The lifetime of relict landforms may be considerable in the case of fossilised forms that were later exhumed; they may date back even to the Precambrian. Such very old palaeoforms were formed under very different geographic conditions than exist at the present, and their interpretation should take into consideration the general geologic evolution of the earth, especially the primordial fact which is the progress of life on its surface.

Natural interruptions in the morphoclimatic equilibrium of the geologic past are not the only ones which affect climatic geomorphology. Through the extent of his clearings or his brush fires man is a significant morphogenic factor, important enough to give rise to a special morphogenic system, namely, *anthropic erosion*, which brutally disturbs the natural morphogenic systems. All geomorphology concerned with reality must allow for the effects of man; concern with these effects draws our discipline, which is becoming increasingly objective and scientific, towards human geography, from which it was formerly severed; the intervention of man in the evolution of landforms opens up to the geomorphologist a whole new realm of practical applications in connection with the problems of land use and economic development.

In this chapter we discuss successively: the concept of morphoclimatic equilibrium; interruptions in the morphoclimatic equilibrium of the geologic past; and interruptions in the morphoclimatic equilibrium due to man.

The concept of morphoclimatic equilibrium

The fundamental unity of nature and the narrow interdependence of its various elements are at the base of the concept of morphoclimatic equilibrium. Each major climatic zone has its biochore, whether forest, savanna, grassland or desert, which transforms the superficial rocks into zonal soils. The morphoclimatic system of each of these environments is dependent not only on climate but on vegetation and soils (*sensu lato*). It is on the inter-relationships of these three factors that the concept of morphoclimatic equilibrium is based.

The concept of morphoclimatic equilibrium, actually, is less a balance than a permanent state, or steady state, in the sense of physicists. Indeed if a stream course or a slope has a constant profile but is nevertheless subjected to slow erosion, there is a slow displacement of the profile but no change in form. The fact that stream water always holds more foreign matter in suspension or in solution than rainwater or the water of melting snow suffices to demonstrate that there is always erosion, no matter how little, or, in other words, that watercourses and slopes shift, no matter how infinitesimally, even if their profiles remain constant. The permanence of forms during a given period is, however, so frequent and so important that we conform to usage and refer to it in terms of an *equilibrium*.

Definition of morphoclimatic equilibrium

To define the concept of morphoclimatic equilibrium involves two successive steps: first, to define the concept of equilibrium and, second, to apply it to climatic geomorphology.

THE CONCEPT OF EQUILIBRIUM IN GEOMORPHOLOGY

The concept of equilibrium is of current usage in geomorphology where it was first applied to the profile of stream beds. The concept, without the term, was already understood by Italian hydraulic engineers of the seventeenth century and by Surell. Davis integrated it into the theory of normal erosion and the cyclic concept. Baulig (*Essais*, pp. 43–77) has traced its history in detail and has later applied the same concept to the evolution of slopes (*ibid.*, pp. 125–47). He gives the following definition of the concept of equilibrium: 'An *actual* profile of equilibrium is always *provisional*; it is never *definitive* nor *final*. All that can be said is that, given an evolution free of interruptions, the profile proceeds *increasingly slowly* towards a *limit* which, by definition, it cannot attain' (p. 57).

This definition contrasts with the one given by de Martonne, who follows Philippson. For these two authors the profile of equilibrium of streams is reached when the gradient has been reduced to such an extent that there is no further change (Philippson's *Erosionsterminante*). For example, the torrent which has reached its profile of equilibrium is a 'dead' torrent which has ceased corrading. This definitive stability is in fact unknown in nature; to set it up as a norm would be to adopt an idealistic position, a philosophical apriority. Rather than stability, a modern geomorphologist is forced to conclude from a careful observation of stream courses that minor or major alterations continuously take place, whether swiftly or extremely slowly. Immobility, however, does not exist.

Baulig has had the merit to point out this unstable nature of stream profiles. Moreover Davis himself had clearly noted the characteristics of an interruption in the geomorphic equilibrium. In his article on the geographical cycle he even drew a curve of it; changes caused by an interruption of the equilibrium at first rapidly increase, then pass through a maximum, and finally progressively decrease. The curve of their intensity becomes tangent to 0°. It is the application of the concept which is wrong in this case, for the interruption of equilibrium analysed is exclusively tectonic. Outside of faults, which are often related to rapid movements, tectonic deformations are generally rather slow; even fault movements are usually reflected by a series of small quakes, for throws exceeding one metre in a single quake are exceptional. The average rate of uplift of fold mountains is generally only two or three times the average rate of denudation, and often less. Tectonic movements, consequently, do not generally cause very sudden interruptions in the geomorphic equilibrium. The figurative curve drawn by Davis in his article on the geographical cycle (1899) shows an exaggerated maximum of this fact, for the slope which denotes it is too steep. An interruption in the equilibrium caused by tectonic deformations is normally less sudden and more progressive and should be shown by a flatter curve. The most sudden interruptions of equilibrium are of climatic origin. In less than 20 000 years continental glaciers have retreated from northern Germany, northern Poland, and northwest Russia to their present position in the highlands of Scandinavia. And the glacio-isostatic reactions which followed the disappearance of the ice sheet are amongst the most rapid deformations on the globe. This is not a theoretical speculation but an observed, even *measured*, reality, a rare fact in our discipline.

The equilibrium concept as elaborated by classical geomorphology may therefore be said to be valid, at least as expounded by Baulig. The realisation of the state of equilibrium in Davisian geomorphology was characterised by the so-called stage of 'maturity'.

The concept of equilibrium may, however, be defined in a more objective manner. The statistical study of slopes indicates a tendency of average slopes to group themselves round a certain value in each environment that is uniform from the lithologic and climatic points of view. This frequency maximum is a function of morphogenic mechanisms. In the high Alps, where there is intense fragmentation, the average slope is between 40° and 55°. In the moderately frost-liable limestones of Lorraine it is only 7° to 12°, in keeping with the predominant action of Pleistocene periglacial solifluction. With statistical studies of this kind it is possible to relate the concept of equilibrium to a classification index of the frequential distribution of slope gradients. For example, a resumption of stream erosion is first reflected in a deepening of the stream courses which gradually spreads to higher levels. In the beginning this process results in a steepening of the hillsides and therefore in a flattening of the frequency curve of the slope values. Slopes whose gradients have been increased by the resumption of stream erosion coexist with others where this action has not as yet made itself felt and which preserve a gentler gradient. Such a juxtaposition of slopes of different gradients at two distinct levels produces the well-known contrast between an old erosion surface with slightly undulating forms and the gorges which incise it, or between a slightly rolling plain and the gullies produced by anthropic erosion which scar its margins. The continuation of such resumed erosion, which augments the dissection, increases the number of steep slopes and at the same time reveals a rising frequency maximum. These facts are observed regardless of the cause of the resumption of erosion, whether tectonic, as is most often the case in ancient massifs, or morphoclimatic. Studies of resumptions of erosion may therefore make possible a general quantitative definition of the concept of equilibrium.

APPLICATION OF THE CONCEPT OF EQUILIBRIUM TO CLIMATIC GEOMORPHOLOGY

Because the concept of equilibrium is valid, it is not surprising to find it in other branches of the natural sciences, particularly those that are most closely related to geomorphology, namely plant geography and pedology. In these sciences the concept of equilibrium is called *climax* or *climactic formation*, which, at least in France, increasingly tends to take the place of *final climactic formation*, still used in vol. 3 of de Martonne's *Traité*. The first mentioned term is indeed much to be preferred as there is no guarantee that the present state is the final one.

The concept of *climax* is valid for vegetation, soils, and fauna.

Let us consider a complex vegetation sample occupying a surface of several square metres and belonging to a certain natural environment. If the surrounding natural conditions do not change except in a minor way during a sufficiently large number of years the sample remains essentially unchanged. In other words, the climax is the state of a complex sample which remains unchanged except for minor fluctuations if the surrounding natural conditions remain the same, except for fluctuations, during a sufficiently large number of years. But in the long run life evolves, its climax changes, and this change in turn has repercussions on the physicochemical conditions (climate, etc.); furthermore it is not excluded that the evolution of the latter is appreciable. The climax therefore changes in the long run.

A climactic soil is a soil whose profile after a prolonged action of the zonal factors of a region remains similar to itself. This is the case, for example, of the brown forest soils of western Europe or the podzols of Scandinavia. The climactic soil helps the establishment of a climactic vegetation and vice versa. The realisation of the climactic soil indeed favours the plant species which are best adapted to it (for example podzols favour conifers), but it handicaps the others. The acid humus of conifers in turn favours podzolisation, which tends progressively to eliminate deciduous trees, thereby producing a renewed acceleration of podzolisation, and so on. The acidification of soils related to the destruction of the natural vegetation by man on the coasts of northwestern Europe has encouraged the development of moorlands, a type of anthropic climactic formation, or disclimax,[1] adapted to the present conditions. Similarly, the repeated fires which destroy the West African savanna and concurrently favour the cuirassing process cause the development of highly siliceous, treeless grasslands, very poor in species, which also, under these conditions, represent a disclimax. Climactic soils and climactic plant formations are therefore closely related; they represent two forms of adjustment to one and the same environment, forms which, of course, react upon one another. The adjustment of the soil is realised in a few millennia; the adaptation of the vegetation in a few centuries. The latter, therefore, occurs on a time scale which is different from that of the general evolution of plant species. The appearance of a new species occurs, on the average, every several hundred thousand years; it is one or several hundred times longer than the appearance of a new vegetational climax. The existence of plant climaxes, therefore, does not hinder the progressive modification of life forms on the earth's surface. It belongs to another order of magnitude. The contradiction is only apparent. But it does forbid the use of the adjective 'final' in the definition of climax, for it is in absolute contradiction with the concept of evolution on the geological time scale.

[1] Or plagioclimax in British literature (KdeJ).

The biogeographic and pedologic concepts of climax are of special importance to climatic geomorphology because of the influence of plant formations and soils on the morphogenic systems. The realisation of the biogeographic climax indeed produces relatively stable conditions which control the morphogenic processes. The stability allows the formation of zonal soils which provide a specific environment to the geomorphic agents. The climactic vegetation determines the type of living screen interposed between the elements of the weather and the ground surface. It also provides the morphogenic processes with certain specific conditions which may be sufficiently *durable* to let the relief adjust itself to them.

It should however be stressed that biogeographic climaxes and morphoclimatic equilibria are not the same. The coincidence is only partial in space as well as in time. Morphoclimatic equilibria may be realised in regions where the influence of life on the evolution of relief is negligible, as in glacial, periglacial, and desert regions. The geographic extent of the concept of morphoclimatic equilibrium is larger than that of the concept of climax. Moreover, the realisation of a climax does not necessarily bring about that of a morphoclimatic equilibrium. A greater length of time is generally required for the relief to adjust itself to a given biogeographic environment than is required for the realisation of a climax. If the climatic

FIG. 4.1. Dissected relief due to runoff, Willunga Hills, near Adelaide, South Australia. Example of morphoclimatic equilibrium

Slightly metamorphosed shales weather into clays. Runoff remains diffuse in spite of steep slopes because the grass is an obstacle to its concentration. The relief is composed of steep ridges with convex crests and concave slopes at the base where the waste removed from the steeper slopes of the median part of the hillsides is deposited (colluviation).

oscillations are too short, the biogeographic climax may be realised whereas the morphoclimatic equilibrium is not. For example, leaving the intervention of man out of account, the postglacial biogeographic climax is fully realised in Fennoscandia. Yet in spite of its realisation most of the relief is composed of relict glacial landforms which are hardly modified in most cases. In the morphoclimatic zones of dense plant cover the realisation of the biogeographic climax is a condition for the realisation of the morphoclimatic equilibrium, but it is not the only condition. Conversely, the destruction of the biogeographic climax, especially by man, has immediate consequences on the morphoclimatic equilibrium; it causes changes in the landforms which may be extremely rapid; decametric and hectometric forms may develop in a few decades (such as gullies, landslides, and rockfalls). For this reason man-induced erosion is properly called *accelerated erosion*.

Taking into account the facts and mechanisms described above, it is now possible to propose the following *definition of morphoclimatic equilibrium*: 'A morphoclimatic equilibrium is a morphoclimatic adjustment that is realised in a given region when the land forms are *predominantly* determined by a morphogenic system that is dependent on the climatic factors.' Interruption of the morphoclimatic equilibrium may, therefore, either result from a change in climate, causing one zonal morphoclimatic system to be replaced by another, or from the action of man who destroys the plant cover and thereby introduces predominantly polyzonal or azonal processes (such as wind or runoff).

Factors of the morphoclimatic equilibrium

When a morphoclimatic equilibrium is realised, erosional and depositional forms are adjusted to the processes that form them. The crescent form and asymmetry of a barchan result from the rapid migration of a limited amount of sand under the effect of the wind. The form of the barchan is characterised by aerodynamic curves. The profile of a slope formed by rockfalls also reflects the processes that affect it; the upper part is a bare escarpment, more or less abrupt, more or less regular depending on the homogeneity and the resistance of the strata that compose it but whose gradient always exceeds 40 to 45° on the average. The lower part of the slope is composed of a rectilinear scree, reposing at an angle of 30 to 35°, and built by falling or rolling rocks. The largest of them roll farther than the others; sometimes they even roll up the opposite slope for a short distance.

There are corresponding forms of equilibrium (or of a permanent regime) for all the other morphogenic processes and mechanisms; forms of equilibrium are not a discovery of climatic geomorphology. The profile of sea cliffs and the long-profile of rivers also have their forms of equilibrium; the profile of equilibrium of stream courses, or the graded profile, is one of the oldest concepts in geomorphology. It is necessary, therefore, to show what the particular applications of this concept to climatic geomorphology are.

A first fact, on which it is essential to insist, is its general value. *Forms of equilibrium*, which are the reflection of the processes and mechanisms that create them, tend to become established in all landscapes. Practically all distinctive landforms are forms of equilibrium: glacial rockbars, terminal moraines, cirques, and glacial troughs are all forms of equilibrium due to a glacier. A terminal moraine represents a certain momentary equilibrium in the regime of a glacier, between its alimentation and its ablation, and between the mass of debris it carries and the fraction of this debris that the melt waters can carry away. The moraine is indeed composed of the debris left behind. The concave long-profile of midlatitude rivers is a profile of equilibrium. But contrary to generalisations based on the concept of 'normal erosion', the broken long-profile of intertropical rivers, formed of quiet reaches alternating with rapids, is also a profile of equilibrium. Nothing permits them to reduce their breaks in gradient under present conditions. The problem is posed in terms closely related to those of glacial rockbars. A form of equilibrium is therefore not necessarily a regularised form except under certain circumstances: when the processes tend to reduce the irregularities, as in the case of the cutting of stream courses in the temperate zone.

A second fact, *a certain tendency of forms to persist*, is revealed by the concept of morphoclimatic equilibrium. The most typical case is that of inselbergs. In the Sahara and in the whole of intertropical Africa, the most perfect surfaces of erosion are studded with inselbergs whose slopes are always more or less equally steep no matter what their dimensions. Average inselberg slopes are approximately 30° to 40° in the savanna, steppe, and desert zones. This value is true of high as well as of low inselbergs, of inselbergs that are completely isolated in the midst of enormous plains and of inselbergs grouped together in dense archipelagos. The shape of inselbergs is the same on recent surfaces of erosion, such as the Mio-Pliocene surface of Inchiri in southwestern Mauritania, and on old surfaces. This uniformity of slope steepness on all kinds of inselbergs forces the conclusion that their slopes retreat *parallel to themselves* without decrease in gradient. The mountain volume is reduced, but the form persists: the largest inselbergs in the south of Mauritania have a tabular top formed by a plateau which is either a structural surface (on the tabular sandstones of the Palaeozoic cover) or a remnant of an old erosion surface (on folded Precambrian rocks); the smaller inselbergs have a conical form because the retreat of slopes caused the disappearance of the last remnants of the plateau surface by intersection. Slope steepness is everywhere the same in identical rocks. This permanence of forms is also found in other morphoclimatic environments: the breaks in slope of glacial rockbars (riegels) do not seem to suffer reduction under the effect of glacial erosion; in the wet tropics flat bottomlands occupied by dense rainforest do not seem to be reduced any more than sugarloaves during the course of their evolution. When the morphoclimatic equilibrium is realised, the relief forms seem to be characterised by a certain permanence which precisely explains the differences of relief in the various morphoclimatic zones.

The parallel retreat of slopes introduces *the concept of the minimum slope*. In a given, adjusted, morphoclimatic environment slopes preserve a certain minimum value in spite of continued evolution. Statistical studies clearly reveal this. The flanks of inselbergs are a typical example. Their slopes are not steep enough to allow them to be formed by rockfalls, but they are not much below the minimum gradient of scree slopes. Thus the debris which covers their flanks is easily evacuated by slope wash which lacks energy and necessitates a steep slope to remove the debris. In this case one is led to think that the minimum slope is a function both of shower intensity and the nature of the fragmentation (dimension of the debris). The degree of slope would in the last analysis be determined by the combination of lithologic and climatic factors. This would explain its uniformity in a climatic region where the rocks are identical. During the course of time the slope gradient tends towards the minimum slope. In statistics it is reflected in a gradual crowding of the slope values round the average slope. Histograms, however, always remain asymmetrical. The falling of the curve towards values lower than the mode is rapid, whereas it is slower and of concave form towards the higher values. This shape is significant; often there is a progressive slowing down in the rate of evolution, for a given mechanism works increasingly slowly towards the realisation of its ultimate form of equilibrium as the actual slope comes closer and closer to the minimum slope. For example, in the case of cliffs which produce screes, the rate of slope development depends on the rate of fragmentation and on the vertical dimension and steepness of the original rock escarpment. The rate of fragmentation in turn depends on the climate and the nature of the rocks. If the cliffs are in morphoclimatic equilibrium, the talus is at first abundantly supplied because the bedrock is widely exposed on the face of the cliff. But eventually the scree, as it keeps on accumulating, progressively buries the base of the cliff, which then ceases to be undermined and becomes protected against further fragmentation. The average slope of the cliff then decreases little by little. When it reaches 40° to 45°, the detached blocks, except in rare cases, cease to fall and accumulate in place, and the slope is gradually stabilised. If the climate permits the development of vegetation, plants will occupy it and contribute to the cessation of rockfalls. A histogram of slope values reflects this evolution. Slopes inferior to the minimum slope or in some cases to the average slope (e.g. of rockfalls) are those which, initially too gentle, are not affected or are little affected, by the dominant mass wasting process, as for example, the flanks of low hills between two confluent streams. Other slopes, fewer in number but with gradients clearly higher than the mode, are abrupt residual slopes which have successfully resisted the erosive processes either because of their position or because of their lithology as, for example, the abrupt conglomerate escarpments in the valleys of the Vosgian sandstone whose flanks have been regularised by periglacial rockfalls. Of course, if several different mass wasting processes attack the slopes of an area, the slope frequency curves will be polymodal.

The *concept of stability*, common in soil mechanics, is thus related to that of the minimum slope. A rock slope which has been eroded to a gradient equal or inferior to that of the minimum slope on which the causative mass wasting process can function is said to be stable as far as that process is concerned. Thus the limestone slopes of Lorraine, averaging 10° to 12°, are immunised as far as rockfalls are concerned; their slopes are clearly inferior to the minimum slope of gravity rockfalls, which is about 45°. The plant cover plays a predominant role in this stability. A particular slope which is perfectly stable as far as runoff is concerned because it is grass covered will be savagely gullied if the grass cover is destroyed because the gradient is superior to the minimum slope of runoff.[1] Minimum slope and stability are therefore related phenomena, but should not be confused. The concept of the minimum slope is essentially topographic and mechanical: it applies to mechanical morphogenic processes and has nothing to do with the plant cover. The concept of stability expresses a steady state: a particular slope is stable as far as a particular process is concerned, either because its slope is not steep enough for the process to operate or because the plant cover protects it effectively. This has enormous consequences: if a particular slope is stable in relation to a specific process because its gradient is equal to or lower than the corresponding minimum slope, a change in morphoclimatic conditions can start the process only in an indirect manner, as for example by previously modifying the weathered mantle. For instance, a limestone slope stable in relation to solifluction in a temperate climate becomes subjected to periglacial conditions; after having removed the loamy and clayey soil, solifluction will only continue if frost-weathering is able to provide it with friable debris rich enough in fines. On the other hand, if the stability of the slope is due to the resistance of the plant cover, a modification of the latter is enough to set the process in motion. This, for example, is the case of gullying in a temperate or tropical forest environment. As soon as the soil is cleared, gullying is unleashed on slopes which were previously stable in relation to it because of the plant cover. Gullying vigorously dissects slopes of only 4° or 5° when they are composed of friable and rather impermeable materials. This reflection has an important practical application: it is now possible to measure the potential danger of erosion on a slope about to be cleared for agricultural development. The concept of stability of slopes

---

[1] This minimum-slope of runoff is also a function of the length of the slope, runoff being a cumulative phenomenon. The discharge from above adding itself to that accumulated in place, the total discharge increases toward the base of the slope as soon as a continuous runoff is organised. The waters then can attack the soil on increasingly gentle slopes. Under given conditions of climate (shower intensity), soil, and plant cover the minimum-slope of runoff decreases with the length of the slope according to formulas which, in a particular case, may be established empirically.

therefore is at the base of every effort of rational agricultural land development. Before deforestation one should determine which slopes, once cleared, would bring about a sufficiently serious erosion to make them rapidly uncultivable. On these slopes the forest should be maintained or replaced by a dense grass exclusively used for grazing and carefully managed to prevent damage from overgrazing, thus insuring that the protective role of the grass is being fulfilled.

Having explained the concept of morphoclimatic equilibrium, we can now study the consequences of interruptions in the morphoclimatic equilibrium. These interruptions are caused either by climatic changes or by man and are at the origin of morphoclimatic adjustments. We begin with geologic interruptions due to changes in climate which conjointly produce changes in the morphogenic processes, the plant cover, and the conditions of soil formation.

Geologic interruptions in the morphoclimatic equilibrium: the influence of palaeoclimates

The influence of past climates, or palaeoclimates, on present geomorphology is particularly great because of the importance of climatic changes during the recent geologic past. The Pleistocene was characterised by exceptional climatic instability. The world climatic map has changed considerably during the last 20 000 years; only 20 000 years ago there were continental ice sheets and periglacial landscapes in the midlatitudes, pluvials in the dry regions, and dry climates on the margins of the humid tropics. The climate changed several times in most climatic zones during the Pleistocene, and the morphoclimatic systems sometimes underwent complete changes with, for example, the formation of continental ice sheets and the appearance of periglacial conditions in the midlatitudes. Successive adjustments and readjustments have taken place during each of the major climatic oscillations. Only a few areas in the world have experienced stable morphoclimatic conditions and have evolved within the framework of one and the same morphoclimatic system since the end of the Tertiary. The inverse is much more frequent; in that case regions are polygenic from the morphoclimatic point of view and preserve traces of different and successive morphogenic systems; these traces are more or less apparent and reflected in survivals of unequal importance.

The influences of palaeoclimates are not limited to the Quaternary. Recent climatic oscillations, however important they may have been, should not make us lose sight of the general evolution of the earth. Some relict landforms date back to the Tertiary and even to the Cretaceous, whereas exhumed forms may be dated back to an even more remote past, including the Precambrian. Most peneplains are very ancient forms, and one should therefore inquire whether their formation took place under morphoclimatic conditions similar to those of the present, and whether they are the end result of the work of erosion in a so-called 'normal' environment.

Recent climatic oscillations and morphoclimatic readjustments

Because of the short period of time since the last radical morphoclimatic changes of the Pleistocene, geomorphology must give the greatest attention to palaeoclimatology. Davis has systematically minimised this factor, and this is one of the causes of the failure of classical geomorphology. Knowledge of palaeoclimatic conditions is indispensable to the geomorphologist as well as to the biogeographer and the pedologist. It is basic knowledge to the sciences of nature which, in return, constantly add new data to palaeoclimatology through the dialectic process which is common to all scientific inquiry. The residual savannas of Lower Ivory Coast, for example, imply a lesser northward extent of the forest several thousand years ago; elsewhere, pollen analysis of peat or humic soils make it possible to reconstruct the main features of the plant cover and, therefore, to define the corresponding morphoclimatic environment.

Morphoclimatic zones during the pleistocene

General palaeoclimatic reconstructions have been attempted especially by German geologists and geomorphologists, and this is not surprising, for being little influenced by Davisian theories, they posed the problem of palaeoclimates earlier.

The most recent synthesis is that of Büdel (1951, 1953), which we will use as a starting point.

1. *Büdel's theory* is based on a climatic cooling which permits the growth of glaciers. Their extension is responsible for increased cold due to thermal autocatalysis. The cooling, however, is unequal depending on the climatic zones. It is maximum in the temperate regions, somewhat less in the high latitudes, and minor in the latitude of the Tropics of Cancer and Capricorn. There the principal change is thought to have been a significant increase in precipitation which would have caused a poleward migration of the savannas. The pluvials of the desert regions and the glacials and periglacials of the midlatitudes are believed to have occurred simultaneously.

In Büdel's theory there is a progressive weakening of the Pleistocene climatic oscillations towards the low latitudes. The intertropical zone would not have incurred any changes, while radical changes affected the temperate zone. The land surface of the earth is thought to have been affected by climatic changes in the following proportions:

15 per cent of the total was covered by Würm (Weichselian) ice
53 per cent of the total experienced an important climatic change
32 per cent of the total was not appreciably affected.

In general the glacial and periglacial climates of the midlatitudes were definitely drier than the present climates.

The Mediterranean zone experienced a cold temperate climate with a

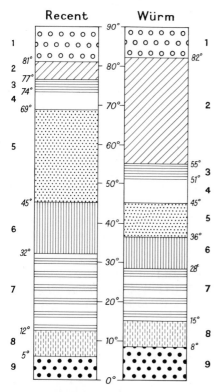

FIG. 4.2. Morphoclimatic zones during the Würm (Weichselian) and the Recent, according to Büdel

1. Sea ice
2. Ice sheets
3. Frost deserts
4. Tundras
5. Midlatitude forests
6. Mediterranean zone
7. Tropical deserts and steppes
8. Savannas
9. Intertropical forest

The diagram applies to a latitudinal strip bounded by the Prime meridian and the meridian of 15° east longitude. The numbers indicate degrees of north latitude.

periglacial environment in the highlands. At sea-level solifluction and frost-weathering occurred in the most favourable rocks as far south as Leghorn. The upper limit of the boreal forest, which included birches, is estimated to have been situated at an elevation close to 500m (1 600ft) in Corsica and 1 200m (4 000ft) in southern Italy.

The dry zone was reduced in size due to the appearance of conditions said to be *pluvial*. The climate was more humid than at present, permitting a greater development of stream courses and producing lakes, such as Lake Bonneville in the western United States. In the Sahara the arid zone extended only from 20° to 28° instead of from 18° to 32° as it does today.

In the humid tropical zone changes were minimal, according to Büdel. The most important one was an extension of the savannas towards the dry zone; in West Africa they reached 15°N instead of 12°N as they do today. The tropical rainforest did not experience any modification except, possibly, a slightly larger latitudinal extension. It would have reached 10° to 15° north latitude. The snowline was lower on the mountains, which would have implied a 4° to 5°C (7·2° to 9°F) lower mean annual temperature, together with reduced precipitation, but without affecting the plant formation-types, the soil-forming processes, or the morphoclimatic system.

2. *Critique and discussion of Büdel's theory.*　　Büdel's theory is based on extensive studies by the author in the midlatitudes and on a short trip to Africa. It is therefore not of equal solidity in all its parts. As a whole the theory is satisfactory for the temperate zone and the polar regions, but it conflicts with established facts in the other morphoclimatic zones.

North of 35° to 40° latitude the theory is satisfactory and not contradicted by the facts. Except for a few details, its main elements can be accepted; they are as follows:

(a)　The primordial cause of glaciations is a climatic cooling and not a mere increase in precipitation, as some authors have affirmed. Everything that we know about the mechanism of glaciers confirms it.

(b)　The precipitation during the glacial periods was, on the whole, certainly less in the midlatitudes than it is today. At the present Antarctica receives on the average only 100 to 200mm (4 to 8in) of water (in the form of snow). In Europe the low temperatures of the glacial periods reduced the water capacity of the atmosphere and the amount of evaporation, while the glacio-eustatic regression slightly increased the surface of the continents to the detriment of the oceans. Less precipitation did not necessarily produce less streamflow, for absorption by the plant cover was certainly most reduced. It seems, moreover, that the decrease in precipitation during the glacial periods affected the lowlands more than the highlands, as Klüte's studies reveal (1951).

(c)　Büdel's southern limit of Pleistocene periglacial phenomena at sea-level in Europe is in full agreement with our own observations. The Mediterranean region was a transitional morphoclimatic zone where the periglacial processes were, consequently, narrowly influenced by the nature of the rocks, exposure, the geographical position, and elevation. Würm sorted talus and frost-weathered features are therefore much more common at sea-level in the area of the Bas-Rhône, which was subject to a frequent icy mistral, than in the more sheltered Nice area where the lowest evidence we have been able to find is at an elevation of 400 m (1 300 ft).

In the tropics, on the other hand, numerous facts established by various authors contradict Büdel's theory:

First of all, the precise equivalence of the pluvials and the glacials is not justified. It results from a confusion between the northern and southern margins of the Sahara. In the Magreb as well as in the western United States, it is true that the pluvials coincided with the glacials at least in their major outlines. This does not seem to be the case south of the Sahara. In Senegal Tricart (1955) has shown that the periods of glacio-eustatic regression were arid phases during the Riss (Saalian) as well as during the Würm (Weichselian). These phases were characterised by an interruption of stream flow and the formation of dunes, particularly an erg which advanced to the banks of the Siné-Saloum rivers (14°N) during the Riss. At least part of the duration of the glacial periods of the midlatitudes was characterised by a southern extension of the arid zone. Senegal also experienced more humid phases than the present one; the main one goes back to the Riss-Würm interglacial because it coincides with the Eemian (Ouljian) marine level, which is higher than the present one. It was characterised by more rains but without modification of the seasonal rhythm; a marked dry season persisted. It was then that the northern margin of the savanna extended further north.[1]

In the second place, important morphoclimatic changes in the humid tropical zone have been proven by various observations. De Heinzelin (1952) has pointed to the existence of several phases of aeolian sedimentation during the Pleistocene in the Kisangani (Stanleyville) area of the Congo, right at the Equator. Cailleux has more than once noted the great frequency of round-frosted grains worked by the wind in the superficial deposits of this area. We therefore cannot agree with Büdel that the equatorial forest then continued to cover this area and that the morphoclimatic system was not modified. In neighbouring areas, such as Ubangi, Angola, and Gabon, Pleistocene alluvium is in part made up of gravel beds, including cobbles (10 to 15 cm: 4 to 6 in). The same facies is found in French Guiana, lower Ivory Coast, and Indonesia. It is frequent between 0° and 15° latitude in the midst of the equatorial rainforest. Such formations contain placers of gold,

[1] Since this was written Elouard (1967) has shown by radio-carbon dating that the 'raised' beaches flanking the lower Senegal and which Tricart thought to be Eemian, actually are Dunkirkian (no shells older than 5 470 ± 110 years B.P.) Elouard, Faure, and Michel: 'Nouveaux âges absolus (C 14) en Afrique de l'Ouest', *Bull. IFAN*, **29**, ser. A, no. 2, pp. 845–9; also see other articles by Elouard in the same number). It seems therefore that the erg referred to above may be Würm in age and that the last humid phase occurred as recently as the Dunkirkian (KdJ).

diamonds, and tin. Furthermore, on the islands of Banka and Billiton in Indonesia, such a coarse alluvium plunges below the sea-level and spreads over the surface of the continental shelf. Its deposition, which is therefore contemporaneous with a glacio-eustatic regression, is attested by the well-known drowned stream courses of the Java Sea. These coarse Pleistocene deposits thus appear to be contemporaneous with the cold phases of the mid-latitudes. In any case they imply serious morphoclimatic changes in the wet tropics as deposition of gravel beds does not occur in the forest of equatorial Africa at the present time. One has to go as far as the Sahelian zone and the margins of the Sahara to find recently deposited gravel sheets. Arid and semiarid climates ought to be considered as they could therefore have reigned in areas presently occupied by the rainforest during the glacial periods of the Pleistocene. Whether the aridity extended to the entire wet tropical zone or to only part of it is unknown. It is possible, and even probable, that the evergreen forest persisted in certain refuge areas, particularly in coastal locations with an oceanic climate and on certain islands. In West Africa, for example, according to Tricart's observations, it would seem that the present salient of savannas which reaches the Gulf of Guinea in Togo and Ghana was wider and more intensely developed during the Würm, reaching towards the west of Ivory Coast. Conversely, the Monrovia area and the area of Mount Cameroon probably remained forested. From there the reconquest of the present evergreen and semi-evergreen seasonal forest zones would have taken place. This problem does not only interest the geomorphologist but the biogeographer.

Certainly, it is no longer possible to consider the humid tropical zone as monogenic in its entirety from the morphoclimatic point of view. Like the temperate zone, it has experienced major climatic oscillations during the Pleistocene, and very different morphogenic systems have succeeded one another. The Congo Basin went through an alternation of deserts and savannas. The Fouta Djallon was several times subject to semiarid conditions which have left behind thick argillaceous screes, the most ancient of which are cemented into cuirasses. Lower Senegal swung between erg and wooded savanna.

In total, then, it seems that the Quaternary was characterised by important migrations of morphoclimatic zones: it appears, but it is not yet firmly established, that during the glacial periods there was an equatorward migration of all the morphoclimatic zones, and that during the interglacial periods there was a poleward migration of all the morphoclimatic zones.

The amplitude of the oscillations seems to have been greater during the early and middle Quaternary than during the late Quaternary. In the Magreb the size of the piedmont alluvial slopes is larger and the elevation of periglacial phenomena is clearly lower in proportion to age. In Senegal extensive alluvial gravels were deposited during the early Pleistocene, whereas a less voluminous erg developed during the Würm. In central Europe the Riss-Saale glaciation was more extensive than the following one, the Weichselian. Interglacial climates also seem to have been warmer than the present one, particularly between the early and middle Pleistocene. The Mindel-Riss (Hoxmian) is thus characterised by the feretto formation in the Po valley and the red soils of Mediterranean France, which, according to pedologists, imply a warmer climate and a more marked dry season than the present. The tundras of the Russian arctic underwent a forest incursion during the Riss-Würm (Eemian) interglacial. In the Fouta Djallon the early Pleistocene scree was cemented into ferrallitic or ferruginous cuirasses, whereas those of the last semiarid phase remained friable. It seems therefore, on the whole, that the Quaternary climatic oscillations have somewhat decreased in intensity, at least since the Mindel-Riss.

Most of the morphoclimatic zones have therefore experienced climates different from the present ones during the Pleistocene. Divergences were not of the same importance everywhere. In general during the glacial periods the morphoclimatic systems were probably completely different from the present ones, whereas during the interglacials the differences were probably less marked (the Paris Basin, for instance, had a mediterranean climate during the Mindel-Riss). The present lack of palaeoclimatic knowledge makes it impossible to determine with precision which areas of the earth have not suffered major morphoclimatic changes. According to Tricart's researches it appears that Pleistocene climatic oscillations were particularly important in Africa, a continent where zonal climates are little affected by geographical configuration. On other continents, on the contrary, whose climates are considerably dependent on geographical configuration, changes were less. In northeastern Brazil the parallel disposition along the coast of a wet tropical strip of mata, a wet–dry strip of *agreste*, and the semiarid caatinga have persisted with clearly defined limits which have only moved a few kilometres. The coastal desert of Peru and northern Chile maintained itself throughout the Pleistocene; it was only interrupted by 'pluvial' phases during which floods emerged from the neighbouring mountains without ever having introduced a regime of seasonal streams and a steppe plant cover. The main monogenic environments from the morphoclimatic point of view are probably the following:

1. Certain coastal areas, islands, and peninsulas of the intertropical zone probably served as refuges to the rainforest during the glacial periods of the midlatitudes (the gulf of Biafra, the area of Liberia, parts of Indonesia, southwestern Ceylon, and the Guianas) while interior areas were subject to desiccation. The lower temperatures of the higher latitudes and the immobilisation of an enormous mass of ice withdrawn momentarily from the hydrologic cycle, probably caused a general decrease in precipitation in the lower latitudes which affected especially the interior areas.

2. Some very continental deserts probably experienced uninterrupted arid conditions with only minor oscillations throughout the Pleistocene. This seems to have been the case of the Libyan Desert and the Tanezrouft in the

Sahara, which would explain their extreme poverty in underground water; the Pleistocene would not have enabled the creation of aquifers which would have survived down to the present. These regions on account of their continental location were probably out of reach of the pluvials, whether of the northern or of the southern margin of the Sahara. For, indeed, in the same latitudes but in areas closer to the ocean, one finds traces of pluvials, as in the north and centre of the Spanish Sahara and in the south of Mauritania.

3. Some monsoon regions of maritime location in the midlatitudes, such as central and southern Japan, and perhaps southern Korea and the south coast of China, probably also served as refuge areas towards which the monsoon forest retreated. All the periglacial phenomena observed in Japan occur at high elevations. Even taking into account the smaller amount of research, there is here a contrast with western Europe and central and eastern United States at the same latitude. Similarly, in the coastal zone of Morocco, in the area of Casablanca, there are no Pleistocene palaeoforms that contrast with the relief in equilibrium with the present climate; in particular there are no piedmont slopes or calcareous crusts.

4. Some periglacial regions experienced neither Pleistocene ice sheets nor a marked rise in temperature during the interglacials. The continental climates with winter drought and rigorous cold seem to have been particularly favourable to this periglacial morphoclimatic continuity. An uninterrupted periglacial environment probably existed during the course of the Quaternary in the central lowland of Alaska and a part of northern Siberia, both of which have not been glaciated, at least during the late and middle Pleistocene. The same may also apply to the curious dry valleys of McMurdo Sound in Antarctica.

But these regions which are essentially monogenic from the morphoclimatic point of view are the exception and their characteristics and distribution remain to be worked out. On most of the earth successive morphoclimatic adjustments are the rule and present the main problem.

PRINCIPLES OF MORPHOCLIMATIC READJUSTMENTS

Pleistocene climatic oscillations have in many places caused alternations between predominantly mechanical and predominantly biochemical morphogenic systems. Examples in the temperate zone: between periglacial or glacial and forested environments; in the arid tropical zone: between pluvial steppe and interpluvial desert environments; in the humid tropical zone: between desert or semidesert and savanna or forest environments.

These are the most general cases which have common characteristics under different latitudes. We will examine them first and then later mention some interesting, specific morphoclimatic readjustments.

1. *General characteristics of geomorphic alternations between predominantly mechanical and predominantly chemical morphogenic systems.* Schematically, it may be

said that periods with a predominantly chemical morphogenic system produce *in situ* soils and a rather deep weathering of the rocks. The whole constitutes a friable material which later may be moved relatively easily by mechanical processes. Inversely, periods with a predominantly mechanical morphogenic system produce and often deposit considerable amounts of friable debris which later may easily be occupied and fixed by vegetation and partly transformed into thick soils. Erhart has proposed to call *biostasy* the system with a biological and chemical predominance, whereas his

FIG. 4.3. Granitic weathered products, including corestones, reworked in a moraine and subjected to frost-weathering. Mount Kosciuzko, Australia

Weathering into corestones probably dates back to the Tertiary. It is the result of climatic conditions producing a predominantly chemical weathering. Corestones and gruss have both been swept by the glacier, then accumulated in the form of a moraine. The present climate, with frequent frost, causes the corestones to shatter either right through the middle or through peeling. On the surface a layer of frost-riven debris has slid over and covered the moraine. It has been stabilised by the grass mat and dates back to late Würm. The succession of morphogenic systems revealed by the material is also found in the relief.

rhexistasy corresponds in part to a predominantly mechanical system. Landforms undergo greater change when the morphogenic system is predominantly mechanical, especially if erosion is facilitated by the chemical weathering of the preceding period. Conversely, under a dense plant cover the relief is relatively stable as the removal of material in solution only causes very slow modifications in the landforms. This stability explains the considerable number of relict landforms in the present topography. Many of them developed during the last predominantly mechanical morphogenic phase, whether glacial, periglacial, or dry. Examples are: all glacial landforms, whether due to ice sheets or mountain glaciers; the periglacial slopes of the forested temperate zone; the dunes and ergs of the Sahelian zone; and lastly, the monolithic domes, sandstone cliffs, and stabilised screes which can be found as far as the zone of the tropical rainforest.

Such morphogenic alternations have played an important role, not only in the genesis of landforms but also in two closely related domains which

FIG. 4.4. Sheet erosion on a granite weathered onto corestones in an arid climate. Palmer, South Australia

Rainfall: approximately 200 mm (8 in). Weathering into corestones took place under a cover of gruss in a more humid climate. Increased runoff caused by decreased precipitation has permitted the removal of the gruss and the exhumation of the corestones, which break and peel.

may have practical applications. First, in the geology of alluvial mining in which the deposition of alluvial beds containing tin, diamonds or rare minerals is determined by the morphogenic system. The existence of placers depends on the Pleistocene climates during which gravels containing valuable minerals were deposited and concentrated into minable products. The discovery and prospecting of these sites should, more logically, fall to the geomorphologist rather than to the geologist. Secondly, the soil-forming processes have also been subject to morphogenic alternations. Pedology and agricultural development have therefore much to learn from a geomorphic study. It provides precise knowledge on the conditions of soil formation and, consequently, the protective measures to adopt for their intelligent use. In many places cultivable soils have developed under different conditions than presently exist and are now prone to easy erosion. Indeed, past periods with a predominantly mechanical morphogenic system have produced fresh or slightly weathered materials which were derived from the direct disintegration of the bedrock and are, consequently, generally more fertile than the old leached soils which date back to periods of predominantly biochemical weathering. In the Fouta Djallon there is a striking difference, even where the contrast has not been increased by man, between the friable slope materials composed of a now argillaceous scree accumulated during the last

FIG. 4.5. Geomorphic effects of the succession of a wet–dry tropical climate by a semiarid climate coinciding with a renewal of erosion. Greenhills, Western Australia

The surface of the butte is a bowal formed by a ferruginous cuirass covering a late Tertiary surface of erosion. Underneath the cuirass the granite has disintegrated into gruss over a thickness of some 30 m (100 f). With renewed erosion the regolith has been dissected. Today the bowal overlooks the pediplain (in the foreground) which was cut from the regolith underlying the cuirass and which plays the role of a non-resistant bed.

semiarid phase, and the bowals of the plateaux which are old cuirasses of Tertiary or even of Mesozoic age. If cattle herders were the first to take possession of the bowals and have later accentuated their characteristics through fire, the vegetation was originally probably already less dense due to the nature of the regolith. In the Magreb the best agricultural lands, which are often composed of Grimaldian Pleistocene loams resulting from a more pronounced mechanical disintegration than is at present the case, are widely spread throughout the valleys and on the lower slopes. It is hardly necessary to recall the late Würm (Weichselian) loess of the midlatitudes nor the deep weathering due to frost and solifluction of the limestone plateaus of northern France; the contrast of the latter with the Mediterranean garrigues, whose thin weathered mantle was quickly removed as a result of more than a thousand years of overgrazing, is well known. Present agriculture is largely dependent on a true stock of relict soils incapable of reconstituting itself if it were destroyed. The future belongs to those peoples who know how to manage it and not overexploit it causing despoliation. To prevent this danger, association between geomorphologist, pedologist, and agronomist is indispensable: it is the fundament of extensive agricultural developments presently being realised, notably in the Russian steppe or in the Hungarian plain.

From the geomorphological point of view, the danger of erosion incurred through cultivation provides a close enough picture of what would have happened when a period with a predominantly biochemical morphogenic system was followed by a period with a predominantly mechanical system. There would have been phases of *accelerated erosion* owing to the presence of a ready stock of debris which the vegetation progressively ceased to protect. This actually happened at the end of the Tertiary and at the beginning of the Quaternary during the Villafranchian in the midlatitudes when the products of a prolonged chemical weathering under a tropical climate became subject to intense erosion during the first glacial lowering of temperature. The land-forming processes then passed through a paroxysm of momentary intensity. Certain changes of climate can cause a degradation of the plant cover which rapidly worsens under the effect of simultaneous morphogenic actions. Changes in relief are then very rapid, just as those which are caused by anthropic erosion. There are indeed analogies in both cases: a stock of loose material inherited from the preceding period and the elimination of the protection provided by plants which assured its preservation.

Variations in the rate of the land-forming processes depend mainly on the intensity of fragmentation, which in solid rocks determines the quantity of material that may be removed by the processes of transportation. Thus in the compact granites of the French or German ancient massifs, the beginning of the periglacial stages was marked by a much more important acceleration of erosion than in the frost-liable limestone regions. The fragmentation of fresh granite by frost is indeed difficult. At first solifluction rapidly liquidated the gruss and the residual corestones of Tertiary chemical weathering, but later it was slowed down through a dearth of materials. This abatement, however, was progressive as the more porous and mechanically less resistant weathered granite is much more frost-liable than fresh granite, as laboratory experiments have proved. It is easy to see, then, how periglacial actions were able to clear the more or less weathered corestones and to expose the irregular subregolithic granitic relief. In plateau regions of frost-liable limestone, on the other hand, a swift fragmentation by frost action produced important quantities of debris formed at the expense of the fresh rock; accordingly there was no lack of available material after the initial phase of liquidation of the stock of interglacial soils, and solifluction did not abate. In the Riss slope deposits of the Nîmes area, one frequently notices important remnants of characteristically red argillaceous Mindel–Riss soils on the lower slopes, whereas the upper slopes exhibit nothing but broken limestone fragments, indicating that periglacial erosion continued its course after the removal of the soils produced during the previous interglacial.

2. *Examples of specific zonal morphoclimatic readjustments.* Specific zonal morphoclimatic readjustments occur in certain parts of the intertropical zone where oscillations of humid and dry climates produce shifts from forest or heavily wooded savannas to steppe. The Fouta Djallon of Guinea offers a good example. Here the climatic oscillations seem to be responsible for part of the cuirassing process, whether ferruginous or ferrallitic (the other part being caused by ground water). The slopes of the Fouta Djallon display innumerable screes now stabilised by the vegetation. Much of the scree consists of boulders, sometimes very large (up to 1 m), which are fragments of the ferruginous cuirass which breaks loose from the margins of bowal plateaus. Certain screes, probably the most recent, have remained unconsolidated; their boulders and rubble are embedded in an abundant red-brown or chocolate-brown argillaceous matrix. In other screes the clays have completely disappeared, and the whole mass of cuirass fragments has been assembled and reconsolidated into a new ferruginous or ferrallitic cuirass in which the original fragments are sometimes hard to recognise. Such conglomeratic cuirasses are of course found at the base of the slopes and have a geomorphic significance which is very different from that of the plateau cuirasses. They record important Pleistocene climatic oscillations during which mechanical systems of erosion (fragmentation and fall) alternated with periods of intense chemical or biochemical activities (reconsolidation on the slopes). The periods of predominantly mechanical actions accompanied by a steppe plant cover were those during which the inselbergs, the marginal cliffs of the bowals, and certain castellated sandstone reliefs were formed. They gave the slopes an essential part of their present characteristics. They did not, however, allow the watercourses to develop smooth concave long-profiles of the type of the temperate zone, for the climate did not cease to be warm, as is the case of the present climate of the Mauritanian Adrar whose wadis have similar long-profiles. During periods of forest climax pedogenesis and biochemical actions were again predominant; the decomposition of the argillaceous matrix gave rise to latosols, which during the next dry phase became indurated into cuirasses. This would explain why the recent slope deposits which have not yet weathered into clays and not yet endured a new dry phase have generally remained unconsolidated. When there is a beginning of induration, brush fires set off by man, rather than by natural causes, are the cause of it.

In the upper basins of the Senegal and the Gambia (on the north flank of the Fouta Djallon, mainly in Guinea and Mali), Michel (1959) has demonstrated the existence of three stepped Quaternary climatic pediplains and gravel terraces below two older erosion surfaces (Eocene and Mesozoic). Whereas the older surfaces are covered by ferrallitic cuirasses, the younger pediplains, except the lowest one, are covered by ferruginous cuirasses. The gravel terraces merge with the pediplains along their outer margins. A fourth (youngest) gravel bed outcrops in the stream beds. These gravels have been called '*graviers sous berges*' (underbank gravels) by Vogt and date from the Pre-Flandrian regression (Würm). Whereas each down cutting and planation (pediplain and terrace) phase was characterised by a semi-arid climate, each cuirassing phase was characterised by a wet–dry, sudanese type of climate. Therefore, a predominantly mechanical morphogenic system producing gravels and forming pediplains alternated with a predominantly chemical morphogenic system producing cuirasses. The present wet–dry tropical climate has not been sufficiently long to produce a cuirass on the lowest pediplain. At the present soils are rapidly forming and the materials eroded from the hillsides are fine (clays and sands).

Another case, which also deviates somewhat from the general scheme, is that of the cold European steppes and the Mediterranean region. In these environments the plant cover is less dense than in the midlatitude forests, and during the interglacials it provided a lesser protection against mechanical actions. The difference between successive morphogenic systems tends to diminish. Thus in the cold steppes periglacial phenomena were in large part of the *aeolian* type with thick loess deposits more or less reworked by runoff and solifluction. In Bessarabia and Hungary the series of Pleistocene loesses together with intercalated fossil soils exceeds 40 m (130 ft). Because of the less rigorous cold, the interglacials of Bessarabia were characterised by a denser prairie which although it provided a better soil protection was very apt to degradation. During the postglacial, aeolian phenomena have continued at a reduced pace in a discontinuous manner in time as well as in space and have especially affected surfaces where the plant cover was momentarily degraded.

In Mediterranean lands such as the Magreb, where important studies have been made by Dresch (1941), Raynal (1961), Coque (1962) and Joly (1962), the glacial periods (= pluvials) were characterised by the formation of *piedmont plains*. The climate, colder than the present one, greatly increased frost-weathering and provided large quantities of debris which were removed by the waters of rainstorms. Runoff spread them sheet-wise over wide surfaces at the foot of the highlands, thus preventing them from attaining great thicknesses. Such *alluvial aprons* or *bajadas* are absent from the more humid parts of the Magreb where the drainage net enabled easier evacuation of the debris (Rif, Little Kabylia, Kroumirie). At present these piedmont plains are being dissected through a lowering of the stream beds because the agents of fragmentation, particularly frost-weathering, are much less intense on the highlands; clearer waters cut them up. Raynal and Joly have observed that during each flood the wadis of the semiarid regions on the bajadas of southern Morocco first aggrade and then degrade at the same point but usually deeper than they have aggraded. This is easily explained by the decrease of solid load resulting from reduced fragmentation since the last pluvial. At the beginning of a flood the waters collect the debris scattered about the slopes and left behind on the wadi floors during the waning of the preceding flood. The heavily loaded waters then deposit the debris at the foot of the mountain. At the end of the flood the available waste has all been evacuated and the waters attack the alluvial material of the piedmont slopes entraining those elements whose calibre is within the stream's competence. In the area of Aïn Séfra (south of Oran) we also have observed that the Pleistocene piedmont slopes continue to function at a reduced rate at the foot of the mountains; the smallest floods, incapable of going further, abandon their load there. But the main wadis are incised into the alluvial outskirts which they presently dissect; only the largest floods reach them, and for want of sufficient fragmentation they are now less charged with debris than they were during the last pluvial. Further north, in a more humid region, not even a part of the piedmont slopes is still functional, and dissection proceeds on all their parts as the transit of the debris is assured at the foot of the highlands.

In Mediterranean regions which were forested in their natural state, such as peninsular Italy, the Algerian Tell, or Corsica, the role of Pleistocene morphoclimatic oscillations is again somewhat different. Pleistocene periglacial actions were only minor. They left their mark on the relief only in details and locally, not in a generalised way as in the Paris Basin or Charente. Moreover, the climate remained seasonally contrasted, which imparted a torrential character to the morphogenic processes. During the interglacials and the postglacial the same torrential character was preserved so that the changes in the morphogenic regime of the streams are more a question of degree than of a radical modification of the processes. Differentiation into climatic and other terraces is less than in higher latitudes. The rocky slopes incompletely grown over by the present vegetation on account of summer drought also continue to be exposed to frost action at the present time. Frost-weathering, often increased by pastoral degradation of the plant cover, is still important and plays its part in imparting a highly dissected, rough aspect, and a stony character to Mediterranean mountains at medium and even low elevations. Frost does not need to be frequent, for if a single, intense cycle occurs the rock breaks at once if its fissures are filled with water.

The above examples are only a few indications intended to emphasise the considerable role played by recent climatic oscillations on the land-forming processes. A great deal of research is urgently required to discover the mechanisms of morphoclimatic readjustments and to trace their distribution. The present evolved out of the past; to forget it would lead to a misunderstanding of nature. Our readers will find other examples of similar phenomena in the various volumes of this *Treatise* which deal with climatic geomorphology; after study of the dynamics and the present landforms, an important place will always be given there to the influence of palaeoclimates and the palaeoclimatic oscillations particular to each zone.

Present and past causes in climatic geomorphology

So far we have confined ourselves to a problem limited in time, having only considered Quaternary climatic oscillations. They do not extend back much more than a million years, which is little considering the age of the earth (approx. 4·5 thousand million years) or even the most ancient landforms available for study. Important remnants of exhumed Lower Cambrian erosion surfaces (approx. 500 million years old) persist in Scandinavia and Canada. In the Saar, on the margin of the Paris Basin, there are exhumed residual reliefs of the post-Hercynian erosional surface (about 250 million years old). Exhumed topographies of such age are not infrequent, particularly on ancient massifs. In platform areas recently uplifted and composed of resistant rocks, remnants of Tertiary relief are common even if they have not been protected by fossilisation. In the Paris area the Oligo-Miocene surface on the siliceous limestones (*meulière*) of Beauce is very well developed and largely preserved (about 25 to 30 million years old). Remains of Palaeogene surfaces, all exhumed forms, are extensive in the Massif Central, the Morvan, and the Paris Basin (about 30 to 60 million years old).

The reconstruction of such old surface remnants is not always without difficulties; the part played by the original slopes or by later tectonic deformations are not always easy to disentangle. A better understanding of the morphoclimatic conditions which affected them is indispensable, the more so as they are ancient surfaces exhumed by erosion which have served as the starting point for the establishment of the peneplain concept which Davis later incorporated into the theory of the cycle of erosion. According to him such peneplains would be the image of what happens to relief subject to the action of 'normal erosion'; they would represent the final stage of the cycle, the old age stage towards which all the present reliefs of the earth are evolving. Except for some differences, this would also include the glaciated regions and the sandy deserts. The morphoclimatic significance of peneplains must, therefore, be explained.

But because we are concerned with relief forms which may be very ancient, it is not enough to consider only the purely climatic factors. The evolution of living organisms, especially the evolution of plant formation-types on which the morphogenic processes are narrowly dependent, must be taken into consideration.

THE EVOLUTION OF CLIMATES AND OF BIOGEOGRAPHIC ENVIRONMENTS

The history of the major living organisms is known directly thanks to fossils and indirectly thanks to certain lithologic data such as round-frosted sand grains which are indices of intense aeolian action and, therefore, indicators of bare landscapes. The same two orders of criteria may be used to reconstruct past climates. Criteria for warm climates include the following phenomena: bioherms; species of genera presently located in the warmer zones; a great variety of species; bauxite; ferruginous cuirasses; and general red colour of sediments. The main criteria for cold climates are: glacial deposits and ice-scoured rock surfaces, periglacial phenomena, impoverished flora and fauna, and species of genera presently found in the colder zones. Aridity and high temperatures are implied by gypsum and salt deposits; wind by the particular wear of sand grains, stones, or bedrock. Forest and steppe are revealed by species of genera that presently grow in such environments. Because nearly all these criteria are also subject to other influences, they must be managed with prudence. Two pitfalls must be avoided: the manipulation of data without careful discrimination, and a too systematic discrediting of data through not fully understanding it. Thus cautioned and armed with a good method, the following results, in their main outlines, are obtained for various geological periods.

1. *Precambrian.* The Precambrian, which is the longest of the periods of earth history, lasted approximately from 4·5 thousand million to 600 million years before the present; it is also by far the poorest in fossils; the latter (a total of ten species) are without exception aquatic. The land must have been denuded and subject to the violence of atmospheric actions because there is no trace of terrestrial plants or animals. Wind action is attested by sandstone formations with round-frosted grains analogous to those of the present day Sahara. One should not necessarily conclude that the climate was dry; the desert might simply have been due to the absence of land plants. Saline deposits or glossy, rounded sand grains have not been reported; there is therefore no proof of aridity or of prolonged wear in water. There are, furthermore, several magnificent glacial formations. To have persisted to the present day in spite of erosion they must have spread over considerable areas and been preserved in basins; there is therefore every possibility that they were ice sheets, indicating very cold climatic periods. The depth to which the corresponding sea-levels were lowered is unknown, but there are few traces of definitely marine formations in the Precambrian, whereas conglomerates of perhaps terrestrial origin are common. Algal reefs (*Collenias stromatholites*), at least, indicate a warm climate.

An Eo-Cambrian glaciation, which seems to have extended to all latitudes, occurred at the end of the Precambrian and at the beginning of the Cambrian. In South-West Africa, for example, crystalline *roches moutonnées* formed by this glaciation are locally exhumed from sandy proglacial deposits. This remote glaciation has more than a geological interest.

The Precambrian also appears to have been propitious to intense subaerial, pseudodesert morphogenic processes: the heavy rains of the equa-

Sedimentary environmental indicators of the geologic past in various parts of the world.

MILLIONS OF YEARS		PERCENTAGE OF ROUND-FROSTED (WIND) GRAINS				EOLIANITES	PERCENTAGE OF GLOSSY, ROUNDED GRAINS, INDIRECT INDICATION OF A PLANT COVER		SALT AND GYPSUM	BIOHERM REEFS	PRINCIPAL GLACIATIONS
		Europe N & W	USA	Sahara	Other areas		Europe N & W	Other areas			
	Quaternary	60	20	70	+	Europe, USA	+	+	+	+ +	+
1	Villafranchian	30			+	Europe, USA				+	+ +
2	Pliocene	10								+	+ +
13	Miocene	5		30			80	+ +	+	+ +	
25	Eocene	15	5			Congo	80	+ +	+	+ +	
63	Cretaceous	10		4			65	+ +	+	+ +	
135	Jurassic	5				USA	80		+	+ +	
180	Triassic	85		55	+	Congo, USA	0?		+	+ +	
230	Permian	85	40		+	Black Forest, Krakow	10		+	+	+
280	Carboniferous								+	+ +	
345	Devonian	60		+			7		+	+ +	
405	Ord.-Silurian	85	95		+		0?		+	+ +	
500	Cambrian	40	90	70	+	Sweden, USA	0?		+	+ +	
600	Eo-Cambrian					Sweden			0	+	+
700	Pre-Cambrian	70	+		+				0	?	+
4500											

Legend: + = isolated reported incidents; + + = detailed observations torial and temperate zones then fell on a bare ground and might have caused, as in other zones, sheet runoff and an intense peneplanation. And, as a matter of fact, the Cambrian rests on a magnificent peneplain in many parts of the world, such as in Canada or in Sweden.

2. *Cambrian to Triassic (from 600 to 180 million years* BP *approximately).* Marine floras and faunas begin to appear in the Cambrian, but palaeontologists are unanimous in proclaiming that because of their organisation and variety they imply a long past. It is not known whether their absence (or at least their extreme infrequency) in the Precambrian is only caused by their effacement through later metamorphism or is also owing to a different geographic distribution (major marine regressions). Whatever the cause may have been, the Cambrian floras and faunas are exclusively marine, and terrestrial forms only appear later. During the Devonian a few marsh plants appear. The rich Carboniferous floras are all adapted to humid localities judging from comparisons with the present. The first plants that are definitely adapted to drought do not appear before the Permian and the Triassic, such as the coniferous-like Voltzia; Grauvogel's detailed studies also show them to be scattered along riverbeds rather than elsewhere in dense stands. It is therefore certain that during the Cambrian, and perhaps still during the Triassic, a good part of the land masses presented a picture much like that mentioned for the Precambrian: vast spaces devoid of vegetation, at least of rooted plants; there was a great capacity for sheet erosion and peneplanation. Abundant evidence also indicates wind abrasion: faceted pebbles; round-frosted sand grains, which predominate in two-thirds of the formations of these periods; and few glossy, edge-rounded grains, no more than are found on the present beaches of Mauritania. In both cases the frosted aeolian aspect acquired inland persists even after reworking by water.

Warm climates are attested by coral reefs since the Cambrian, and warm and arid climates by deposits of salt and gypsum. A series of glaciations with ice sheets, whose traces are found almost to the Equator, occurred during the Permo-Carboniferous. The amplitude of the corresponding marine regression is unknown, but it is certain that climatic variations were severe. Other climatic oscillations are clearly revealed by aridity indices during the Permian and the Triassic; furthermore, there was a notable impoverishment (in species) of faunas and floras, marine as well as terrestrial, during the Permian. A recovery under new circumstances occurred during the course of the Triassic and reached full development from the Jurassic to the middle Pliocene.

3. *Jurassic, Cretaceous, and Tertiary (upper Pliocene excluded; approximately 180 to 3 million years before the present).* One of the most striking differences of this interval as compared with the preceding interval is the increase and frequent dominance of glossy, edge-rounded sand grains fashioned in water; they imply that vast areas protected by vegetation from the action of the

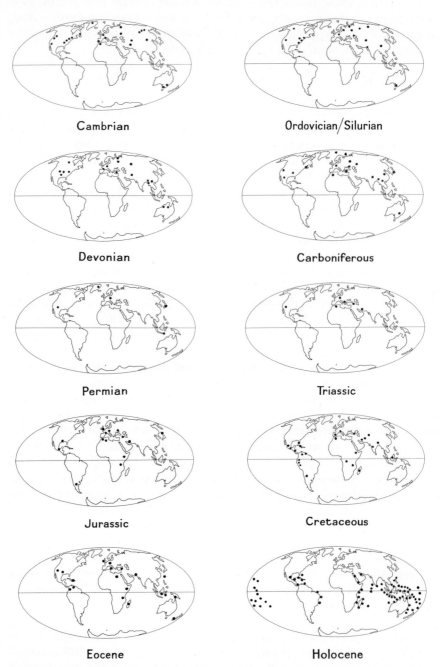

FIG. 4.6. Paleogeographic evolution of the distribution of coral reefs on the earth's surface (after Schwarzbach, 1949, p. 158)

wind existed in the interior of the continents. As a matter of fact, authentic round-frosted grains (in much smaller quantities than during the preceding periods) only remain important in large deserts such as the present Sahara. Elsewhere, in Europe and in North America, a complete plant cover was generally the rule; angiosperms appeared, then predominated starting with the Cretaceous and continuing down to the present. From the Jurassic to the middle Pliocene coral reefs and animal and plant genera indicate an average temperature 5° to 15°C (9° to 27°F) higher than the present, at least in the temperate and cold zones. Whether the equatorial regions then experienced *hyperthermal climates*, that is too hot and unfavourable to plants and animals, is unknown. The question has been posed but not answered.

Saline deposits, which are proof of a certain aridity, are not absent; they form a belt which is clearly north of the present one in the northern hemisphere and reached the latitude of France during certain epochs. It seems, therefore, that more arid phases have existed at certain times and places in areas favourable to planations. This may have been the case in France during the middle and near the end of the Tertiary; the siliceous limestones known as *meulières* which were then formed recall the surface silicifications of the Kalahari desert, the age and origin of which have not even been established. As the data are therefore still insufficient for a reconstruction of the possibly semiarid Tertiary landscapes of France, one should remain prudent. These planations in any case did develop very different landscapes from those of the last major and real peneplain of western Europe which developed a little prior to the Permo-Triassic. Tertiary deposits are indeed definitely less red in France, and authentic round-frosted grains not nearly as common. Climatic geomorphology has much to gain from progress in sedimentology in these matters and from a more detailed knowledge of the palaeogeography of past geologic periods.

4. *Quaternary, including the Villafranchian.* This epoch began approximately 1 to 3 million years ago. It represents such a short period of time that evolution of living organisms is minor although not negligible. The major well-known fact is the climatic cooling which caused glaciations, marine regressions of 100 to 200 m (330 to 660 ft), raised beaches, and stepped terraces. Other more immediate climatic consequences include various types of deposits and forms, some of which were periglacial whereas others were linked to pluvials and interpluvials. Alternations of cold and temperate climates, which were repeated at least four and perhaps ten or eleven times, caused major migrations of floras and faunas and, as a result, modified the plant cover. Sometimes the vegetation was even completely eliminated and later reappeared. During the interglacial periods and even more during the glacials, the types of morphoclimatic effects and their zonation were markedly different from what they were during the period preceding the Quaternary, as we have already noted, and which was one hundred times as long.

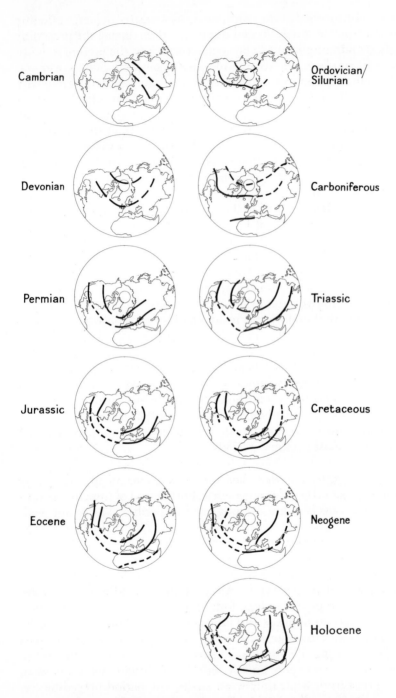

Fɪɢ. 4.7. Alkaline deposits of the northern hemisphere during geologic time, according to Lotze (1938)

The Pleistocene and the Holocene also are rather exceptional epochs from the climatic point of view; they show us a picture of the earth which we have no right to project into a more remote past. A transposition is necessary. In the present relief the geomorphologist must not only make allowance for the present climate but also for those of the past. His task is thereby singularly complicated but all the more fascinating on account of it.

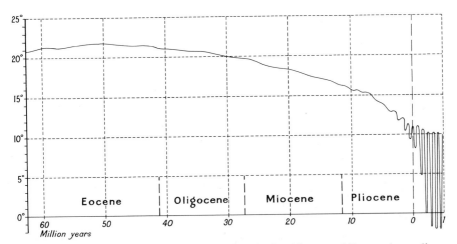

FIG. 4.8. Curve of mean annual temperatures at sea-level in central Europe (according to Woldstedt (1954))
Vertical scale: temperatures in degrees Centigrade. *Horizontal scale:* geological epochs. The time scale of the Quaternary is five times that of the Tertiary.

INFLUENCE OF PAST CLIMATES ON THE LAND-FORMING PROCESSES

Great progress in geology was made in the nineteenth century when Charles Lyell formulated the principle of uniformitarianism: to observe present mechanisms and present causes and with their help to explain past phenomena. He made a clear distinction between *actual* causes and more or less *imaginary* ones such as the universal flood or the universal fusion of rocks; in other words, realist and materialist, Lyell opposed himself resolutely to *idealism* (in the logical, not moral, sense of the term). But how is one to know whether a cause is actual except by trying to observe it in action at the present time? Lyell of course proceeded in this fashion.

The principle of uniformitarianism, it is true, is at the basis of all geologic reconstructions. Nevertheless it must be discussed. Has the geological past known any processes which have ceased to function at the present time? In other words, are there any *past causes* of geologic evolution? The problem has been clearly posed by Lucien Cayeux (1941), and the following example will show it:

The continental glaciations which existed in the Precambrian and in the Permo-Carboniferous seem not to have existed during the Mesozoic and the Tertiary (excluding the Upper Pliocene). For an observer who would have lived at that time, the concept of continental glaciation would have been a past cause because the phenomenon was entirely unknown in the epoch in which he lived. In the laboratory, however, it would have been possible to produce ice, to preserve it long enough to observe its plasticity, and even to make it flow as miniature glaciers. For us, on the contrary, who are contemporaries of continental ice sheets, glaciation is a part of present causes.

The concept of present and past causes in geology must therefore be well understood. A certain number of mechanisms have constantly operated since the most remote time in the history of the earth: for example, the force of gravity. But while rains which have reached the earth's surface always have caused water to flow from high to low points, their abundance has not determined the morphogenic influence of streamflow. This influence is determined by both the amount of evaporation, which depends on temperature, and the amount of evapotranspiration, which depends on the extent and nature of the plant cover.

Elementary physical or chemical processes have not changed since the Precambrian. They are typically present causes. The laws of hydrodynamics, of aerodynamics, and of ice flow have remained strictly the same. Streams have always been able to erode their beds, the wind has always been able to construct dunes, and breaking waves have always been able to move sands and pebbles on shores as long as there were pebbles and sands. Rainwater charged with carbon dioxide while the drops fall through the air has always been able to transform limestone into a soluble bicarbonate and to effect chemical erosion. The crystallisation of water into ice has always been accompanied by an increase in volume capable of breaking certain rocks. Temperature oscillations have always caused expansions, contractions, and desiccations generating mechanical efforts in rocks.

But the *frequency* and *intensity* of these physical and chemical processes has varied. If, as recent research tends to demonstrate, the length of days and nights has increased since the Precambrian, the geomorphic conditions of marine littorals subject to tides must have been affected by them: high tides succeeding one another only every seven or eight hours subject cliffs to a more rapid evolution. Hot, *hyperthermal* climates, the hypothesis of which is likely for certain periods, largely eliminated frost-weathering. On the other hand, they could have facilitated certain purely chemical reactions, such as the dissociation of silicates, while they retarded others, such as the dissolution of limestone to form bicarbonate. All such purely physicochemical processes have not changed. Their mechanisms have remained the same, but their frequency or their intensity may have changed considerably within the framework of known or very probable climatic changes. Therefore their combination into morphogenic systems could have assumed very different forms from the ones that now exist.

Another important factor has modified the morphogenic systems during earth history: the development of life. The multiplication of species, the diversification of faunas and floras and their progressive perfection have enabled a gradual expansion of the *biosphere* on the terrestrial surface and interposed a more and more effective screen between the lithosphere and the atmosphere. In general the result has been a decrease in the direct effects of the weather and of the purely physicochemical phenomena in favour of the biochemical and biomechanical actions. New, more complex processes have appeared at the same time as organic matter was added to the mineral matter to form soils. Modifications of the morphogenic systems during the course of geologic history are implied by changes in floras and faunas. We can affirm their past existence from the present influence of the plant cover on the morphogenic systems: an application justified by the principle of uniformitarianism. We have already affirmed the existence of combinations of processes which do not exist any more today. Thus bauxites are related to the cuirassing process, but nowhere on earth have bauxites been reported to be forming at present. All bauxites are older than the Quaternary (infra-Cretaceous in France, infra-Miocene in Hungary, Pliocene in Surinam, etc.). They correspond to mechanisms that do not function any more today: microbic activities due to species which have since disappeared, or simply climatic conditions without present equivalents.

As a whole, therefore, the principle of uniformitarianism applies best to the most elementary processes and the simplest, mainly physical or chemical, mechanisms. Inversely, past causes were all the more different as plant and animal life participated increasingly in the processes and the complexity of the combinations was greater. And because climatic geomorphology is based on complex combinations, in which physiochemical and biological factors play a fundamental role, it must grant an important place to past causes. To adopt a mechanistic uniformitarian attitude towards past causes would be to deprive oneself of the understanding of a large number of facts. This is precisely what has happened to Davisian geomorphology. Its concept of the present world indeed is a truncated, partial, and artificially impoverished concept: it ignores the basic fact of the existence of the organic world. We have seen how this mutilation of nature has isolated Davisian geomorphology from the other natural sciences and has led it towards errors in its own domain. The concept of normal erosion, however, is not the only one at fault, but also the concept of the cycle of erosion. These concepts are based on a mechanistic application of the principle of uniformitarianism; accordingly, all past peneplains would be the end result of the system of 'normal erosion', which thus would have existed during the Precambrian as well as during all the remaining geological periods and the present time on the near totality of the earth's surface, for the sandy and glacial deserts are only regarded as 'accidents' by the Davisians. This view must therefore be abandoned because the evolution of relief is based, as we have seen, on complex combinations of processes in which the biosphere often plays a determinant role. We must look for another explanation of peneplanation which agrees with what we know about the evolution of the earth.

The largest and most clearcut present planations are found in the semi-arid regions. They are the pediments and pediplains worked by rillwash at the foot of steep reliefs known as inselbergs. The development of pediments is particularly advanced on intrusive and metamorphic rocks, especially on granites and gneisses. Plains of almost perfect planation are indeed found on these rocks and may extend over tens or hundreds of kilometres. Isolated and clearly delimited inselbergs without any transitional slopes stud these plains. Some pediplains are recent such as the one of Inchiri, near Akjoujt, Mauritania, which is partly covered by Plio-Quaternary deposits. Others are more ancient and extend uniformly over hundreds of kilometres. Africa offers innumerable examples of them. The ease with which pediments develop on granular rocks is explained by active granular disintegration in warm climates; the role played by alternations of temperature and of wetting and drying has been demonstrated by Birot. Variations in temperature is thought to be particularly effective on bare rock if it is composed of minerals of different colours and its expansion coefficient is relatively high. Climates in which these mechanisms function are relatively varied, ranging from wet–dry tropical to deserts. They are best developed in savannas with a long dry season where the vegetation does not cover the soil completely. In wooded savannas with a better protected soil and in arid regions where there is not enough humidity, they function at a reduced pace but nevertheless mark the relief with their characteristics. The term *pediplain* was proposed by King to describe these extensive piedmont slopes and plains.

The formation of pediplains, or *pediplanation*, is the most effective planation process presently at work on the earth's surface; it is the only one which definitely ends in the elaboration of peneplains as extensive and level as those which exist on old lands fossilised by Palaeozoic or Mesozoic sedimentary covers. Worthy of note is the fact that past peneplains as well as recent pediplains are particularly well developed in granitoid rocks, whereas inselbergs are frequent in other lithologies. This is notably the case of the Armorican massif, in the Falaise area, where Precambrian schists and granites of the Athis massif are truncated by a perfectly plane surface fossilised under the Lias (Lower Jurassic), whereas the Silurian Armorican sandstone which produces abrupt and clearly circumscribed ridges, in fact veritable inselbergs, has only been buried during the Middle Jurassic, or some 20 million years later. The Sierck area on the Moselle, on the Franco-German border, also displays inselbergs of Upper Devonian Taunus quartzites fossilised by the Triassic Buntsandstein. The magnificent pre-Triassic peneplains on the northern margin of the Massif Central truncate all the granitoid rocks, and the arkoses which have been preserved have a facies very similar to that of the products of recent granular disintegration of African pediplains.

Because of the importance of the preceding analysis, it seems legitimate to discuss in more detail the analogy between peneplains and pediplains. Let us first note that the only recent planations besides pediplains are the periglacial reliefs of equiplanation. Palaeoclimatology, however, indicates that very cold climates have been the exception in the geologic past; accordingly, reliefs due to equiplanation can in all likelihood only be credited with a tiny part of past peneplains, at least since the Cambrian. On the other hand, there are a few indications that the most advanced planations of recent geological history were perhaps elaborated under some kind of semiarid conditions. In France this would have been the case of the rather well developed Oligo-Miocene surface of the Paris Basin, the Massif Central, and the Ardenne. On this surface Tricart (1952) has noted fine ferruginous gravels similar to those which exist on the semiarid pediplain elaborated on the lower Senegal during the early Pleistocene. The Eocene surface of the Massif Central and the Armorican Massif may have formed under similar conditions. However, the scarcity of wind abraded sand grains in the Tertiary (the Stampian and Bartonian have revealed glossy, rounded grains fashioned by water) poses a problem because such grains are the rule in other formations overlapping peneplains, as, for example, in the Triassic from Catalonia all the way to Silesia. A kind of pediplanation without aeolian action during certain phases of the Tertiary may explain certain characteristics of the relief of the ancient massifs. On the African continent where fossilisation of past erosion surfaces is limited, King has remarked that planations are usually stepped: the lower ones abut distinctly against the foot of the higher ones by means of a rather uninterrupted scarp escorted by inselbergs. Dissection of the highest plateaux is hardly begun because streams have corraded little. Stream profiles more or less repeat the stepped profiles of the interfluves with a slight upstream offset at the most. This general disposition is precisely what is found in regard to the planations of the ancient massifs of Europe. True, it cannot display the same sharpness, jarred as it is by tectonic deformations and the morphogenic actions of warm and humid Tertiary climates (depressions of the Morvan and the Vosges), temperate Plio-Quaternary climates, and cold Quaternary climates. There are, however, several particularly typical planations, such as that of the Annonay plateau (southeast of St Etienne) formed during the Neogene at the foot of the Pilat and which Pelletier has, it seems rightly, interpreted as a pediment. Recent research on the French Tertiary suggests fluctuations of dry and humid tropical climates until the end of the Oligocene. One is therefore tempted to compare the Annonay plateau with what has been observed during the Pliocene and the early Pleistocene in the present tropical zone, particularly in West Africa and Brazil; during periods of humid tropical climate the rocks weather deeply, the facies that are more resistant to chemical attack produce inselbergs and monolithic domes. Büdel (1957) has reconstructed a Miocene needle and dome karst of humid tropical type in the Swabian Jura. During drier periods the weathered products are easily truncated by pediplains and in part removed, alimenting accumulations of quartzose sand and kaolinitic clays in neighbouring depressions. Such facies indeed have been found at various stratigraphic levels in the Tertiary (Thanetian, Sparnacian, Cuisian, Lower Bartonian, Stampian, and Burdigalian of the Paris Basin). This interpretation is similar to Erhart's (1956) theory of biorhexistasy which sees in the geologic past an alternation of periods with a high degree of weathering with mostly chemical sedimentation (clays, limestones, iron) and of periods of mechanical attack with detrital sedimentation. Tricart, in his doctoral dissertation, had already proposed analogous concepts in connection with the Paris Basin but without giving them a specific formulation (1949–52).

One may moreover pose a problem of palaeophytogeography. A variety of palm, the sabal, is the principal plant type encountered in Eocene sandstone formations of western France; these formations seem to be alluvial deposits which correlate precisely with pediplains. The sabals, like the rhyzomic grasses of the tropical savannas, are the very type of plants which favour rillwash. They only partially cover the ground allowing splash erosion to take place between them; through their implantation they hinder the concentration of the rills into gullies in such a way that the runoff remains unconcentrated over extensive surfaces even during heavy rainstorms. We have observed this phenomenon more than once in West Africa, Brazil, and Venezuela. Such a plant cover is particularly favourable to pediplanation. It would be extremely interesting to know whether the process of pediplanation was widespread during the not very humid tropical climates of the Tertiary.

More ancient peneplains, particularly those of the Precambrian and the Palaeozoic whose dimensions are often imposing and whose degree of planation is astonishing, seem to have benefited from exceptionally favourable conditions. Indeed plant life has only belatedly occupied the continents between the Devonian and the beginning of the Mesozoic. Moreover, in the beginning it was restricted to marshes and humid depressions which were close to the original aqueous environment. Interfluves remained bare longest, and the first plant formations, poor in species and perhaps composed of rather poorly adapted plants, probably covered the ground to a lesser degree than our present forests and grasslands. Thus the soil only belatedly and progressively escaped the direct effects of the weather. For a long time it experienced conditions which are no longer realised in the natural state today. The soil was no better protected than in the present deserts and sparsely vegetated steppes, i.e. in environments of pediplanation; it was subject to a much more intense action of the weather. Certainly much more than an average 100 to 200 mm (4 to 8 in) of precipitation fell annually over the major part of the landmasses. There is indeed no reason to admit a lower global amount of precipitation for the earth as a whole; the higher temperature caused an increased physical evaporation capable of compensating for the lesser evaporation of plants. For want of protective plant cover the ground suffered particularly high variations of humidity and temperature. Thus the fragmentation of rocks, especially granitoid rocks, was particularly rapid, and the considerable mass of water given to runoff had a high capacity for transport. The sparse vegetation, or its non-existence, and the near absence of correlative soils considerably diminished infiltration and accentuated the torrential character of the runoff. Thus in spite of a greater humidity than in the present semiarid regions, the land-forming processes obeyed analogous laws. But the processes were much more rapid, not unlike what presently happens in the wet tropics after removal of the vegetation when violent showers rapidly wash away the superficial friable products which have been left unprotected.

The importance and characteristics of past peneplains and their correlative formations should now be clear: the latter consist of enormous masses of more or less argillaceous and more or less ferruginous sandstones which are the residue of mechanical disintegration with a minimum of biochemical alterations. Their red colour probably resulted from the dehydration of iron oxides due to evaporation and high temperatures.

Later the progressive occupance of the continents by living organisms little by little modified these geomorphic conditions. Pediplanation gradually became restricted to certain semiarid environments whose extension and localisation on the earth's surface varied according to time as a function of palaeoclimates. This progressive decrease of pediplanation explains the stepped nature of the pediplains on a continent such as Africa; without this decrease the oldest pediplains could neither be the most extensive, which is the case, nor, because of it, have persisted to the present time. In France, too, the decreasing extent of the Tertiary planations as a whole is perhaps partly due to the progressive evolution of the climate. It is not a continuous curve, it is true; and important oscillations are contained within the general trend. During the Eocene and the Oligocene the climate was warm but, it seems, with alternations of wetter and drier periods. The alternate formation of pediplains (at least in rough outline) and depressions, sometimes called 'cuvettes', would thus be explained. Later, climatic oscillations progressively involved cooler, subtropical climates, followed by warm temperate and during the Quaternary by temperate and periglacial climates. At first, periods of planation alternated with valley deepening, then valley deepening alternated with slope development by solifluction. The planations became increasingly less extensive and less perfect, revealing a progressive transition from pediplains to less clearly defined truncations.

The theory of the cycle of erosion therefore does not conform to the observations: regions whose relief evolves towards peneplanation have become progressively more and more restricted, which is due on the one hand to the development of vegetation and on the other hand to the evolution of climate. They are presently limited to certain semiarid regions. A peneplain cannot, therefore, be considered as the end product of the evolution of relief under humid climates in general and even less under humid temperate climates to which some have explicitly restricted the concept of normal erosion. The general significance given to the peneplain concept by Davis is therefore incorrect. Biology teaches that the embryo often retraces in its development the major stages in the evolution of its species; but the death of the organism is not a return to the embryo. The same is true for the relief of the earth: there is an unequal forward march but no return to the point of departure. There is evolution but not cycle.

To understand the present relief it should not be interpreted in relation to a fatal, future end product such as the peneplain of the Davisian cycle of erosion. It must be interpreted as the product of a complex evolution, as a combination of new elements in the course of development, and of relicts in the course of more or less rapid elimination. It is important to specify the part played by each of these categories as it determines the possible action of man on the evolution of relief. To organise the future, at least in the short span of human duration, the present and the past should be known as well as possible. Palaeogeomorphic reconstructions are of prime importance to applied geomorphology. They may help prevent the sudden man-caused interruptions of the morphoclimatic equilibrium, which often result in the destruction of valuable natural resources.

Interruptions in the morphoclimatic equilibrium caused by man: anthropic erosion

The postglacial morphoclimatic equilibrium has been broken in most areas of the earth by the intervention of man, who has often caused waves of erosion as important as those of the Quaternary climatic oscillations. It is in certain cultivated regions of the earth that the morphogenic processes are at present the most active, with enormous rates of erosion and deposition. In certain river basins of Indonesia average erosion is as high as 3 mm per annum, or 3 m (10 ft) per millennium, on surfaces of 100 to 1 000 sq km (40 to 400 square miles). The Loesi, a rather important river in central Java, annually removes a film of 0·87 mm; the Irrawaddy above Prome, 0·52 mm. These values, which are probably world records, exceed by far the rate of deposition of the formations concerned during the geological past. The large amount of suspended matter supplied by the piedmont glaciers of Alaska, which are at present in accelerated retreat, do not represent an erosion exceeding 2 m per millennium (6·6 ft). This is a particularly high value related to an interruption of the morphoclimatic equilibrium. The Massa, as it leaves the Aletsch glacier (the longest glacier of the Alps) has a solid load which only corresponds to 50 cm per millennium (1·6 ft), six times less than the rivers of Indonesia with a slightly larger basin. In his remarkable statistical study on the present intensity of erosion on the earth's surface, Fournier (1960) ends up with mean values which extrapolated to a

Quaternary of 1 million years would indicate an average erosion of 400 m (1 300 ft) on all the landmasses, which is a startling amount. It is known that on the greatest part of the land areas, i.e. the continental plates, Quaternary erosion was modest. Their low average elevation, which is less than 400 m (1 300 ft), would not have allowed pre-Quaternary landforms to subsist under such circumstances. But such forms are extensive; for example, in the Paris Basin, the Saône lowlands, the Aquitanian Basin, the Armorican Massif, and the Ardennes, to mention France only. The serious contradiction between Fournier's results, which are trustworthy, and reality illustrates the pathologic character of anthropic erosion.

Assuming that a surface of erosion is equal to that of deposition, 1 000 m (3 300 ft) of molasse would be deposited in only 333 000 years granting erosion rates of the order of those known to exist in Java, and in 1·2 million to 2 million years if the rates of the Loesi or the Irrawaddy were applied. Yet about five to thirty times as much time was required to arrive at the same result in the peri-Alpine regions during the Oligocene and the Miocene, although the surface of erosion was perhaps even larger than that of deposition (assuming 2 000 to 3 000 metres of molasse and 23 million years). True, Indonesia is a rather extreme example because the intense mechanical erosion which characterises it would be practically non-existent under natural conditions dominated by a biochemical morphogenic system. The interruption in the morphoclimatic equilibrium caused by man results in the mobilisation of a large mass of weathered materials previously prepared by biochemical action. Once the stock of soil and subsoil is liquidated, erosion slows down except where an important accumulation of volcanic ashes continuously renews it. Excluding this particular case, present accelerated erosion represents a peak whose duration, except in friable rocks, is necessarily limited. Such averages could not, of course, be maintained over hundreds of thousands of years, but they do exist and have been important enough to cause a change in the sediments in the span of a few thousand years (even a few hundred), and to give a special character to the relief of many areas. Similarly, in western Europe valley bottoms began to silt up with the major periods of forest clearing, Neolithic in the Paris region, Gallo-Roman in Lorraine, and medieval in Germany. In the Fouta Djallon of Guinea the spread of agriculture caused the superposition of anthropic sandy deposits on top of older blackish ferrallitic clays (coloured by poorly decomposed organic matter) in the bottomlands. In the southern Alps and the Atlas Mountains, gentle marl slopes formed by solifluction alternate with deeply incised gullies between unstable ridges. The minor undulations of the Russian steppe are slashed by abrupt gullies which extended themselves at the rate of several decimetres per year before man arrested their development.

Man-caused interruptions in the morphoclimatic equilibrium, therefore, often are at the origin of specific landform features on the scale of the seventh and even the sixth order of magnitude (i.e. metric or hectometric forms).

Significantly, their study is one of the meeting grounds of geomorphology and human geography. This convergence helps to direct our discipline away from the sterile ivory tower in which it became enclosed and to re-orient it towards practical applications.

In the next section we identify the principal forms of anthropic erosion, discuss their characteristics and indicate their relationships with climate. Our main task is to show what the place of anthropic erosion is in climatic geomorphology as a whole.

Definition and principal causes of anthropic erosion

We start with a definition: 'Man-induced morphogenic processes result from the readjustment of the environment whose natural morphoclimatic equilibrium has been upset by man.'

The definition is broad to include every kind of man-caused interruption in the morphoclimatic equilibrium, whatever its origin. Some interruptions are purely mechanical, such as the case of a road cut in a hillside, which by increasing the slope of the hill, throws it out of balance and causes land-slides; or the case of the sapping action of a stream which has been displaced by man and acts in a manner similar to that of the road cut. Interruptions caused by a modification of the plant cover, even when involuntary, are more subtle and general. An example is that of the Gippsland Lakes in Victoria, Australia, analysed by Bird (1963). These coastal lagoons had a poor communication with the sea and were invaded by fresh water in which *Phragmites communis* and other higher plants eventually prospered. These plants effectively trapped shore sediment conveyed by the streams and caused a rapid sedimentation thanks to the comb-like effect of the plants. The opening of an artificial entrance through the Ninety Mile Beach caused the lakes to be invaded in 1889 by sea water resulting in a brackish environment and the dieback of the *Phragmites*. The sediments which the reeds stabilised were swept away and erosion succeeded deposition on the deltas, 'the eroded material being added anew to spits and forelands that were temporarily insulated by swamp encroachment'.

In most cases the modification of the plant cover creating the morpho-climatic disequilibrium is produced by agricultural development. Often one speaks of *soil erosion*, which is only part of the adjustment to new conditions: anthropic erosion is not restricted to the soils but often proceeds well beyond by causing a weathering of the bedrock which has been exposed as a result of the disappearance of the soil, or by causing gullies which eat their way into the subsoil. The expression 'soil erosion' is attributed to pedolo-gists who have for a long time been the only ones interested in it; it has been generalised abusively because of the default of geomorphologists. The ex-pression 'accelerated erosion', proposed by certain geomorphologists, is not satisfactory either; it should refer to a certain state of nature which is not necessarily the result of the intervention of man but may also be caused by a natural interruption in the morphoclimatic equilibrium. Furthermore, although erosion is caused by man, its rate may be constant.

It is therefore necessary to break down the above mentioned definition into several kinds of man-induced morphogenic processes.[1] We will classify them according to cause.

INTERRUPTIONS IN THE MORPHOCLIMATIC EQUILIBRIUM CAUSED BY ENGINEERING WORKS

The best-known breaches of the morphoclimatic equilibrium caused by engineering works are those that are linked to the development of an advanced industrial civilisation. They are, for this reason, very unequally distributed over the earth's surface. They do not only affect climatic geomorphology but also the azonal or polyzonal processes and occasionally assume a pseudotectonic aspect. A few examples will show their variety.

Pseudotectonic disturbances occur in connection with mining; for example, in the Belgian Borinage progressive settling of the ground above mining galleries causes surface saggings which finally result in a notable subsidence of the floor of the Meuse Valley. In places the subsidence exceeds several metres and must be counteracted by the construction of embankments and drains. If the area were sparsely populated such costly works would not be undertaken, and a local alluvial filling would result in the manner of a subsiding basin. Intense pumping of water in the mine galleries causes a considerable acceleration of underground water circulation which is probably not without its effect upon subterranean erosion although the question has not been studied. Rising spoil dumps cause an overloading of the ground, resulting in compaction. Marlière (1951) has shown that there is a certain amount of bulging around the tailings due to the flux of friable subjacent sedimentary strata (sands, clays, marls, and silts). Faults form in the spoil dumps themselves; their throw may reach 12 m (40 ft). In a similar way typical but intensive exploitation of aquifers in friable sediments may cause the ground to sag. In some fifty years sagging has reached 2 m (6·6 ft) on the former lake bottom in Mexico City. Identical phenomena have been reported in Japan. In Long Beach, California, petroleum exploitation has caused a subsidence of several metres, resulting in the flooding of the lowlands close to the harbour.

Littoral morphogenic processes may be severely perturbed due to harbour engineering. The construction of jetties may stop the drifting of sands and pebbles and cause beach degradation downwind. A well-known example is that of the beach at Santa Barbara, California: on a part of the shore situated

---

[1] It is better to say 'morphogenic process' rather than 'erosion', for this phenomenon not only includes erosion but also correlative deposition, and this should never be forgotten: the intense erosion of mountains often causes lowlands to be buried by alluvium which may cause as much havoc as erosion. The filling of irrigation reservoirs, catastrophic floods and the burial of cultivated soils below sterile beds of sand and gravel are the fate of the best agricultural land in areas of intense anthropic erosion.

in the lee of the jetty, in the main direction of drifting, waves now exert an important erosion, whereas before the construction of the jetty this part was only affected by transportation. The incessant dredging of the outlets of navigable rivers stops the build-up of deposits in the estuaries and removes considerable quantities of sediments (often hundreds of thousands of tons per annum) from the natural geomorphic system. The fixation of coastal dunes by blocking the migration of sand towards the interior has repercussions on the evolution of the beaches. Thus in the Bas-Champs area of Picardy the resumed migration of dunes resulting from the destruction of the pine forests by the Germans during the second world war seems to have caused an acceleration in the retreat of the coast. It is easily explained: the sands are no longer detained on the backshore, so that deflation is accelerated.

Fluvial regimes are also profoundly modified in many countries through the canalisation of navigable rivers, the construction of hydroelectric dams, or the creation of extensive irrigation systems. Such uses replace the natural fluvial regime with an artificial one. In dry regions there is, on the whole, a decrease in stream flow due to irrigation. In humid regions navigable waterways and hydroelectric installations cause a reduction in the variation of discharge. To analyse the geomorphic consequences of such interference and to study the modifications in the fluvial dynamics caused by engineering works, as has been done by Meynier with regard to the Vézère in the Brive Basin (French Massif Central), is of the highest interest. Regularisation of the discharge of this river consequent upon the construction of dams has stopped the evolution of meanders by eliminating the large floods whose mechanical action undermined the concave banks. Clays rich in organic matter (debris of leaves and branches) are now deposited on the concave as well as on the convex banks. Immediately above an artificial reservoir the raising of the local base-level may completely modify fluvial dynamics. A good example is that of the Colorado River above Havasu Lake. The formerly cultivable and fertile plain of Needles has been transformed into a net of braiding watercourses dividing the former cropland into marshes and alluvial embankments. Having failed to foresee this course of events and having constructed artificial embankments to contain the course of the Colorado, thousands of acres which might have supported irrigated crops have been lost.

Large artificial reservoirs create new littoral phenomena which, too often, have not been foreseen by engineers. The reservoir of Tsimlyanskaya on the Don, Lake Mead, and the Serre-Ponçon reservoir on the Durance erode cliffs in the friable terrains of their banks. The seasonal variations in level caused by the operation of the works vertically shift the line of attack during the course of a rapid series of transgressions and regressions, thereby increasing their morphogenic efficacy. Cut and built terraces develop and precipitous cliffs appear. In less than a year one to two metre cliffs developed in the black earth bordering the Serre-Ponçon reservoir near Savines. On the Tsimlyanskaya reservoir ten metre (33 ft) high cliffs appeared within eight years due to a 100 to 150 m (330 to 500 ft) retreat of the lake shore. Jones, Embody, and Peterson (1961) have studied and described spectacular landslides in the Pleistocene lake beds containing the reservoirs of the Columbia River valley. All these phenomena reduce the capacity of the reservoirs and menace constructions such as settlements, public thoroughfares, and cultivated land on their margins.

The morphology of slopes may be influenced by engineering works. The road from Dublineau to Oran, in the Beni-Chougran, Algeria, is unfortunately a typical example: leaving Dublineau, the road crosses an area of marly-clay hills cultivated by the native population which European colonisation has evicted from the lowlands. The natural relief is characterised by broad rolling forms developed by solifluction during Pleistocene pluvial phases. The slopes are rather steep (10 to 20°) and the climate is noted for its intense showers. The asphalt road, about 10 m (33 ft) wide, constitutes an excellent catchment area capable of concentrating a large discharge in its gutters during heavy storms. Such masses of water are directed by the inadequate gutters towards gullies into which they often fall from a height of several metres. The runoff, which is much more intense (practically undiminished) along the road than on the neighbouring land, causes an intense spurt of erosion. Gullies with crumbling vertical walls are cut from the bottom of the ravines and progressively encroach on the cultivated land whose ruin they prepare. It would have been easy either to improve the gutters in order to diminish the dangerous concentration of waters, or to extend their revetment to the ravine heads. Such phenomena, unfortunately, are extremely frequent, and there are innumerable cases of gullies formed by the concentration of waters due to roads, airports, and settlements. In the Swiss Alps prolonged measurements in the Sperbalgraben have shown that 10 to 40 per cent of the runoff during heavy rains came from the road net alone, even though this net is much less dense than in urban areas. The construction of forest trails or simply the tracks of machinery used during periodic tree felling can trigger a dangerous erosion of which foresters are well aware. In mountains the laceration of the soil and the grass tier by logs dragged down in the direction of the steepest slope may cause the appearance of gullies which incise themselves to the bedrock.

The spread of engineering works and of mining activities often threatens the natural morphoclimatic equilibrium.[1] The wealth obtained from nature should not have as a corollary, on a planet which may become too small for

---

[1] Excellent examples are to be found in the opencast or strip-mining that is becoming more and more widespread. In the United States, for example, some 18 million acres are affected, for one third of all U.S. coal is now obtained by strip-mining. In the Appalachian area there are some 32,000 miles of strip mines, about 20,000 miles of which are on steep and formerly forested slopes. In Kentucky the erosion of certain areas of strip-mine spoil heaps produced 27,000 tons of sediment per square mile in one year—as against 25 tons per square mile from an equivalent area with its forests intact. (Editor's note.)

humanity, massive losses such as those caused by the harnessing of the Colorado River. Engineers alone cannot envisage all the consequences of their intervention in nature, therefore geomorphologists must help. This is the principle of applied geomorphology as it has been defined by Tricart (1962).

MAN-INDUCED INTERRUPTIONS IN THE BIOGEOGRAPHIC EQUILIBRIUM

Exploitation of the plant cover by man, even without agriculture, modifies the biogeographic equilibrium and may influence the morphogenic system.

We have already spoken about the brush fires of certain grazing lands of the wet–dry tropics. They cause the retreat of the wooded savannas and their progressive replacement by grass savannas. These fires free large quantities of silica which, later, may be partly swept up by the wind and form loams that are easily washed away. This morphogenic process helps the alluviation of bottomlands. The cuirassing process is concurrently accelerated and perhaps even made possible in regions where cuirasses would otherwise not form. Runoff becomes abundant but without morphogenic efficacy on the cuirasses which mould the relief and arrest its evolution. The brush fire is an important factor in the geomorphology of the wet–dry tropics. It affects ever larger areas and becomes increasingly important as the degradation of savannas begun by man and his animals proceeds apace. The geography of brush fires, however, is an aspect of human geography; its importance depends on the technological level of the peoples and the density of their herds. Brush fires are not fatal: education made possible by a higher standard of living will help to restrict their extension.

Grazing is a factor in the degradation of the plant cover. In the Ukraine and in southern Russia, even before the clearings of the last centuries, grazing has caused a considerable extension of the steppe at the expense of the wooded steppe, which originally was composed of alternating groves and clearings. In the Mediterranean region, over a period of several thousand years, sheep and goats have reduced the original woodlands to sparse shrubs which incompletely protect the soil. Runoff has consequently removed a good part of the regolith and exposed rock knobs. *Overgrazing* is characterised by an intense degradation of the pastured plant formations; they become impoverished in species as well as in numbers and fail to play their natural protective role in regard to the mechanical morphogenic processes. The overpopulation of Corsica in the nineteenth century and the overgrazing which resulted from it have largely ruined the intermediate mountain level. There the bedrock has often been exposed around abandoned sheepfolds as, for example, in the Monte d'Oro, and intense fragmentation, the result of climate, has produced a progressive burying of the landscape below stone fields. Finally, screes have developed and extended themselves as wedges into the forests, burying the trees and making them retreat little by little. The construction of stock tanks in the Senegal valley and the drilling of deep wells in the waterless Ferlo region cause large herds to congregate in restricted areas. The vegetation is quickly browsed by the animals and shrubs and trees are burnt by the cattlemen. The sand of the Pleistocene dunes becomes exposed and disturbed by treading, and the wind starts to move it once again. As seen from the air, each well is marked by a desolate and denuded expanse veiled in dust.

In the Canberra area of eastern Australia, the pasturing of sheep without apparent overgrazing has caused the appearance of deep gullies with crumbling walls in broad, slightly concave, grass-covered vales carved from

FIG. 4.9. Formation of gullies due to the concentration of runoff near Canberra, Australia

Under natural conditions there is overland flow because the grass is an obstacle to the concentration of the strings of water. This type of runoff which produces sheet-erosion has accumulated thin beds of fine materials (clays, silts, and sands) on the surface of the nearly flat vale in the foreground. A short period of agricultural development and grazing, reducing the comb-like effect of the grass, has facilitated the concentration of the runoff. Today, the larger maximum discharge has been able to incise a gully in the floor of the vale. The gully is of the *lavaka*[1] type with precipitous walls and a poorly drained floor because the groundwater is close to the surface. The retreat of the gully head is partly promoted by the effect of subsurface drainage and the undermining it causes.

weathered granite. There is no doubt that under the original woodland runoff was much slower and its concentration retarded; because of it the maximum stream flow was smaller under a similar climate of violent but short thunderstorms, and the running water had less force and did not

[1] A large type of gully with abrupt and crumbling walls incised in deep tropical regolith, as found in Madagascar (KdeJ).

corrade. Today, the swifter flow of maximum discharges first incises the bedrock at the bottom of the steepest vales, after which the growing incision migrates upward and downward. Upstream the gully ends in a sudden step which retreats vertically in the manner of a small waterfall.

Steppe regions, where the plant cover is fragile and the morphogenic potential of the climate high, are particularly sensitive to the consequences of overgrazing.

There are other causes of man-induced interruptions in the biogeographic equilibrium. The introduction of rabbits into Australia was rapidly followed by an overcrowding which had morphogenic repercussions. The stirring of the soil by these innumerable burrowers exposed the friable materials brought up from the lower soil horizons to the action of the wind. It was probably reflected by an increased importance of aeolian phenomena, particularly dust storms. Research concerning this matter would be of interest. On the Kerguelen Islands Aubert de la Rue has reported an intense honeycombing of stabilised pumice dunes through the actions of rabbits. These animals have started to swarm here as in Australia since their unfortunate introduction in 1874. The exploitation of forests also has its influence on the morphogenic processes. Trees uprooted by the wind are not as frequent in cultivated forests as in natural forests where trees have the time to become worm-eaten. The evacuation of logs is, however, an important factor of erosion.

INTERRUPTIONS IN THE MORPHOCLIMATIC EQUILIBRIUM CAUSED BY AGRICULTURE

Of all the man-induced interruptions in the morphoclimatic equilibrium those caused by agriculture are commonly recognised and, because of it, more familiar than the preceding types. Here are their principal characteristics.

Agriculture means the replacement of the natural vegetation by an artificial plant cover consisting of cultivated plants. It causes a double modification of the natural conditions:

(*a*) Because of harvesting an important part of the mineral matter is not returned to the soil. Under natural conditions, on the contrary, plant debris is returned to the soil upon death. In the case of crops the plants are removed completely or in part (fruit, stems, tubers, etc.). An important mass of mineral products is thus taken out of the fields, which is equivalent to a veritable chemical erosion. This removal is more or less compensated for by the addition of manure, fertilisers, and other improvements.

(*b*) Because of clearing the protection of the soil from the elements of the weather is modified. Crops only cover the soil more or less completely: cotton protects it very poorly and corn hardly any better, whereas the small cereals provide a superior cover. Measurements made in the United States have shown that the part of runoff during a shower of 50 mm (2 in) per hour is as high as 50 per cent on corn and cotton plots, as against only 25 per cent on wheat and oat plots. In Texas, on a slope of 5° (8·75 per cent) a 20 cm (8 in) layer of earth would be removed in twenty-one to forty-six years on a cotton field, whereas it would take sixty-seven years under a varied plant cover. Another important factor of the mechanical processes in a cultivated environment is the seasonal rhythm of the crops. A crop such as the vine, which grows on bare soil and whose roots poorly anchor the topsoil, allows a much more intense erosion than the small cereals which form a dense mat during part of the year. Clean tilled crops such as sugar beets, groundnuts, and manioc are a very poor protection.

Recalling such elementary facts is enough to show the complexity of the problem. The climatic factors, which certainly do play a role in field erosion, are, however, not the only ones. They are but one element in a combination of factors which include the following:

1. The techniques of agriculture are determined by the development of material civilisation and financial resources, which in turn are the result of a particular type of economic organisation. The struggle against erosion, for example, is hardly compatible with a speculative economy as it raises the cost of production and is, consequently, a liability in a competitive economy. The practice of soil conservation safeguards the future but compromises immediate profits. For this reason successive generations of peasants living in a subsistence economy but mindful of the future have terraced the slopes of the Mediterranean region and of certain parts of the island of Luzon, whereas certain settlers from Europe who first of all tried to make a rapid fortune in the vast open spaces of the New World ploughed in the direction of the slope and later abandoned the ruined land to find new land further west (United States, Brazil). On the molasse hills of the Aquitaine Basin, peasants driven from the Magreb, little mindful of conservation but rather armed with a speculative mentality, started to cultivate corn on lands which their tractors ploughed in the direction of the slope. The deplorable consequences have become evident after only two or three years: gullies formed and considerable quantities of soil were removed to the valley bottoms, whose drainage has progressively worsened. Unless precautions are taken, a whole process of regional degradation will ensue.

2. The cropping system determines the characteristics of the plant cover and, consequently, the protection it offers to the soil. This system depends on the economic conditions and the kind of economic organisation. Monoculture is more apt to produce erosion than diversified cropping, for runoff can easily become concentrated and assume increased force on slopes planted in the same crop, whereas with diversified cropping each field has its own regime, depending on the crop. Runoff is more here and less there, and, in the last analysis, damage is less. Runoff removes soil on a freshly ploughed field, but lower down as the water reaches a wheat field, for example, it becomes subject to a comb-like effect and drops its load. In this

207

way colluviation tends to take place on the scale of each individual field, and the concentration of runoff is impeded. Intercropping, like the Florentine *coltura promiscua*, in which cereal strips, rows of vines, and tree crops alternate, is also an empirical method of fighting soil erosion.

3. Social structure and land tenure determine the division of the land and the means available to peasants or farmers to adopt a certain farming technique. They even influence the type of economy. Large estates are necessarily a source of revenue and, because of it, tend towards a speculative type of farming. Water or wind erosion is greater on large than on small fields if the latter are in different crops. The stepped terrain linchets due to ploughing on the slopes of Picardy had the advantage of impeding erosion. An inconsiderate consolidation of parcels has led to the elimination of many, leading to a serious spurt of erosion, plainly visible on aerial photographs as white tones of chalk in the midst of darker loams. The consolidation of parcels and the regrouping of farms demanded by economic evolution and technical progress are thus often dangerous and reflected by an increase in erosion, which is sometimes catastrophic. Large parcels consisting of a single field occupying an entire slope are ideal for the concentration of runoff previously impeded by the diversity of crops: shoestring rills appear on kolkhozes as well as on large capitalistic farms. Often, too, old devices which prevented erosion have been eliminated, as the hedges of Vendée. Each one of them served as a barrier on the slopes and retained the slope wash. Their removal accelerates runoff and eliminates the obstacles they formed. Consolidation measures made in a purely legal spirit, as in the case of France, for example, may have disastrous effects. The whole agricultural development of the land, and especially its problems of conservation, should be taken into account if one were determined to practise a really rational and progressive policy.

All the elements of human agrarian geography are therefore combined with the physical factors to determine the particularities of agricultural soil erosion. The data to be considered are much more varied than in the case of geomorphic interruptions in equilibrium caused by engineering works or by overgrazing. Physical and human geography are more intertwined. We are now at the very heart of the geographic disciplines.

Having reviewed the principal forms of man-induced morphogenic processes, we will now proceed to describe the succession of events which result from man-induced interruptions in the morphoclimatic equilibrium.

Characteristics of anthropic erosion

By its very nature man-induced morphogenic processes only partly reflect climatic characteristics. Like the morphology of stream beds, they belong to the category of polyzonal mechanisms, and their actions have important points in common on the various continents. We will therefore study them first, before going into their distribution which is unrelated to the major natural morphoclimatic zones.

STAGES IN MAN-INDUCED MORPHOGENIC PROCESSES

We will deal primarily with anthropic landforms that are related to pastoral and agricultural activities, as only they affect sufficiently large and varied areas. Engineering works cause more localised perturbations of nature, related to a predominant process, such as littoral in the case of harbours or fluvial in the case of dams. They are special aspects of these morphogenic processes, whereas the agricultural and pastoral morphogenic breaches in the natural equilibrium assume a pseudoclimatic character. The effects of man-induced morphogenic processes succeed one another in a number of characteristic stages:

1. *Reinforcement of the natural processes.* The first stage consists of a reinforcement of the natural processes which begins with a decreased resistance of the plant cover. From the chemical point of view this stage is principally characterised by an impoverishment of the soils in soluble nutritive products for the plants; it is equivalent to a leaching. The removal of mineral matter extracted from the soil by the plants is not compensated for by manure or fertilisers. The fertility of the soil drops progressively. This mechanism is very important in the process of overgrazing; it contributes, together with the trampling and mutilation of plants by the teeth of the animals, to the gradual impoverishment of the plant cover. There is first of all a decrease in the number of species, the most fragile or the most demanding being eliminated, followed by a decrease in their total numbers, which means in the degree of plant cover. Bare spaces appear, and the processes of mechanical erosion take over. This also occurs, with similar effects, in cultivated regions because of the impoverishment of the soils.

The mechanical processes thus become more active than in the natural environment, but their characteristics remain about the same, at least they do not give rise to new landforms. In steppe regions, for example, overgrazing allows the wind to pick up an increased load of silt, but its action is insufficient to dig yardangs or to honeycomb friable formations; nor do live dunes appear. In the mountains of the temperate zone the replacement of many forests by short grasses, called alps (*alpes* or *alpages*), increases runoff but without gullying the slopes as high mountain grasses provide a very effective soil protection. The regime of torrents, however, is changed: their freshets are more sudden and voluminous, and thus have an increased capacity and competence. But their network remains the same: no gullies form on the slopes, at least not as long as the grass cover prevents them.

On the whole this stage is characterised by a certain preservation of soil profiles. Soil erosion is not yet generalised, but it has begun. The soil does not reconstitute itself at the base at the rate at which it is lost at the top. This is the criterion of the beginning of accelerated erosion, of the *incipient phase* of man-induced morphogenic processes. The importance of this stage is obvious because it indicates the moment when it is still possible to preserve the soil and the plant cover from destruction. It signals the moment when arresting action becomes indispensable.

It must be emphasised that this incipient phase of anthropic erosion is not in any way different from the purely natural modifications brought about by a climatic deterioration. The effects of either cause seem to be similar. Both cases reflect the beginning of an accelerated morphogenic crisis. But in the case of an artificial cause the beginning is usually more sudden, the process more rapid, and the plant formations resist less well than under a natural change of climate.

2. *The disappearance of the soils.* A second stage appears with the disappearance of the soils, here by erosion, there by burial under a massive influx of sediments. It marks the culmination of the accelerated morphogenic crisis caused by man. It is the *paroxysmal phase.*

This phase can best be observed on ploughed fields where the plant cover is completely absent for weeks, or even months, on a soil artificially softened up before seeding. Conditions for erosion are therefore optimum; the main natural obstacle is the structure of the soil itself, particularly its cohesion,

FIG. 4.10. Increased runoff due to the cultivation of a slope underlain by shale in the Rif, east of Tamchecht, Morocco

The destruction of the obstacle of vegetation facilitates the concentration of runoff, which occurs rapidly beyond a very small maximal surface of unconcentrated runoff. The shoestring rills easily deepen to form gullies, which are oriented according to the maximum slope.

which is due to the humus which cements the colloidal particles into aggregates. The aggregates make the soil more permeable and thus facilitate infiltration, which is also favoured by certain methods of cultivation. But as soon as overgrazing or cultivation without sufficient restitution cause soil degradation, the impact of the raindrops scatters the aggregates which then break and pulverise, thereby facilitating erosion. Erosion first rapidly increases and removes the superficial layers which were rich in humus in their natural state, then reaches down into the deeper layers poor in aggregates, and finally attains the C horizon devoid of aggregates. This horizon

FIG. 4.11. Agricultural colluviation in the Dombes (Les Caronnières de Priay, Ain, France)

Furrows on a soil poor in structure are causing a damaging concentration of runoff. Small fans form at the lower end of the furrows. Ploughing wipes out such microforms but does not stop the transport of material. Colluvial accumulations result at the foot of a slope or in front of obstacles such as hedges, low walls, or, as here, at field margins.

may be friable, however, if leaching has impoverished it of colloids; erosion then simply attacks a weathered product which may be close to having an optimal texture for easy removal.

Therefore, in proportion as the soil is removed, erosion encounters increasingly favourable conditions which produce a gradual acceleration of the erosional process. The paroxysm is approached and is often realised when the C horizon is being exposed. The complete removal of the humus eliminates the retarding effect which the aggregates oppose to erosion. The prior weathering of the bedrock has prepared easily removable materials. Poverty in organic matter and in colloids impedes the re-establishment of the vegetation which then approaches the ultimate stage of degradation.

This paroxysmal phase is characterised by the appearance of specific processes which did not exist in the previous state of morphoclimatic equilibrium. They are sheetwash and gullying which develop progressively during the course of accelerated erosion.

Sheetwash appears during very severe storms; it consists of a layer of water which sweeps the entire slope; it evens irregularities and produces a form of colluviation; bulges are truncated, exposing the deeper soil horizons, and hollows are filled. Generally, the higher parts of the slopes are subjected

FIG. 4.12. Anthropic erosion in a wet tropical climate, El Salvador, Central America

Basalts are weathered into red clayey soils. Rainfall is approximately 1 800 mm (72 in). The forest has been destroyed and the soils exhausted. Intense runoff, quickly concentrated, has developed, incising gullies straight down the slopes. The gully walls crumble, widening the gashes and forming niches. Slumps such as shown in Fig. 4.13 also exist.

to erosion while sheets of waste are spread at their foot, encroaching on the valley bottoms or even filling them with one or two metres of debris (as in Beauce, near Etampes, France). Sheetwash even develops below the plant cover if the soil is insufficiently protected. It is common in overgrazed areas where the grasses are reduced to tufts alternating with bare patches of soil, as the degraded tropical savannas (the sandstone plateaux of the Kayes area in Mali, for example), the esparto (alfa) grass steppes of the High Plains of the Magreb and the steppes around the Caspian Sea. Sheetwash is, of course, common on cultivated fields where after ploughing it is reflected by differences in colour due to the truncation of different soil horizons. This stage is the last in which anti-erosive action can still be taken at not too great a cost. It is therefore important to be able to identify it.

Gullying appears as soon as runoff becomes concentrated; inequalities in the topography and the least undulation can concentrate the water; the streamflow then erodes more than the neighbouring sheetwash because of its greater velocity. It first forms a rill, then, if conditions are favourable, a gully. In friable materials where weathering has generally permitted the development of an important loamy binding material such gullies often have abrupt walls, which crumble as they are undermined. Their head

FIG. 4.13. A man-induced slump caused by forest clearing near Tres Rios, Minas Geraes, Brazil

Deeply weathered gneiss forming convex hills typical of the forested tropics. Destruction of the forest causes the desiccation of the soil; the soil cracks which then form are filled with water during storms; they are the cause of mass movements including irregular creep, which forms terracettes, and large slumps, which form niches.

retreats progressively by sections, encroaching on the surfaces modelled by sheetwash or by shallow rills. A network of gullies thus gradually spreads at a relatively fast rate; average retreats of the order of one metre per year are not exceptional. The gullies develop at breaks in slope where the topographic surface is steep enough to allow the concentration of running water. They gash the steeper slopes with their parallel and about equally spaced incisions, while the more gentle slopes above and below are the locus of rilling or sheetwash. Frequent crumbling of the abrupt walls produces a rapid widening of the gullies, which thus encroach upon the generally broad valley bottoms in which they establish themselves. The crumbling provides the water with a heavy solid load which gradually reaches more important watercourses and modifies their depositional regime. Alluvial fans, the homologues of colluvial slopes produced by sheetwash, develop at the mouths of affluent gullies. The more heavily loaded main streams begin to meander, sapping their banks and giving them abrupt profiles, which leads to more gullies.

The transition from sheetwash to streamflow is counteracted by man on cultivated land. Every time a field is ploughed the rills are destroyed and their definitive formation is therefore retarded. The progressive removal of the topsoil, however, favours erosion, whereas the loss of soil decreases crop yields, and in this unequal struggle there comes a moment when the farmer grows weary and gives up, either at the end of an unusually violent storm which caused havoc beyond repair, or after an economic depression which compromised his profit in relation to falling yields, or simply because of age.

3. *The end of the crisis.* The maximum intensity of the morphogenic process is reached when erosion has succeeded in destroying the last subsoil remnants and in creating an environment unfavourable to plant life.

The duration of this climax depends on two factors:

1. The thickness of the subsoil that may be removed by the phase of accelerated anthropic erosion. It naturally varies according to the rocks. In a terrain of *consolidated* rocks erosion rapidly reaches the bedrock on which, later, the water runs off without incision. Such rock knobs are common in the mountains of the Mediterranean region (Algeria and Catalonia, for example) which have been rapidly denuded of their thin regolith. From then on the rate of mechanical erosion depends on the processes of fragmentation, that is on a combination of lithology and climate. Mechanical erosion may continue as long as fragmentation produces enough debris. This final stage is reached more or less quickly depending on the thickness of the original regolith. The duration of the paroxysmal phase is therefore a function of both climate and palaeoclimate, on which the thickness of the regolith depends. The paroxysmal phase has the characteristics of a zonal feature.

2. The other factor which influences the length of the paroxysmal phase is the presence of *poorly consolidated* rock formations; the problem then is different as the available material is practically unlimited on the human time scale. The question is whether after a certain period of time the processes triggered by man are going to decrease progressively in intensity? Such a slowing down of the processes presupposes that the forms evolve towards minimum slopes. This is the case of certain gullies; for example, those which attack the edge of a low plateau above a rather stable stream. After a certain period of time the gullies attain a profile of equilibrium, and a progressive stabilisation occurs which progresses upstream. As the gullying process is slowed down, the base-level of the gully walls in turn becomes stabilised and the evacuation of the waste at their foot becomes more

difficult, which changes the profile as debris falls produce concave scree slopes where there formerly were crumbling walls. The upper part of the gully walls is rounded off and gradually becomes convex. A ravine with less steep, convexo-concave forms thus takes the place of the gully. Evolution having sufficiently slowed down, a second factor becomes increasingly important: the possibilities of a reoccupation by vegetation. As soon as the intensity of erosion decreases, conditions become more favourable—or less unfavourable—to the development of plants. Small surfaces remain untouched by erosion for several months enabling the establishment of not very demanding annuals. A process antagonistic to mechanical erosion thus appears. Under the most favourable conditions such plants are able to resist, to form colonies, and to protect their seeds, which germinate in place and begin to reconstitute the plant cover. A little humus develops and begins the pedologic evolution, which in turn allows the development of new, more demanding species and the enrichment of the plant association. The soil is fixed by degrees, and the intensity of mechanical erosion decreases little by little.

The fixation of the soil may, of course, under certain conditions and with appropriate methods, be begun artificially. It is one of the great merits of French foresters to have been the first to succeed in arresting a wave of accelerated erosion caused by forest clearing in the mountains of France: in the Massif Central, the Alps, and the Pyrenees. An important reforestation was undertaken in the second half of the nineteenth century, notably in the Aigoual massif (Cévennes). The principle of the method, like the fixation of coastal dunes by Brémontier, another pioneer, consists of momentarily slowing down the intensity of mechanical erosion with the help of small engineering works, such as dry wall barrages in gullies and torrents or fascines and gabions on slopes and in rills, and to take advantage of the respite by planting and growing a plant cover. One must, of course, choose a succession of species: at first the least demanding plants, such as lyme-grass on dunes, which afford a protection and start the formation of a soil; later, more varied and useful species must be planted, and finally an artificial climactic association should be created. Another method useful in the friable formations of lowlands has produced good results in the United States: gullies are filled in with the help of bulldozers or, at least, they are considerably reduced; the vegetation can then take hold and protect the filled in gullies against erosion.

In short, such possibilities of reoccupation by vegetation depend on the relationships between two inverse factors: the intensity of mechanical erosion and the ecologic demands of the vegetation. Climate determines the first and more or less satisfies the second.

In conclusion, can one speak of a cycle in connection with crises of accelerated morphogenic processes? Can the concept of the cycle of erosion, which is inadequate on the geologic time scale and in relation to the natural conditions of land evolution, be applied to anthropic erosion?

First it is necessary to distinguish between zones of deposition and zones of erosion.

In the zone of deposition the crisis is reflected by massive deposits which bury the former soil if it has not been eroded. When the intensity of the crisis decreases the supply of sediment diminishes, and the vegetation can progressively occupy the surface and allow the development of a new soil. True, the latter will not be exactly like the former because the materials from which it has been derived differ; the products of accelerated erosion are not the same as those of natural erosion: the material is less weathered at the beginning; sorting is less during transport; the climatic conditions, however, are assumed to be the same, and so the soil tends towards a climax similar to that of the former soil. Considering the general evolution of the earth, the cyclic concept is inadequate; in this particular case we are close enough to it to make use of it without hurting the facts. In other cases the soil particles are transported by streams towards the sea where they are deposited. In this way the littoral of Charente (La Rochelle area) has been alluviated by muds called *bris* which is very hard to plough. Here the cycle is of course not closed.

In the zone of erosion the conditions are quite different. There the heritage of past periods is lost forever. The palaeosols and the ancient regolith, which could play a considerable role in the development of vegetation, are removed and scored. Some of them do not form again under the present climate. For example, a loess-covered slope may be entirely despoiled by erosion and the subjacent limestone exposed; if the limestone is invaded by plants after a period of frost-wedging and other weathering processes, and a new soil develops after a number of centuries, the past conditions will not therefore again be realised. The soil will be a poor rendzina rather than a fertile loam. Of course if one insists on being abstract one could consider the concept soil rather than the actual soil and admit that there is a cycle because one soil has been replaced by another. But such an abstraction would be inaccurate: it would sever the facts from observation, confuse good and bad soils, and conceal what man has lost in the change. And what has been lost is essential: it is the possibility of having food, firewood, or clothing which affects the development of future generations. The cyclic concept, as in natural geomorphology, is, here again, a narrow view which withdraws our discipline from reality and excludes it from contributing to the general progress of humanity.

THE MAJOR REGIONS OF MAN-INDUCED MORPHOGENESIS

The distribution of man-induced morphogenic processes on the land areas of the earth is determined by that of humanity itself and its economic activities; their characteristics, however, are influenced by climate. Climate indeed determines the mass of regolith that may be removed as well as the possibilities of reoccupation by plants and, therefore, in the final analysis, the extent and duration of the geomorphic crisis.

The human geographical factors responsible for the geomorphic crisis depend on the type of social and economic organisation. This organisation also determines what kind of measures can be taken to limit the wound inflicted on nature by agriculture and grazing in the development of the land. It decides to a large extent the farming methods used, especially the application of fertiliser and manure which preserve soil structure, improve it through cultivation, and enable it to resist erosion better. A speculative, extensive type of farming is the worst form of agrarian economy as far as soils are concerned. It is practised in countries where land is cheap, which encourages its abandonment in favour of virgin land when profits are jeopardised by falling yields. It also aims at a maximum profit through reduction of costs by mechanisation or utilisation of underpaid labour and by a minimum use of fertilisers which are considered to be too expensive. The degradation of soils is therefore rapid, and erosion is easily unleashed. Sometimes in order to obtain large profits in a short period of time, the land is deliberately plundered by practising the devastating methods of mine and forest exploitation, or frenzied gathering, activities which in German are called *Raubwirtschaft* (pillage economy). Convergence of a certain degree of overpopulation and a low living standard may produce similar effects: poverty precludes the use of fertilisers, and overpopulation compels the use of lands which should not have been cleared; the people are then resigned to the appearance of erosion, which aggravates the problems even more. Elsewhere, on the contrary, as in the Netherlands, the very high density of population coincides with an extreme care of the soils, which is the result of a concurrence of enough labour and capital. The wealth obtained from several centuries of colonial trade has partly been invested in the reclamation and improvement of land (polders). The high density of population makes possible the care of this new land, whose value is high because of the investments which have been put into it.

The morphoclimatic factors determine the importance of the damage inflicted by anthropic erosion on a specific type of farm. In zones of seasonally contrasted climate the erosive potential is particularly high because the agents of the weather are violent and the ecologic conditions mediocre. In humid climates, clays or marls are rapidly reoccupied by vegetation: in the Jura or the Alps of Savoy the plant cover reconstitutes itself in four or five years after a landslide. On the other hand in climates with dry seasons, such as in the southern Alps or the Mediterranean region, the reoccupation by vegetation is more difficult: drought cracks the hardened ground, which hinders the development of saplings and favours the concentration of runoff with the first rain; the surface which tends to remain bare becomes gullied and transformed into badlands. Steppes and savannas are also fragile environments; palaeosols (loess in particular) play an important role there, and the vegetation struggles against difficult ecologic conditions; the climatic erosive potential is high because of high temperature variations, heavy rains, and high air turbulence on the ground due to the importance of ground radiation. As in all transitional zones, the equilibrium is particularly unstable and easily upset. Lastly, in the wet tropical zone very active chemical weathering has accumulated an enormous mass of movable debris the thickness of which exceeds 10 m (33 ft). Once the vegetation is destroyed, the high daily temperature variations and the violent and abundant rains cause an active erosion. The reconstitution of the vegetation, however, is rapid.

The part played by the various climatic factors and the relief are well reflected in Fournier's (1960) global statistical analyses. Evaluating the intensity of erosion from the suspended stream load for the entire world, he in fact examines an image in which man-induced morphogenic processes play a predominant part. The values obtained are grouped around four distinct straight lines summarised by the following equations:

$Y = 6{\cdot}14x - 49{\cdot}78$ valid for lowlands with regular rainfall
$Y = 27{\cdot}12x - 475{\cdot}41$ valid for areas of low relief and irregular rainfall
$Y = 52{\cdot}49x - 513{\cdot}21$ valid for mountain regions with a rather humid climate
$Y = 91{\cdot}78x - 737{\cdot}62$ valid for semiarid regions

These straight lines clearly bring out the influence of two factors: the vigour of the relief (for each type of climate the straight lines are steeper for mountain regions) and the irregularity of the precipitation which is represented by x. This x parameter is equal to p^2/P in which p is the total precipitation of the wettest month and P the total annual precipitation (in millimetres); it provides a rough but quickly calculated representation of the irregularity of precipitation from the readily obtained but perfunctory data of climatic statistics. Mediterranean, monsoonal, and savanna climates, all with well-marked dry seasons, are characterised by high values for x (above 20). Continental climates also show higher values than do the maritime climates, which is satisfactory. By means of this simple parameter one obtains, on the one hand, the resistance of the plant cover reduced by seasonal drought and, on the other hand, the intensity of precipitation, usually higher in climates with very contrasted seasons. It therefore faithfully reflects reality.

Taking into account both the role of the human geographical factors and the characteristics of the physical environment, one may differentiate the following major areas of man-induced morphogenic processes on the face of the earth:

1. *The new countries of the nineteenth century* characterised by a speculative economy, cheap land, extensive methods of agriculture and, consequently, generally large fields and the practice of monoculture The economic principle in these regions was the systematic application (even if it was unconscious) of a robber economy. The development of the land was followed by a wave of accelerated morphogenic processes of unheard of

violence. In one hundred years 25 per cent of the surface of the cultivated land of the United States was ruined, often in an irremediable way. Millions of tons of topsoil were accumulated in valley bottoms or lost to the ocean. In the prairie states dunes invaded former fields and gullies frequently gashed the hills. Having become aware of the often irreparable damage, the unlimited optimism of the past century has tended to give way to a Malthusian attitude. The belated struggle to reclaim this land is costly; it has had some success, but the proposed Missouri Valley project, inspired by that of the Tennessee Valley, may never be realised because of the high cost of investment.

If the United States is the most typical of the countries of this group, it is not the only one. The Canadian prairie, a region of speculative economy par excellence, is gravely menaced by anthropic erosion, whereas French Canada, characterised by a more traditional economy, is less so; it is more like European countries. South Africa, Australia, Argentina, and southern Brazil are characterised by the same problems as the United States, with differences in degree only. In Brazil the coffee boom rode on a wave of intense man-induced morphogenic processes leaving behind ruined soils and gullied slopes; Jean Goguel, who has flown over these regions, has noted the close association of agriculture and intensive gullying.

2. *The overpopulated lands of traditional agriculture*, which include such regions as the Apennines, Mexico, India, and the Algerian Tell. In these lands poverty maintains such archaic techniques that they prevent a successful fight against erosion. Overpopulation is the result of a lack of employment due to a retarded economy where industrial development remains in a beginning stage; unemployment is chronic. In these circumstances a maximum amount of land is used to feed the hungry, even at the risk of destroying the soil of tomorrow. The grip of misery excludes all concern for the future; for example, certain native Algerians very well realise the danger of erosion and manifest some reactions, but they are too poor and too hard pressed really to engage in the struggle. This situation was aggravated in certain colonial territories where part of the land was taken away to satisfy settlers who planted speculative export crops. The Oran area of Algeria was a notable case. In tropical Africa, Central America, and Southeast Asia the introduction of export crops, sometimes in the form of forced agriculture, increased the danger of erosion. The products were sold at low prices, whereas the imported foodstuffs were of the highest value. The price difference forced export crops to expand over vast areas so that they could help pay for the foodstuffs whose importation became necessary by their very extension, for they often took the place of subsistence crops. Thus there was an extension of agricultural land without a concomitant improvement in cultivation methods. The fallow period, during which the soil reconstitutes itself and resists erosion, became shorter and shorter. In the Fouta Djallon,

for example, the increase of population without a concurrent improvement of farming techniques has forced a fallow period of ten to twenty years to be reduced to one or two years only; in ten or twenty years the forest more or less reconstituted itself, and soil protection was sufficient, so that only the incipient phase of the man-induced morphogenic processes could develop; in one or two years, on the other hand, only a meagre steppe can grow on soils that are rapidly being exhausted. The paroxysmal phase is reached: the bottomlands silt up and the streams start to meander; the solid load starts the mechanical corrasion of the falls and rapids, and potholes form; gullies score cultivated hillsides; and innumerable corestones exposed by erosion make their appearance.

Elsewhere the spread of shifting agriculture with a more rapid rotation of crops and overgrazing accelerates the development of sterile cuirasses, particularly in the African savannas. And while the Algerian Tell is gashed with gullies, in Senegal the extension of groundnut cultivation on Pleistocene sands once more allows the migration of the sand. An equally lamentable situation exists in Latin America. If some ancient Indian fields, particularly in the Peruvian Andes, are still subject to traditional conservation practices (terraced fields dating back to the pre-Columbian period), nearly everywhere half-starved peasants and absentee landholders do nothing to stop an alarming erosion. While shifting cultivation with corn crops encroaches on steeper and steeper slopes and the forest fallow becomes shorter and shorter erosion rages. In countries which still possess vast vacant spaces, such as Brazil or Venezuela, the peasants migrate in search of new land and rapidly spread the extent of the damage. Sometimes a policy based on education, expert supervision, and intelligently distributed subsidies remedies the evil. This is the case of a small area in the Venezuelan Andes not far from Mérida. Here the administration encourages the formation of co-operative conservation units which adopt a programme established by technicians. All the members participate in the work by exchanging services; they construct terraces which the owners pledge to maintain and to raise progressively with the use of stones removed from the fields. A modest subsidy, much lower than the same works would have cost had they been built by wage-earners, is granted for each metre constructed. For such partly occupied peasants it means a small capital immediately invested in improvements realised with the help of materials supplied at low cost by the administration, and completed, in turn, by subsidies; the ameliorations include such items as improvement of living conditions, water supply, or sprinkler irrigation system. A process of development which may have a chain reaction has been set in motion. It assures an intensification of agricultural production and soil conservation, actions which must be undertaken jointly as the investments required for conservation are impractical within the framework of a production whose yields are too low. A low productivity which unceasingly causes the extension of the cultivated surface in proportion as the population increases and the yields decrease, sets in motion a mechanism of destruction which has tragic effects in underdeveloped countries, particularly in those with a rapidly growing population.

3. *The monsoon lands of southeast Asia* have experienced, until recently, the same economic regime as the countries of the preceding group, and in spite of the introduction of a new agricultural policy in China the old regime has had its particular effect on geomorphology.

The action man has had on the relief is particularly important due to the convergence of a certain number of factors:

(*a*) The ancient character of peasant farming which goes back some five or six thousand years in China and India. These countries are, after the Middle East and with Europe, those which have known for the longest period of time a continuous agricultural occupation; the cumulative effect of man-induced morphogenic processes is therefore considerable.

(*b*) The high density of rural population in the areas which have been cultivated since ancient times. The effect of such crowding is to create conditions of overpopulation of the type we have described above. Particularly in India these conditions have been intensified by the colonial regime which encouraged the development of export crops (jute, cotton, and wheat) at the cost of subsistence crops and without sufficiently improving yields.

(*c*) The considerable morphodynamic potential of the climate in the form of violent showers concentrated in a short, extremely wet season alternating with a dry season. Such a climate with two very contrasted seasons is morphologically very effective: the first monsoon storms, often the most torrential, drench a desiccated and overheated landscape, causing enormous damage. They initiate gullying which continues throughout the rainy season, and produce landslides in waterlogged soils.

The joint action of such divers factors explains the extremely rapid erosion of parts of southeast Asia. In the drier northeast the intensity of the man induced morphogenic processes only decreases slightly because the lesser precipitation is partly offset by increased fragmentation due to the presence of frost, which is severe in continental climates. The action of the wind is added to the action of frost in the lowlands, especially in northern and eastern China.

In the most anciently peopled areas erosion long ago has removed the reserve of friable regolith from the steep and thoroughly deforested hills and mountains. Agriculture has often become impossible on them, and this factor certainly must have played a role in the concentration of farming on the alluvial plains, so characteristic of southern China and Vietnam, for example. The persistence of farming in the lowlands is assured by the fertilising silts provided by the erosion of the uplands, but the increased torrential nature of the rivers, their high solid load, and rapid deposition are grave dangers made evident in floods and catastrophic changes in rivers' courses, which are particularly frequent in China where the impact of man is at a maximum. Shifting agriculture still exists in mountains not yet severely affected by gullying. In Indonesia, mainly, an evolved agriculture is found on mountain slopes, especially in Java and Sumatra. It produces an enormous amount of erosion and can only continue to exist thanks to thick and friable rock formations such as volcanic ash and shales.

4. *The old European farmlands* are, after southeast Asia, the second most important locus of intense farm life. For one or two thousand years, sometimes more, agriculture has maintained high densities of population here. This region is characterised by several traits:

(*a*) A smaller climatic morphogenic potential than in monsoon Asia and even on part of its area than in North America. The most extensive damage is caused by winter rains falling on frozen or snow covered ground; the soil aggregates then have been partially destroyed by frost enabling a thin, slowly flowing film of water to remove an appreciable quantity of material. Severe summer thunderstorms are generally less effective except in Mediterranean Europe. The predominant role of these two types of weather conditions explains the steadily increasing intensity of the man induced morphogenic processes as one proceeds from west to east: unimportant in Normandy it is already more noticeable in Alsace; it reaches a maximum intensity in the steppes of southeast Russia. On the whole, the impact of man on morphogenic processes is more important in the Mediterranean region than in the maritime countries of northwestern Europe.

(*b*) Another trait is the age-old empirical traditions of fighting soil erosion. They are visible everywhere in the use of the land: the linchets of the hills and plateaux of western and central non-Mediterranean Europe; contour ploughing; rows of cypresses and hedges of the Mediterranean lands and the enclosures (*bocages*) of oceanic regions which, among other things, constitute an effective defence against the wind; preservation of the stubble during the winter between the harvesting of the winter wheat and the sowing of the spring cereals which reduces the damage done by rain during periods of thaw in regions of a three field system. These precautions have assured the preservation of soil fertility down to the present generation.

The old European farmlands also bear the mark of periodic past crises which were caused by more or less serious upheavals of the traditional economic balance. They deserve more attention from rural sociologists, for their consequences are often important. Vogt (1957) has studied an example in the Paris Basin and neighbouring areas, where renewed clearings, a redistribution of land parcels, and an elimination of the grass fallow due to the beginning of speculative agriculture occurred during the eighteenth century; the result was a resumption of soil erosion. A rural exodus which no longer enables the maintenance of the age-old protective devices may have the same consequences: thus in the Mediterranean region of France terraces crumble and slopes become the prey of gullying and often catastrophic landslides as at Menton. Speculative agriculture has similar effects: the substitution of a diversified agriculture by vineyards in Languedoc, the introduction of beetsugar on the large estates of Prussia, or the extension of wheat cultivation for export on the steppes of the Ukraine and Russia all produced an important acceleration of anthropic erosion.

In conclusion, the problem of man-induced morphogenic processes is a fascinating field of research for both geographers and geologists. They permit them to grasp the unity of their discipline and the interaction of physical and human factors. They enable them to contribute to the development of productive enterprises and, as a result, to the material progress of humanity.

Bibliographic orientation

On palaeoclimatology

A basic work with valuable data is of the *Geol. Rundsch.* (Klimaheft) **40** (1952), 200 p. Also consult: 'Das Klima der Vorzeit als Tagungsthema der Hauptversammlung', 1951 in *Geol. Rundsch.* **37** (1949), pp. 139–40.

CAILLEUX, A. (1952) *La Géologie*, Paris, A. Colin Coll. Que sais-je? 128 p.

BORISOV, A. A. (1959) 'Sur les conditions paléoclimatiques de la formation des principaux centres barométriques du climate actuel de la Terre', *Ivz. Vsiess. geogr. Obchtch.* **91**, 259–65 (French trans. by Bur. Rech. Géol. Min. no. 2487).

BROOKS, C. E. P. (1949) *Climate Through the Ages*, 2nd edn, London, Benn, 395 p.
Presents the explanatory theories of palaeoclimates and reviews them in connection with each continent. References at the end of each chapter. The work tries to demonstrate the merits of Wegener's theory and sometimes assumes the character of an argument rather than of an objective study.

FURON, R. (1941) *La Paléogéographie. Essai sur l'évolution des continents et des océans*, Paris, 530 p.

FURON, R. (1954) 'Biogéographie et paléogéographie', *Rev. Sc.* **61**, pp. 158–69.

MARKOV, K. K. (1956) 'Origin of contemporaneous geographical landscapes', *Geogr. Essays, USSR Akad. Sc.* pp. 42–53.

MARKOV, K. K. (1960) *Paleogeography*, 2nd edn, Univ. of Moscow, 267 p.

SCHWARZBACH, M. (1963) *The Climate of the Past; an Introduction to Paleoclimatology*, trans. and ed. by Richard O. Muir, London and Princeton, Van Nostrand, 328 p.

TERMIER, H. and TERMIER, G. (1952) *Traîté de géologie*, vol. 1: *Histoire géologique de la biosphère* (*La vie et les sédiments dans les géographies successives*), Paris, Masson, 721 p. 35 maps in colour.
Certain authors, including J. Tricart and A. Cailleux, do not agree on some of the facts and interpretations of the text. But this does not reduce the value of the maps.

On the mechanisms of *climatic oscillations* and on the present climatic oscillation one should consult:

'Solar variations, climatic change and related geophysical problems', *Ann. NY Acad. Sc.* **95** (1961) 1–740.

ALBRECHT, F. (1962) 'Paläoklimatologie und Wärmehaushalt', *Meteorol. Rundsch.*, no. 2, 38–44.

BERGSTEN, K. E. (1950) 'Some characteristics of the dispersion of the annual precipitation in Sweden during the period 1881–1940', *Lund Stud. Geogr.*, ser. A, *Phys. Geogr.* no. 1, 18 p.

BERNARD, E. A. (1962) 'Théorie astronomique des pluviaux et interpluviaux du Quaternaire africain', *Acad. Roy. Sc. Outre-mer, Cl. Sc. Nat. Medic.* **12**, no. 1, Brussels.

BROOKS, C. E. P. (1950) 'Selective annotated bibliography on climatic changes', *Amer. Meteorol. Soc., Meteorol. Abstr. Bibliogr.*, July, vol. 1, no. 7.
Presents a useful critical bibliography, a basic work to which we refer.

BUTZER, K. W. (1957) 'The recent climatic fluctuation in lower latitudes and the general circulation of the Pleistocene', *Geogr. Annaler*, **39**, 105–13.

BUTZER, K. W. (1958) 'Russian climate and the hydrological budget of the Caspian Sea', *Rev. Can. Géogr.* **12**, 129–39.

BUTZER, K. W. (1958) 'Die Ursachen des Landschaftswandels der Sahara und Levante seit dem klassischen Altertum', *Akad. Wiss. Litt., Abh. math. Naturwiss.* Kl., no. 1, pp. 2–19.

BUTZER, K. W. (1961) 'Climatic changes in arid regions since the Pliocene', UNESCO, *Arid Zone Res.*

EMILIANI, C. and GEISS, J. (1957) 'On glaciations and their causes', *Geol. Rundsch.* **46**, 576–601.

ESTIENNE, P. (1952) 'Le problème des variations climatiques en pays tempérés', *Rev. Géogr. Alpine*, **40**, no. 2.

EWING, M. and DONN, W. L. (1958) 'A theory of ice ages II; the theory that certain local terrestrial conditions caused Pleistocene glaciations is discussed further', *Science*, **127**, pp. 1159–62.

FAIRBRIDGE, R. F. (1961) 'Convergence of evidence on climatic change and ice ages', *Ann. NY Acad. Sc.* **95**, 542–79.

FAIRBRIDGE, R. F. (1961) 'Radiation solaire et variations cycliques du niveau marin', *Rev. Géogr. Phys. Géol. Dyn.* n.s. **4**, pp. 2–14.

GREGORY, S. (1954) 'Climatic classification and climatic change', *Erdkunde*, **8**, 246–52.

KRAUS, E. B. (1955) 'Secular changes of tropical rainfall regimes', *Q. J. Roy. Meteorol. Soc.* **81**, 198.

LAMB, H. H. and JOHNSON, A. I. (1959) 'Climatic variation and observed changes in the general circulation', *Geogr. Annaler*, **41**, 94–134.

LAMB, H. H. and JOHNSON, A. I. (1961) 'Climatic variation and observed changes in the general circulation', *Geogr. Annaler*, **43**, 363–400.

LAMB, H. H. (1961) 'Climatic changes with historical time as seen in circulation maps and diagrams', *Ann. NY Acad. Sc.* **95**, 124–61.

LYSGAARD, L. (1952) 'Recent climatic fluctuations', *Folia Geogr. Danica*, vol. 5, Copenhagen. Review of climatic oscillations since 1798; is especially concerned with data obtained in Europe and North America.

MALKIN, N. P. (1961) 'On the influence of sea transgressions and straits on the Quaternary glaciation', *Izv. Vsiess. Geogr. Obchtch.* pp. 122–35.

MANLEY, G. (1951) 'Climatic fluctuation, a review', *Geogr. Rev.* pp. 656–60.
Completes Brooks (1950); more elementary and accessible.

SCHEIDEGGER, A. E. (1961) *Theoretical Geomorphology*, Berlin, Springer, 333 p. 2nd ed. 1970, 435 p.

SHALEM, N. (1952) 'La stabilité du climat en Palestine', *Rev. Biblique*, **58**, 54–74. Reproduced with modifications in *Desert Research*, UNESCO, pp. 153–75.
Good example of method; ingenious use of ancient texts.

SOKOLOV, A. A. (1955) 'Decrease in the durability of ice in relation with the climatic amelioration', *Priroda*, no. 7, 96–8.

THAMBYAHPILLAI, G. (1958) 'Rainfall fluctuations in Ceylon', *Ceylon Geogr.* **12**, 51–74.

VERYARD, R. G. (1963) 'A review of studies on climatic fluctuations during the period of the meteorological record', UNESCO, *Arid Zone Res.*

WILLETT, H. C. (1949) 'Long period fluctuations of the general circulation of the atmosphere', *J. Meteorol.* **6**, no. 1, pp. 34–50.

WILLETT, H. C. (1953) *Atmospheric and Oceanic Circulation as Factors in Glacial–Interglacial Changes of Climates*, Cambridge, Harlow-Sharphey.

On the methods of studying palaeoclimates

CAILLEUX, A. (1950) 'Paléoclimats et optimums physiologiques', *C.R. Somm. Sc. Soc. Biogéogr.* **230**, 23–6.

CAILLEUX, A. (1951) 'Reliefs coralliens et paléoclimats', *C.R. Somm. Sc. Soc. Biogéogr.*, no. **239**, pp. 22–4.

HEINZELEN, J. DE (1952) *Sols, paléosols et désertifications anciennes dans le secteur nord-oriental du Bassin du Congo*, Brussels, Publ. INEAC, 168 p.

EMILIANI, C. (1958) 'Paleotemperature analysis of core 280 and Pleistocene correlations', *J. Geol.* **66**, 264–75.

EMILIANI, C. (1958) 'Ancient temperatures', *Sc. Amer.* Feb. 11 p.

KOCH, H. G. (1952) 'Bericht über eine Exkursion durch das Nadelwaldgebiet in Lappland', *Polarforsch.* Kiel, p. 193.
Analysis of a study by Regel published in *Experientia*, **8** (1952), no. 1, p. 34. Trees 430, 280, 125, and 50 years old . . . are found close to the tree-line; they indicate years in which germination was successful; during the intervals they failed.

MARTIN, P., SABELS, B. and SHUTLER, D. (1961) 'Rampart Cave coprolite and ecology of the Shasta ground sloth', *Amer. J. Sc.* **259**, no. 2, 102–27.

MITCHELL, J. M. (1965) 'Theoretical paleoclimatology', in *The Quaternary of the United States* (a review volume for the VII Congress of the Int. Assoc. Quatern. Res.) ed. H. E. Wright, Jr. and David G. Frey, pp. 881–901.
Consult bibliography.

ÖPIK, E. J. (1950) 'Secular changes of stellar structures and the ice ages', *Mon. Not. Roy. Astron. Soc.* London, **110**, 49–68.
The climatic variations during the course of time were caused by variation of solar energy. It was also the idea of Gerard and Ebba de Geer. Critical analysis of the other hypotheses.

RUDAKOV, B. E. (1952) 'Méthode d'utilisation des anneaux annuels des arbres pour la mise en évidence des variations du climat d'après leur épaisseur', *Dokl. Akad. Nauk. SSSR*, **84**, no. 1, 169–71 (French trans. by BRGM, no. 787).

SABAN. (1951) 'Thermomètres géologiques', *Bull. Trim. d'Inf. CEDP*, **3**, no. 12, 3–13.

SCHOVE, D. J. (1954) 'Summer temperatures and tree-rings in north Scandinavia, A.D. 1461–1950', *Geogr. Annaler*, **36**, 40–80.

SOUBIES, L., GADET, R. and MAURY, P. (1960) 'Le climat de la région toulousaine et son influence sur les récoltes de blé et de maïs', *C.R. Acad. Agric.* (France), pp. 185–95.

THEOBALD, N. (1952) 'Les climats de l'Europe occidentale au cours des temps tertiaires d'après l'étude des insectes fossiles', *Geol. Rundsch.* **40**, 89–92.

WENTWORTH, C. K. and DICKEY, R. I. (1935) 'Ventifact localities in the United States', *J. Geol.* **43**, 97–104.

On Quaternary climates

BÜDEL, J. (1951) 'Die Klimazonen des Eiszeitalters', *Eiszeitalter und Gegenwart*, Ohringen (Württ.), **1**, 16–26.

BÜDEL, J. (1953) 'Die "periglaziale" morphologische Wirkungen des Eiszeitklimas auf der ganzen Erde', *Erdkunde*, **7**, 249–65.
These two studies by Büdel constitute an attempt to reconstruct the variations of the climatic zones of Europe and Africa during the glacial periods. Numerous references.

BÜDEL, J. (1956) 'The Ice Age in the Tropics', *Universitas.* Stuttgart, **1**.

DEEVEY, E. S. (1949) 'Biogeography of the Pleistocene', *Bull. Geol. Soc. Amer.* **60**, 1315–1416.
Study of the biogeographic changes of the Pleistocene accompanied by useful maps and diagrams. Extensive bibliography.

FIRBAS, F. (1951) 'Die Quartäre Vegetationsentwicklung zwischen den Alpen u. d. Nord- und Ostsee', *Erdkunde*, **4**, Heft 1, 6–15.

FLINT, R. F. (1957) *Glacial and Pleistocene Geology*, New York, Wiley, 553 p.

FLOHN, H. (1953) 'Studien über die atmosphärische Zirkulation in der letzten Eiszeit', *Erdkunde*, **7**, 266–75.
Attempt, based on Büdel's theories, to reconstruct meteorological mechanisms during the last glacial.

FURON, R. (1950) 'Les problèmes de paléoclimatologie et de paléobiologie posés par la géologie de l'Arctide', *C.R. Somm. Sc. Soc. Biogéogr.* **230**, 13–23.

GUILLIEN, Y. (1955) 'La couverture végétale de l'Europe Pleistocène', *Ann. Géogr.* **64**, 241–76.

GRITCHUK, B. M. (1952) 'Basic results of a micro-palaeontological study of the Quaternary deposits of the Russian plain', *Materials for the Quat. of the USSR*, **3**.

LANDSBERG, H. (1949) 'Climatology of the Pleistocene', *Bull. Geol. Soc. Amer.*, pp. 1437–42.

LEMEE, G. (1948) 'La méthode de l'analyse pollinique et ses rapports à la connaissance des temps quaternaires', *Année Biol.* **24**, 49–75.
Complete and lucid account of the method; extensive bibliography to which we refer.

MORTENSEN, H. (1957) 'Temperaturgradient und Eiszeitklima am Beispiel der Pleistozänen Schneegrenzdepression in den Rand- und Subtropen', *Z. Geomorph.* n.s. **1**, 44–56.

WOLDSTEDT, P. (1954) *Das Eiszeitalter. Grundlinien einer Geologie des Quartärs*, vol. 1 : *Die allgemeine Erscheinungen des Eiszeitalters*, 2nd edn, Stuttgart, Enke, 374 p.
Excellent basic study with abundant bibliographies.

WRIGHT, H. E. and FREY, D. G., eds. (1965) *The Quaternary of the United States* (a review volume for the 7th INQUA Congress), Princeton Univ. Press, 922 p.
Consult especially part II on biogeography.

Concerning the climatic oscillations of the various latitudinal zones one should refer to the chapters especially devoted to palaeoclimates in the volumes of the present *Treatise of Geomorphology* which are concerned with the relief of these zones. For example, for the actions of the cold Quaternary climate to the volume devoted to periglacial relief, for the pluvials to the volume on dry regions, etc.

On Pre-Quaternary climates

'Klimaschwankungen und Krustenbewegungen in Afrika, südlich des Äquators von der Kreidezeit bis zum Diluvium', *Geogr. Ges. Hannover*, **3** (1950).

ANTEVS, E. (1952) 'Cenozoic climates of the Great Basin', *Geol. Rundsch.* **40**, no. 1, 94–107.

BEHRMANN, W. (1944) 'Das Klima der Präglazialzeit auf der Erde', *Geol. Rundsch.* **34**, 763–76.
Account of Tertiary palaeoclimatic zones. Conclusion: there is no continental drift or shifting of the poles. One of many studies demonstrating that Wegener's theory is contrary to the facts.

DERRUAU, M. (1960) 'Quel était le climat du Massif Central français pendant la seconde moitié de l'ère tertiaire?' *Rev. d'Auvergne*, **74**, 179–86.

EMILIANI, C. (1954) 'Temperature of Pacific bottom waters and polar superficial waters during the Tertiary', *Science*, **119**, 853–5.

EMILIANI, C. (1956) 'Oligocene and Miocene temperatures of the equatorial and sub-tropical Atlantic Ocean', *J. Geol.* **64**, 281–8.

EMILIANI, C. (1961) 'Cenozoic climatic changes as indicated by the stratigraphy and chronology of deep-sea cores of Globigerina-ooze facies', *Ann. NY Acad. Sc.* **95**, 521–36.

ERHART, H. (1956) 'La vie végétale continentale aux époques pré-dévoniennes vue sous l'angle de la théorie bio-rhéxistasique et des dernières découvertes palynologiques', *Bull. Soc. Géol. France*, 6ᵉ ser., **6**, 445–50.

ERHART, H. (1962) 'Témoins pédogénétiques de l'époque permo-carbonifère', *C.R. Somm. Soc. Biogéogr.* **355/7**, 21–53.

LOTZE, F. (1938) *Steinsalz und Kalisalze*, Berlin.

TYCZYNSKA, M. (1957) 'Climate in Poland during the Tertiary and the Quaternary', *Czasopismo Geogr.* **28**, 131–70.

WOLDSTEDT, P. (1954) 'Die Klimakurve des Tertiärs und Quartärs in Mitteleuropa', *Eiszeitalter und Gegenwart*, **4–5**, 5–9.

WUNDT, W. (1948) 'Eiszeiten und Warmzeiten in der Erdgeschichte', *Deut. Geographentag*, Munich, Amt für Landeskunde, **27**, 114–19.

On the influence of palaeoclimatic oscillations on relief

For examples one should also refer to the various volumes of this treatise devoted to the several morphoclimatic zones in which there will always be a chapter about the role of climatic oscillations. Here we will only list references concerned with specific general problems.

The essential factors concerning the problem of *peneplanation and pediplanation* may be found in the following publications:

BAULIG, H. (1952) 'Surfaces d'applanissement', *Ann. Géogr.* **61**, 161–83; 245–62.
Describes the conditions which lead to the realisation of planation surfaces under various climates. Recognizes, like ourselves, the influence of biotic palaeodeserts.

BAULIG, H. (1957) 'Les méthodes de la géomorphologie d'après M. Pierre Birot', second article, *Ann. Géogr.* **66**, 221–35.

BÜDEL, J. (1957) 'Die Flächenbildung in den feuchten Tropen und die Rolle fossiler solcher Flächen in anderen Klimazonen', *Deut. Geographentag*, Würzburg, pp. 89–121.

BÜDEL, J. (1957) 'Die "doppelten Einebenungsflächen" in den feuchten Tropen', *Z. Geomorph.*, n.s. **1**, 201–28.

CAILLEUX, A. (1950) 'Ecoulements liquides en nappes et applanissements', *Rev. Géomorph. Dyn.* **1**, 243–70.
Clearly poses the problem of the relationship of climate and peneplanation.

COTTON, C. A. (1961) 'The theory of savanna planation', *Geography*, **46**, 89–101.

DRESCH, J. (1947) 'Pénéplaines africaines', *Ann. Géogr.* **56**, 125–30.

DREYFUSS, M. (1960) 'Sur quelques problèmes de géomorphologie dynamique', *Ann. Sc. Univ. Besançon*, 2ᵉ sér. Géol. **12**, 93–106.

DYLIK, J. (1957) 'Tentative comparison of planation surfaces occurring under warm and under cold semiarid conditions', *Biul. Perygl.* **5**, 175–86.

HOLMES, C. D. (1955) 'Geomorphic development in humid and arid regions: a synthesis', *Amer. J. Sc.* **253**, 377–90.

KING, L. C. (1950) 'The study of the world's plainlands: a new approach in geomorphology', *Q. J. Geol. Soc. London*, **106**, 101–31.

KING, L. C. (1953) 'Canons of landscape evolution', *Bull. Geol. Soc. Amer.* **64**, 721–52.
These two publications by L. C. King give a general view of the position of the author, his critical account of the Davisian theory of peneplanation, and the bases of his concept of pediplanation.

LOUIS, H. (1957) 'The Davisian cycle of erosion and climatic geomorphology', *Proc. Int. Geogr. Union, reg. conf. Japan*, pp. 164–6.

LOUIS, H. (1957) 'Rumpfflächen Problem, Erosionszyklus und Klimageomorphologie', *Machatschek Festschrift*, pp. 9–26.

MELTON, M. A. (1958) 'Correlation structure of morphometric properties of drainage systems and their controlling agents', *J. Geol.* **66**, 442–60.

MICHEL, P. (1959) 'L'évolution géomorphologique des bassins du Sénégal et de la Gambie, ses rapports avec la prospection minière', *Rev. Géomorph. Dyn.* **10**, 117–43.

PIOTROVSKII, M. (1948) 'Concerning the theory of the fluvial cycle of erosion', *Proc. Inst. Geogr. USSR Akad. Sc.* **39**, 119–33.

RUHE, R. V. (1954) *Erosion Surfaces of Central African Interior High Plateaus*, Publ. INEAC (Yangambi), Sér. Sc., 59, 40 p.

Morphoclimatic equilibrium and interruptions in the morphoclimatic equilibrium

RUHE, R. V. (1956) *Landscape Evolution in the High Ituri, Belgian Congo*, Publ. INEAC, Sér. Sc., no. 66, 91 p.

TRICART, J. and CAILLEUX, A. (1952) 'Causes actuelles et causes anciennes dans la genèse des pénéplaines', *Proc. Int. Geogr. Congr., Washington*, pp. 396–9.

WISSMANN, H. VON (1951) *Über seitliche Erosion. Beiträge zu ihrer Beobachtung, Theorie und Systematik im Gesamtaushalt fluviatiler Formenbildung*, Bonn, Dümmler, 71 p.

Discusses the problem of planation by lateral erosion. Extensive bibliography to which we refer.

On the effect of man-induced breaches of the morphoclimatic equilibrium

Fundamental ideas and examples are found in the following publications voluntarily restricted in number:

General works

ARMAND, D. L. (1955) 'Water erosion and its importance for us', *Priroda*, no. 6, pp. 31–41.

AUSTIN, R. and BAISINGER, D. (1955) 'Some effects of burning on forest soils of western Oregon and Washington', *J. Forestry*, **3**, 275–80.

BAILEY, R. W. (1939) 'A new epicycle of erosion', *J. Forestry*, **35**, 999.

BAILEY, R. W. (1941) 'Land erosion—normal and accelerated in the semiarid West', *Trans. Amer. Geophys. Union*, pp. 240–50.

Basic study posing the problem of the rate of anthropic erosion.

BENNETT, H. H. (1955) *Elements of soil conservation*, New York, McGraw-Hill, 358 p.

BORCHERT, H. (1961) 'Einfluss der Bodenerosion auf die Bodenstruktur und Methoden ihrer Kennzeichnung', *Geol. Jahrb.* **78**, 439–502.

BOUDY, P. (1951) 'Considérations sur le problème de la défense et de la restauration des sols (Etats-Unis, Afr. du Nord)', *Soc. Forest. Franche-Comté*, pp. 174–211.

DEMONTZEV, P. (1882) *Traité pratique du reboisement et du gazonnement des montagnes*, Paris, Rothschild, vol. 1.

Shows the remarkable conclusions at which nineteenth century French foresters had arrived and to what degree contemporary theoretical and abstract geomorphology was behind in this matter.

FOURNIER, F. (1949) 'Les facteurs climatiques de l'érosion du sol', *Bull. Ass. Géogr. Franç.*, pp. 97–103.

FOURNIER, F. (1960) *Climat et érosion*, Paris, PUF, 201 p.

HEMPEL, L. (1957) 'Soil erosion and water runoff on open ground and underneath wood', *Int. Union Geod. Geophys. Int. Ass. Sc. Hydrol.*, Toronto meeting, vol. 1, 108–14.

KITTLER, A. G. (1955) 'Merkmale, Verbreitung und Ausmass der Bodenerosion', *Peterm. Geogr. Mitt.* **99**, 269–73.

KURON, H. (1953) 'Berücksichtigung des Bodenschutzes bei Beratung und Umlegung', *Mitt. Raumforsch*, Bonn, **20**, pp. 1–14.

MARLIÈRE, R. (1951) 'Terrils en marche', *Publ. Ass. Ingénieurs Fac. Polytechn. Mons*, Belgium, Mons.

MENSCHING, H. (1951) 'Une accumulation post-glaciaire provoquée par des défrichements', *Rev. Géomorph. Dyn.* **2**, 145–56.

Study of the silt deposits of German valley bottoms caused by anthropic erosion.

MISZCZAK, A. (1960) 'Amalgamation of landholdings—an agent of increasing soil erosion', *Czasopismo Geogr.* **31**, 179–90.

PACKER, P. E. (1953) 'Effect of trampling disturbance on watershed condition, runoff and erosion', *J. Forestry*, **51**, 28–31.

Erosion caused by forest exploitation.

PONCET, J. (1963) 'Défense des sols et restauration', *La Pensée*, **111**, 43–54.

PONCET, J. (no date) 'Les rapports entre les modes d'exploitation agricole et l'érosion des sols en Tunisie', *Secr. Agric., Et. et Mém.* no. 2, Tunis, 173 p., 23 photogr. pl.

QUANTIN, P. and COMBEAU, A. (1962) 'Relation entre érosion et stabilité structurale du sol', *C.R. Acad. Sc.* **254**, 1855–7.

SMITH, D. D. (1957) 'Factors affecting rainfall erosion and their evaluation', *Int. Union Geod. Geophys., Int. Ass. Sc. Hydrol.*, Toronto meeting, vol. 1, 97–107.

STALLINGS, J. H. (1957) *Soil Conservation*, New York, Prentice-Hall, 575 p.

STRAHLER, A. N. (1956) 'The nature of induced erosion and aggradation', in W. L. Jr. (Ed.) Thomas, *Man's Role in Changing the Face of the Earth*, University of Chicago Press.

TRICART, J. (1953) 'La géomorphologie et les hommes', *Rev. Géomorph. Dyn.* **4**, 153–6.

TRICART, J. (1962) *L'Epiderme de la terre, esquisse d'une géomorphologie appliquée*, Paris, Masson, Coll. Evolution des Sciences, 167 p.

VOGT, J. (1956) 'Culture sur brûlis et érosion des sols', *Bull. Sect. Géogr., Comité Trav. Hist. et Sc.* **69**, 339–42.

VOGT, J. (1957a) 'La dégradation des terroirs lorrains au milieu du XVIIIᵉ siècle', *Bull. Comité Trav. Hist. et Sc., Sect. Géogr.* **70**, 111–16.

VOGT, J. (1957b) 'Protection des sols et modes de tenure dans l'agriculture ancienne', *Bull. Sect. Géogr., Comité Trav. Hist. et Sc.* **70**, 117–29.

ZVONKOVA, T. V. (1959) *Etude du relief à des fins practiques*, Moscow, Geografguiz, 304 p. (French trans. by SIG, Bur. Rech. Géol. Min., no. 3188).

Monographs listed by way of examples

'Research studies on soil erosion in Poland', *Wiadomosci Inst. Melioracji i Uz Ziel*, **1** (1960), 11–26.

AUBERT, G. and MONJAUZE, A. (1946) 'Observations sur quelques sols de l'Oranie nord-occidentale. Influence du déboisement, de l'érosion sur leur évolution', *C.R. Somm. Soc. Biogréogr.* **23**, 44–51.

BEAUDET, G. (1962) 'Types d'évolution actuelle des versants dans le Rif occidental', *Rev. Géogr. Maroc*, **1–2**, 41–6.

BENCHETRIT, M. (1954) 'L'érosion accélérée dans les chaînes telliennes d'Oranie', *Rev. Géomorph. Dyn.* **5**, 145–67.

BENNETT, H. H. (1960) 'Soil erosion in Spain', *Geogr. Rev.* **50**, 59–72.

BESAIRIE, H. (1948) 'Deux examples d'érosion accélérée à Madagascar', *Conf. Afr. des sols*, Goma, communic. no. 21.

BIRD, E. C. F. (1963) 'The physiography of the Gippsland Lakes, Australia', *Z. Geomorph.*, n.s. **7**, 232–45.

BOTELHO DA COSTA, S. V. and LOBO AZEVEDO, A. 'Aspectos da erosao do solo em Angola', *Agros*, **33**, no. 1–2, 15–22.

BRUNET, R. (1957) 'L'érosion accélérée dans le terrefort toulousin (1ᵉʳ examen)', *Rev. Géomorph. Dyn.* **8**, no. 3–4, pp. 33–40.

CALEMBERT, L. 'Le sous-sol. Etude de l'influence des facteurs géologiques et miniers sur les déformations du sol de la région liégeoise. Plan d'aménagement de la région liégeoise', I, *l'Enquête*, pp. 57–76.

CAVAILLE, A. (1958) 'Sols et séquences de sols en coteaux de Terrefort aquitain', *Ass. Franç. Etude Sol*, no. 94, pp. 4–22.

COOK, S. F. (1949) 'Soil Erosion and Population in Central Mexico', *Ibero-Americana*, Berkeley, no. 34, 86 p.

FOOD AND AGRICULTURE ORGANISATION (1954) 'Soil erosion survey of Latin America', *J. Soil Water Conserv.* no. 4, 158–68.

FLEGEL, R. (1958) 'Die Verbreitung der Bodenerosion in der Deutschen Demokratischen Republik', *Bodenkunde und Bodenkultur*, no. 6, 104 p., 1 map.

FLOHR, E. F. (1962) 'Bodenzerstörungen durch Frühjahrsstarkregen im nordöstlichen Niedersachsen', *Göttinger Geogr. Abh.*, no. 28, 119 p. 29 photos, 5 maps.

GEGENWART, W. (1952) 'Die ergiebigen Stark- und Dauerregen im Rhein-Main Gebiet und die Gefährdung der landwirtschaftlichen Nutzflächen durch die Bodenzerstörung', *Rhein-Main Forsch.* no. 36, 52 p. 17 maps, 2 pl.

GEGENWART, W. and RUPPERT, K. (1955) 'Ein Versuch zur Feststellung der winterlichen Bodenzerstörung', *Peterm. Geogr. Mitt.* **99**, pp. 21–3.

GREENHALL, A. F. (1951) 'Soil erosion and its control in the south half of the North Island of New Zealand', *Int. Union Geod. Geophys., Int. Ass. Sc. Hydrol.* Brussels meeting, vol. 2, 106–14.

GROSSE, B. (1953) 'Untersuchungen über die Winderosion in Niedersachsen', *Mitt. Inst. Raumforsch.* Bonn, **20**, 137–45.

GROSSE, B. (1955) 'Die Bodenerosion in Westdeutschland, Ergebnisse einiger Kartierungen', *Mitt. Inst. Raumforsch.* Bonn, **11**, 35 p. 5 sep. maps.

HAASE, H. (1948) 'Ergebnisse der Abholzung des Harzes auf fliessendes Wasser und Erosion', *Gas- und Wasserwirtschaft*, **89**, no. 9, 265–9.

HEMPEL, L. (1956) 'Über Alter und Herkunfstgebiet von Auelehm im Leinetal', *Eiszeitalter und Gegenwart*, **7**, 35–42.

JONES, F. O., EMBODY, D. R., and PETERSON, W. L. (1961) 'Landslides along the Columbia River Valley, northeastern Washington', *USGS, Prof. Pap.* 367, 98 p.

KAYSER, B. (1961) 'Recherches sur les sols et l'érosion en Italie méridionale, Lucanie', Paris, SEDES, 127 p.

KUHN, W. (1953) 'Hecken, Terrassen und Bodenzerstörung im hohen Vogelsberg', *Rhein-Mainische Forsch.* no. 39.

KURON, H. (1954) 'Ergebnisse 15-jährigen Untersuchungen über Bodenerosion durch Wasser in Deutschland', *Int. Union Geod. Geophys., Int. Ass. Sc. Hydrol.*, Rome meeting, vol. 1, 220–7.

KURON, H. (1954) 'Landwirtschaft und Bodenerosion Untersuchungen typischer Schadensgebiete. I, der Rossbacherhof bei Erbach im Odenwald', *Mitt. Inst. Raumforsch.*, Bonn, no. 23, 48 p.

LISSITCHEK, E. N. (1950) 'Analyse quantitative des processus actuels d'érosion dans les régions semi-sèches du Tian-Chan', *Probl. Phys. Geogr., USSR Akad. Sc.*, **16**, 191–2 (French trans. by BRGM, no. 1312).

MAACK, R. (1956) 'Über Waldverwüstung und Bodenerosion im Staate Parana', *Erde*, pp. 191–228.

MARKER, M. E. (1959) 'Soil erosion in relation to the development of landforms in the Dundas area of western Victoria, Australia', *Proc. Roy. Soc. Victoria*, **71**, 125–36.

MARTHELOT, P. (1957) 'L'érosion dans la montagne Kroumir', *Rev. Géogr. Alpine*, **45**, 273–88.

MENSCHING, H. (1957) 'Soil erosion and formation of haugh-loam in Germany', *Int. Union Geod. Geophys., Int. Ass. Sc. Hydrol.*, Toronto meeting, vol. 1, 174–80.

MINTIER, T. (1941) 'Soil erosion in China', *Geogr. Rev.* **31**, pp. 570–90.

MULLER-MINY, H. (1954) 'Bodenabtragung und Erosion im südbergischen Bergland. Ein Beitrag zur Frage der Bodenzerstörung und zur quantitativen Morphologie', *Ber. Deut. Landeskunde*, Remagen, **12**, 277–92.

NICOD, J. (1951) 'Sur le rôle de l'homme dans la dégradation des sols et du tapis végétal en Basse-Provence calcaire', *Rev. Géogr. Alpine*, **39**, no. 4., pp. 709–48.

PITOT, A. and MASSON, H. (1951) 'Quelques données sur la température au cours des feux de brousse aux environs de Dakar', *Bull. IFAN*, **13**, no. 3.

POLOVITSKAIA, M. E. (1953) 'Soil erosion in the United States', *Priroda*, no. 6, pp. 38–43. Excellent summations.

POUQUET, J. (1956) 'Le Plateau du Labé (Guinée française, AOF). Remarques sur le caractère dramatique des phénomèmes d'érosion des sols et sur les remèdes proposés', *Bull. IFAN*, **18**, sér. A, 1–16.

RAYNAL, R. (1957) 'Bodenerosion in Marokko', *Wiss. Z.*, Martin Luther Univ., Halle, Math-Nat. Kl. **6**, 885–94.

RIQUIER, J. (1958) 'Les "lavaka" de Madagascar', *Bull. Soc. Géogr. Aix-Marseille*, **69**, 181–91.

RUXTON, B. and BERRY, L. (1957) 'Weathering of granite and associated erosional features in Hong-Kong', *Bull. Geol. Soc. Amer.* **68**, 1263–92.

SCHMIDT, W. (1952) 'Art und Entwicklung der Bodenerosion in Südrussland', *Mitt. Inst. Raumforsch.*, Bonn, no. 13, 50 p.

SCHMITT, O. (1952) 'Grundlagen und Verbreitung der Bodenzerstörung im Rhein-Mainischen Gebiet mit einer Untersuchung über Bodenzerstörung durch starkregen im Vorspessart', *Rhein-Mainische Forsch.* no. 33, 130 p.

SCHULTZE, J. (1952) 'Bodenerosion in Thüringen, Wessen, Stärke, Abwehrmöglichkeiten', *Peterm. Geogr. Mitt. Ergänzungsheft*, **247**, 186 p.

SCHULTZE, J. (1953) 'Neuere theoretische und praktische Ergebnisse der Bodenerosionsforschung in Deutschland', *Forsch. und Fortschritte*, **27**, no. 1, pp. 1–7.

SEGALEN, P. (1948) 'L'érosion des sols à Madagascar', *Conf. Afr. des sols, Goma*, pp. 1127–37.

STEINMETZ, H. J. (1956) *Ausmass des Bodenabtrages in einem Teilgebiet der Wetteran und Vorschläge zur Verhütung der durch ihn entstandenen Schäden*, Hessischen Ministerpräsidenten, Wiesbaden, 105 p. 10 sep. maps.

THORARINSSON, S. (1962) 'L'érosion éolienne en Islande à la lumière des études téphrochronologiques', *Rev. Géomorph. Dyn.* **13**, 107–24.

TIKHONOV, A. V. (1959) 'Complex problems of soil conservation in the Volga heights', *Izv. Akad. Nauk. SSSR, Geogr. ser.* no. 4, pp. 55–66.

TIVY, J (1957) 'Influence des facteurs biologiques sur l'érosion dans les Southern Uplands écossais', *Rev. Géomorph. Dyn.* no. 12, 9–19.

TRICART, J. (1953) 'Erosion naturelle et érosion anthropogène à Madagascar', *Rev. Géomorph. Dyn.* **4**, 225–30.

UNO (1954) 'Soil erosion survey in Latin America, II: South America', *J. Soil Water Conserv.* **9**, 214–19, 223–9, and 237.

VOGT, J. (1953) 'Erosion des sols et techniques de culture en climat tempéré maritime de transition (France et Allemagne)', *Rev. Géomorph. Dyn.* **4**, 157–83.

WALLWORK, K. L. (1956) 'Subsidence in the Mid-Cheshire industrial area', *Geogr. J.* **122**, 40–53.

WOLFF, W. (1950–51) 'Bodenerosion in Deutschland', *Erde*, Heft 3–4, pp. 215–28.

5

The major morphoclimatic zones of the earth

The delimitation of the morphoclimatic zones of the landmasses must, of course, first be based on presently existing phenomena. But it should not neglect palaeoclimatic influences, for relict landforms are often numerous, and through failure of taking them into account one risks an incomplete understanding of the relief and an erroneous interpretation of its origin. We examine the criteria for a division of the land areas of the earth into morphoclimatic zones, and briefly indicate the main characteristics and extension of each.

Criteria for the division

For three reasons the division of the land masses into morphoclimatic zones presents certain problems:

1. There are not enough systematic studies dealing with this question. Attempts to classify morphoclimatic types have been made, but they have not been accompanied by precise maps. In many cases the limits ought to be charted directly in the field, but this has not been done. For example, the exact extent of the various morphogenic systems of central Asia, South America, or even Africa is inadequately known. Particular problems of delimitation should be studied and mapped in the field prior to making even small scale maps. Proceeding in this way we have noted important differences in the morphogenic systems of the Sahelian zone and the Sudanese savannas of West Africa, where, apparently, these differences had not been clearly distinguished before. Broad interpretations such as those of Tricart (1958) on the eastern seaboard of Brazil defining and mapping the present morphogenic systems as well as recognising the part played by palaeoclimatic oscillations are still exceptional. But it is precisely such studies that are needed to delimit the major morphoclimatic zones of the earth.

2. Because of both the direct and the indirect influence of climate on relief, we cannot base a morphoclimatic classification of the landmasses exclusively on climatic criteria, even if conceived in an ecological sense, as are Thornthwaite's climatic divisions; neither can it be based on a map of plant

formations although plant formations may have common boundaries with morphoclimatic divisions. Climate and plant formations are also classified into zones, but their number and extent, here too, do not necessarily coincide. Similarly there do not exist thirty-two different morphogenic systems, each one coinciding with one of Thornthwaite's thirty-two different types of climate; their number, like that of the plant formations, is definitely less. One should therefore not adopt on the morphoclimatic plane the differentiation that exists on the biogeographic plane between the pine and the hardwood forests of the southeast of the United States. And conversely we cannot group together the 'orchard steppes' of the Sahel with the steppes of the Magreb or those of central Asia; neither can the Scandinavian coniferous forests be lumped together with the boreal forest of central Siberia. A map of morphoclimatic zones must show originality in so far as the morphoclimatic facts constitute a particular category of natural phenomena. This explains the serious consequences of the want of research in this domain.

3. The frequent importance of subtle variations and of transitional elements is still another difficulty in the morphoclimatic division of the earth. Morphoclimatic limits are sharp only where there are marked orographic contrasts. For example, in South America there is a clear contrast between the arid Pacific coast and the Amazon forest, separated by the montane Andean morphogenic system with different morphoclimatic characteristics. But often transition zones are spread over hundreds, even thousands, of kilometres. Divisions on each side of a narrow line are artificial. This is the case, for example, in central Europe with the transition from a maritime region with mild winters to a continental region with very cold winters. It is also the case in the Arctic in the transition between the barren grounds and the tundra and between the tundra and the forest. There are imbrications, enclaves and exclaves, depending on exposure, lithology, and human and animal depredations.

In the present circumstances, therefore, it is not possible to make a satisfactory map of the morphoclimatic zones of the earth, but only a provisional sketch. One may divide the earth into several major morphoclimatic zones, but their subdivisions eventually will suffer modifications because of the progress of geographic knowledge. The most advanced attempt at such a classification is that of Büdel (1948) who distinguishes the following divisions and subdivisions:

1. *Submarine forms:* ocean and sea floors
2. *Subglacial forms:* the glacial zone
   Subaerial forms, namely:
3. Zone of frost splitting (*Frostschuttzone*)
4. Tundra zone
5. Zone of mature (*in situ*) non-tropical soils (*Nichttropische Ortsbodenzone*)
   (*a*) maritime temperate
   (*b*) subpolar without permafrost

    (*c*) subpolar with permafrost
    (*d*) continental
    (*e*) steppes
6. Mediterranean transitional zone (*Etesische Ubergangszone*)
7. Arid zone of rock waste (*Trockenschuttzone*)
    (*a*) tropical deserts and semideserts
    (*b*) midlatitude deserts
    (*c*) highland deserts
8. Zone of sheetwash (*Flächenspülzone*)
    (*a*) tropical
    (*b*) subtropical
9. Zone of mature (*in situ*) equatorial soils (*Innertropische Ortsbodenzone*)

This classification is not consistent because it puts submarine relief, which is azonal, on the same footing as subglacial relief and the various morphoclimatic zones of subaerial relief. It does not take into account, except once (7*c*), the special extrazonal conditions of highlands. It includes subdivisions that are difficult to justify; for example, permafrost regions instead of being united are spread over three categories (3, 4, and 5*c*) whereas the Siberian taigas with permafrost are grouped together with the maritime temperate forests and the steppes in one and the same group (5). On the other hand the 'Mediterranean' regions are considered to be a separate zone with its own originality; but there are in fact fewer differences between the degraded 'Mediterranean' regions and the steppes, or between the 'Mediterranean' regions which remained forested and the midlatitude forests, than there are between the forested regions with a maritime temperate climate (5*a*) and the 'zone of mature (*in situ*) non-tropical soils with permafrost' (5*c*). Lastly, there is an inconsistency as far as the 'zone of sheetwash' is concerned because, no less than the midlatitude and equatorial zones, it is also a zone of mature (*in situ*) soils (*Ortsbodenzone*).

We therefore cannot adopt Büdel's classification, in spite of its interest. The classification we propose is based on the following two types of criteria:

1. the major climatic and biogeographic zones, which provide the principal divisions;
2. certain climatic or biogeographic subdivisions and palaeoclimatic factors, as a basis for the morphoclimatic subdivisions within each of the major preceding zones.

We then distinguish the following zones (applicable only to areas of low elevation where there is no vertical zonation):
1. The *cold zone*, characterised by the predominant importance of frost. Depending on the nature of the latter, this zone may be subdivided into:
    (*a*) a glacial zone where runoff is mainly in the solid form of glaciers;
    (*b*) a periglacial zone with seasonal liquid runoff, but where the formation of ground ice plays the predominant role in the development of interfluves.

2. The *forested midlatitude zone*, more or less profoundly transformed by man and where relict forms, principally Pleistocene glacial and periglacial forms, play a considerable role. The variable importance of winter frost (whose geomorphic role is considerable), together with the influence of palaeoclimates, determines the following morphoclimatic subdivisions:
   (*a*) a maritime zone with mild winters, a minor role of frost, and important Pleistocene glacial and periglacial relict forms;
   (*b*) a continental zone with very cold winters, a high influence of present and Pleistocene frost, which may include the preservation of Pleistocene permafrost;
   (*c*) a 'Mediterranean' zone, with dry summers, where the influence of Quaternary periglacial relict forms is minor.

3. The *dry* (arid and semiarid) *zone* of the middle and lower latitudes. It is characterised by a meagre vegetation of steppes, xerophytic brush and desert, and an intermittent, local runoff. Two sets of subdivisions must be recognised:
   (*a*) one based on the degree of drought, that is on the intensity of the fundamental criterion, which leads into a division of steppes, xerophytic brush and desert;
   (*b*) the other based on winter temperature, which is at the origin of certain important processes (frost-splitting, duration of the snow cover); it leads into a division of midlatitude and tropical–subtropical regions.

4. The *humid tropical zone (humid tropics)*, where temperatures are always high and moisture abundant enough to allow stream flow. Depending on the seasonal distribution of rainfall, its annual total, and the density of the plant cover, it may be subdivided into savannas and forests.
   (*a*) The savannas (wet–dry tropics) have a seasonal rainfall and a moderate plant cover subject to brush fires kindled by man. They are characterised by considerable overland flow and intense chemical weathering during the rainy season. The campos cerrados and deciduous seasonal forests correspond to the savannas in regions which have not been periodically ravaged by brush fires for centuries.
   (*b*) The forests (wet tropics) receive rain throughout most of the year and have an imposing plant cover not subject to fire. The vegetation is most exuberant here, and chemical and biochemical weathering are at their peak. A minor distinction should be made between the evergreen forest and the semi-evergreen seasonal forest characterised by a partial seasonal leaf fall which influences runoff.

With our present knowledge any region of the earth well known from the geomorphological point of view may be included in one or another of the above mentioned zones. Some of them, however, may be modified in the future as our knowledge progresses (opposition of insular and continental climates; possible role of relict forms in the realm of the tropical forest . . .).

We will now treat the characteristics, extent, and subdivisions of each of the zones here distinguished.

The cold zone

Frost is the determining climatic factor in the cold zone; it acts either directly (e.g. by preventing vegetation or by producing contraction cracks in the soil) or through the transformation of water into ice. Frost is a major morphogenic agent; it not only produces specific processes but profoundly modifies the action of the azonal processes: the actions of waves, wind, and running water. For this reason the subdivisions of the cold zone must be based on the attributes of frost. Thus the forest stops at the 10·5°C (51°F) isotherm for the warmest month; notwithstanding very cold winters (Siberia, Canada) its growth is favoured by continental climates.

The *glacial zone* exists in those regions of the earth where there is insufficient heat to permit the thawing of snow from one year to the next: the precipitation accumulates in solid form and produces runoff in the form of ice. The result is a distinctive sculpturing of the subglacial relief under conditions which differ greatly from those of subaerial morphology. Taking into account the peculiar mechanical laws of this sculpturing, the study of the zonal and extrazonal phenomena of high latitude and high altitude glaciers may to a certain extent be compared. Moraines, roches moutonnées, and striations related to analogous mechanical conditions are attributes of both types. But the similarity is not complete; considerable differences separate high latitude glaciers from high altitude glaciers. First there is the all important difference in scale: ice sheets extend over millions of square kilometres, whereas mountain glaciers only cover a few dozen or hundreds of square kilometres. The thickness of the first are of the order of several thousands of metres; that of the second of a few tens or hundreds of metres at the maximum. The climatic conditions are also different. Ice sheets are located in rather dry and very cold regions; the greater part of the ice has a temperature of several degrees or of tens of degrees below zero Centigrade. The mountain glaciers of the midlatitudes or the tropics are much better alimented, and their ice, at the base, is close to the melting point. Their conditions vary considerably with latitude: within the tropics, ablation, like temperature, follows an essentially daily rhythm: the snowline does not fluctuate throughout the year. Within the midlatitudes, on the contrary, the rhythm of ablation and the snowline are subject to important seasonal variations. During the summer the surface of the glaciers is wet, and the water which runs down from it plays an important morphologic role in the ice and on the bottom. Many errors have been committed in glacial geomorphology through extrapolation to all glaciers of observations which are valid only on the mountain glaciers of the midlatitudes. In the latter, the massive liberation of meltwaters facilitates the removal at the extremity of the glacier of the debris entrained by the ice. Frontal moraines mainly develop during phases of glacial advance, whereas fluvioglacial deposits accumulate during phases of retreat. On glaciers situated close to the Equator, on the contrary, in the Venezuelan Andes, for example, fluvioglacial deposits are rather limited and enormous terminal moraines accumulate at the extremities of even small glacial tongues. Terminal 'push moraines', formed by frozen materials, do not exist at the extremities of midlatitude mountain glaciers but are, on the contrary, frequent on the margins of ice sheets. Vertical and latitudinal zonation thus introduce marked differences in the glacial realm.

The limits of the glacial zone are clear: they coincide with the limits of glaciers. They are not strictly climatic for ice flow has its own inertia; except in Antarctica, glacial tongues extend beyond regions of permanent snow, protruding more or less according to the mass of moving ice, the slope gradient, and the seasonal variations of climate. Inversely, periglacial enclaves exist in the midst of icesheets in the form of peaks, called nunataks, emerging from the ice and too abrupt to be covered by snow, or in the Antarctic in the form of dry valleys where there is so little snowfall that it is swept by the wind or melted and evaporated in the course of a very brief summer. The limits of the glacial zone do not exactly correspond to the present climate because of ice surviving from the Pleistocene. At a depth of 100 to 200 m (330 to 600 ft) the temperature of the Greenland ice, far from increasing, decreases several degrees Centigrade, which indicates an imbalance with the present climate. If the ice were to disappear, the rock bottom would be below the snow-line and the ice-sheet would probably not form again.

The *periglacial zone* is characterised by the action of freezing temperatures on the ground, by periodically freezing weather and by more rapid freeze–thaw alternations; furthermore the snow cover does not persist throughout the year. Nearly everywhere thaw generates a liquid runoff in the form of streams which combine with the predominant influence of freeze–thaw alternations in the sculpturing of the relief. The two principal geomorphic mechanisms are frost-splitting of rock outcrops and geliturbation of friable materials, the latter producing solifluction or gelifluction on humid slopes sufficiently rich in clay.

The periglacial zone includes several subdivisions. They are based on three different criteria: (1) the periodicity of frost as a function of the present climate and to a certain extent of past climate; (2) the resistance offered by the vegetation; and (3) the total precipitation.

The periodicity of frost varies according to climate. C. Troll has distinguished two principal types: daily freeze–thaw alternations and seasonal alternations. The daily rhythm, which is also common in tropical highlands, occurs in certain maritime climates, as those of Iceland or the Kerguelen Islands; it affects only a small part of the landmasses. The seasonal rhythm is most widespread (Greenland, Alaska, Canada, Siberia, and the midlatitude mountains). In truth, it is more important to distinguish between differences in the duration of *ground* frost, whether daily, seasonal or permanent. They, in fact, determine the morphogenic processes. Whereas a few days of frost only affect a thin ground layer, seasonal frost influences a depth of several decimetres, or perhaps one or two metres (up to 6 ft); as to permafrost, it completely modifies the conditions of runoff, thereby producing a maximum morphogenic influence. This distinction also takes into account the role of the snow cover, which, when it is thick and durable enough, is an obstacle to a large number of periglacial morphogenic actions. Many areas with heavy snowfall and very cold winters are free of permafrost. Permafrost only occurs in regions with a mean annual temperature below $-2°$ or $-3°$C ($28·4$ or $26·6°$F) but which do not have glaciers, either because there is not enough precipitation (dry valleys of Antarctica), or, more often, because of its distribution: falling during summer, it occurs in the form of rain and not snow (continental climates of Canada and Siberia). Permafrost furthermore has a high inertia and may be residual from the Pleistocene.

The resistance offered by vegetation varies according to its own characteristics, which, in turn, are related to climate. Here appropriate use can be made of Büdel's distinction between frost desert (barren grounds), where mechanical actions operate without hindrance, and tundra. In the latter there are soils, however thin and little evolved and, above all, bogs which react to frost in a different way than do mineral formations. Special forms develop on them as in string bogs. But as a whole tundra impedes geliturbation and gelifluction and is characterised by distinctive microforms such as gelifluction lobes, tundra hummocks, and palsen. The same is true of mountain meadows. Lastly, vegetation is an obstacle to the wind; the latter, on the contrary, plays a notable role in the geomorphology of the barren grounds.

On the basis of climatic, biotic, and palaeoclimatic criteria, one may distinguish the following five provinces in the periglacial morphoclimatic zone:

1. A *hyperperiglacial province* as exemplified by the barren land areas of Antarctica and Peary Land. Cold is intense; positive temperatures are exceptional in the atmosphere, more frequent on the ground (up to $33°$C: $93°$F). Frost-weathering is limited by drought but helped by high daily temperature ranges (up to $50°$C: $122°$F on the ground). Screes are less developed than in midlatitude mountains. Gelifluction is non-existent or minor, enabling contraction cracks to exist on slopes of $30°$. Wind action is the principal agent of transport, mainly producing nivaeolian cover sands; silts are blown out to sea. Snowfall and runoff are minor.

2. A *mesoperiglacial province* which includes all the remaining barren grounds, mainly those of North America and Eurasia. Permafrost is almost general except in Europe; summer thaw is important, but plant life is negligible. The processes of frost-weathering, geliturbation, solifluction or gelifluction are very active. Most of the province is dry and continental in climate, with intense winter cold, foggy summers, a single freeze–thaw cycle (at a depth of a few centimetres), and minor aeolian activity. Patterned

ground, gelifluction slopes, and boulder streams are common. Altiplanation terraces also occur. Streams characterised by torrential floods producing ice jams are important morphogenic agents. Alluvial lowlands are characterised by cryokarst, thaw lakes (some oriented), giant polygons, and pingos. Part of the province is exposed to maritime influences such as parts of Spitzbergen. This influence is weakened, however, by the presence of pack-ice during winter. Freeze–thaw alternations are numerous, producing extensive screes. Snowfall is important, producing avalanche chutes and meltwater rills.

3. A *tundra province* in which vegetation plays an antagonistic role to runoff, to the development of a thick active layer, to gelifluction, and to the effects of the wind (the main effect is indirect: no tundra on exposed sites). Tundra hummocks and gelifluction lobes are common. In areas of little snowfall permafrost is deep, especially in formerly unglaciated areas (Alaska), and the tundra is characterised by ice-wedge polygons, cryokarst, oriented lakes, and pingos. In formerly glaciated areas with heavy snowfall (eastern Canada) there is sporadic permafrost and much frost-weathering mainly in connection with nivation phenomena. An oceanic environment, such as that of the Kerguelen Islands, is characterised by many freeze–thaw cycles, much frost-weathering, an absence of permafrost, solifluction (rather than gelifluction), and considerable creep due to needle-ice. In southwest Iceland stream flow is also important.

4. A *steppe periglacial province* characterised by the importance of wind action producing loess. Frost-weathering, favoured by a sparse plant cover, is reduced by drought. This province is transitional between the grass steppes of the dry zone which reach the zone of permafrost and the periglacial zone in Alberta and Mongolia. It is also found in northeast Iceland (without permafrost), well known for its aeolian features.

5. A *taigan province* on residual (Pleistocene) permafrost. Gelifluction seems to be limited by the rapidity of thawing in the spring. After a few weeks the thickness of the active layer permits a subsurface drainage which favours the formation of hydrolaccoliths in the fall. There seems to be two varieties in this province, one developed on continuous permafrost, the other on discontinuous or sporadic permafrost. Although a part of the boreal forest, it is preferable to study this environment with the periglacial zone because of the large influence of permafrost on the morphogenic processes.[1]

Mountain regions with cold climates above the tree line are also subject to the morphogenic action of frost and defrost which may generate forms that are in part similar to those produced in the latitudinal periglacial zone. The processes are the same, but the morphogenic system is different as it is highly influenced by altitude and latitude. One cannot, for example, equate the morphogenic conditions of the montane periglacial regions of

[1] The last five paragraphs were written by the translator to update the author's text according to his new book (Tricart, 1967) on the geomorphology of periglacial regions (KdeJ).

East Africa and the Andes of Peru with those of the Alps nor, *a fortiori*, with those of the permafrost regions of Siberia.

The limit of the cold zone and the forested midlatitude zone cannot be traced with precision. Without doubt all the tundras belong to the cold zone. But the tundra-boreal forest limit is irregular and often uncertain. There is interpenetration if looked at in detail. This transitional zone corresponds to Rousseau's periarctic zone; it is very clear in Newfoundland and in the north of Quebec. Frost action remains the predominant morphogenic agent in the sparse forests bordering the tundra. The limit is therefore only an indication: it is a transition zone rather than a true limit. Forests growing on permafrost are extensive enough to be mapped separately. Unfortunately, though their extent is well known in Siberia, where it has been systematically surveyed, the same is not true in North America in spite of an enormous recent effort: they seem to exist at least in Alaska and in the northwest of Canada, but it is impossible to state exactly in what areas. As to mapping the exact limits between the tundra and the barren grounds, one should not expect too much, because, like the forest and the tundras, these two types of environment overlap or merge into each other and often alternate on short distances. It is only in the very high latitudes that the barren grounds are extensive enough to appear on a world map.

The forested midlatitude zone

Between the tundras and the dry zone there is a wide zone occupied in the natural state by forest formations. It is centred on the midlatitudes. As a result of the configuration of the continental masses, it is broadest in the northern hemisphere. It forms a long strip in Eurasia from the Atlantic to Lake Baikal, beyond which it appears again in the Amur Basin, Korea, and Japan. It is interrupted between Lake Baikal and the headwaters of the Amur River because here the taiga on permafrost (partly residual) reaches the steppes of Mongolia. In North America the midlatitude forests cover the eastern half of the continent from Texas to Labrador and from Florida to the Yukon Valley. In the latter area it seems to pass into the taiga on permafrost, but the extent of the permafrost has not been determined. On the west coast the forest extends from northern California to the Alaska Peninsula in a mountainous region. In the southern hemisphere the midlatitude forests only cover restricted areas: the Pacific coast south of Santiago de Chile, again in a mountainous region, the coast of Natal, the eastern coast of Australia all the way to the Tropic of Capricorn, Tasmania, and New Zealand.

In a certain number of regions, especially on the eastern faces of the continents, there are forests that are transitional between the tropical and the midlatitude forests. These correspond to climates that are humid enough to interrupt the dry zone. This is the case in southern China, southern Japan, southeastern United States, and eastern Australia. These *subtropical forests*, characterised by a mixture of species, occupy areas where Pleistocene periglacial actions have only been felt at higher elevations. They are transitional between the wet tropical forest environment and the frankly midlatitude forests of, for example, the northeastern United States, northern Japan, or the Soviet Far East. Because of warm and humid summers there is a deep regolith not very different from that of the wet tropics, as the studies of Lautensach (1950) and Krebs (1966) have shown in Korea, although Korea is less typical than southern Japan.

In the forested midlatitude zone the basic morphoclimatic characteristics consist of combinations of a relatively dense plant cover, opposing important resistance to the mechanical processes, and a moderately high mean temperature, which decreases the intensity of chemical actions. Under natural conditions it is a zone of low morphogenic intensity where the evolution of the relief is markedly slower than in the two contiguous zones, the cold and the dry. Very steep slopes are possible under forest cover, as in the Apuan Alps north of Pisa, in Japan, or in the fjords of British Columbia or southern Chile. The relative abundance of plant debris provided by the forest produces a thick layer of litter because the mineralisation of humus is slow. This litter and a good soil structure considerably decrease overland flow, which is often negligible. It is more important under conifers, which produce an acid humus that flocculates the soil and decreases its resistance. Creep is often more important than overland flow. As a whole, however, this morphoclimatic environment is characterised by a weak intensity of the morphogenic processes, physical and mechanical as well as chemical and biochemical.

Slow morphogenic processes are favourable to an important preservation of Pleistocene palaeoforms. Much of the relief of this zone is relict (i.e. it owes its forms to past processes). The activity of the mechanical processes here depends on the steepness of the slopes and their lithologic composition. Steep slopes are formed by rockfalls and landslides, by lobate solifluction, and by the incision of first order streams. Gentler slopes bear deep soils which reflect the slowness of sheet erosion. The slow downward movement of particles subject to the forces of compaction caused by the removal of matter in solution, the work of burrowing animals, and the effects of trees uprooted by the wind is the main type of mass movement. Biochemical processes are not as rapid as in the tropics: the average temperature is moderate, humus decomposes slowly, persists, and facilitates the migration of colloids. The soil minerals are less weathered than in the humid tropics and the regolith is not as thick (about one metre). The removal of dissolved matter is not as great. In areas of low relief the total amount of solids transported seems to be very small under natural conditions.

There are differences, however, within the forested midlatitude zone; the climates are varied enough to influence the intensity of the individual morphogenic processes. One can therefore divide this forested zone into three subdivisions:

241

(*a*) A *maritime zone*, humid, but lacking a large temperature range and a large seasonal variation of humidity. It is particularly well developed in western Europe from Norway to the Pyrenees, and it penetrates rather far into the interior, reaching Poland. It also exists in British Columbia, southern Chile, Tasmania, and New Zealand. Frost is moderate and does not last. Its penetration into the ground rarely reaches the bedrock. There is not much desiccation as the summers are rainy. Variations in stream flow are moderate, and mechanical processes, whose principal agents are living organisms, limited. Chemical erosion clearly predominates and is active the year round. It is only limited by moderate temperatures, which slow down the chemical reactions. Humus accumulates on the ground and tends to become acid, thus facilitating the chemical attack of certain rocks such as granite.

(*b*) A *continental zone*, which, like the continental climates of the midlatitudes, reaches the east coasts of Asia and North America. It is characterised by a much more contrasted climate: heavy showers, even with small annual precipitation totals, and intense winter cold. Mechanical processes are more important than in the maritime zone: frost penetrates more deeply into the ground and often reaches the bedrock; it influences runoff in a major way during several months of the year. Runoff during periods of thaw is a common phenomenon: it is repeated each spring and even several times during winter, for a span of a few days, as a result of momentary thaws that reach as far as western Siberia. The morphogenic effect of this runoff is considerable, reinforced as it is by the waters of melting snow. It produces sheet erosion and gullying; dissection is quite different from that of the maritime realm. Thunderstorms, though morphologically less important, also contribute to the maintenance of forms produced by runoff due to thaw. Chemical erosion is restricted because of winter frost and the runoff of large quantities of water which escape infiltration; the chemical actions are consequently less active than in the maritime realm; they are replaced by mechanical processes as soon as the slopes are steep enough. The mechanical processes, however, are obstructed by the forests which may occupy rather steep slopes, at least on sufficiently consolidated rocks. These factors partly explain the considerable difference between the erosional landforms of northern Japan or Korea and those of northwest Europe. Slopes are very steep (20° to 35° on the average) and rectilinear. Other differences are caused by palaeoclimates, which we discuss later.

(*c*) A *warm temperate* or *subtropical zone*, most typical in regions of mediterranean climate. Frost at low elevations is uncommon and light; only advanced destruction of the plant cover by man can make it an important morphogenic process; under present natural conditions its action is limited to rock outcrops. The predominant role is played by alternations of drought and humidity that cause variations in the volume of argillaceous soils, thereby increasing the possibility of landslides. The seasonal precipitation also causes streams to have a torrential regime that facilitates the evacuation of coarse debris. Showers are usually violent, which favours runoff in small streams and coincides with a dense drainage net and a more rapid dissection. Erosion, of course, is a function of the total annual precipitation: in exceptionally wet regions, such as the Apuan Alps or the Ligurian Apennines, slopes maintained steep (30° to 40°) by vegetation have been deeply marked by a network of secondary stream courses of an exceptional density, more or less similar in both regions, though one is formed of schists and the other of limestones and marbles. Such intensity of runoff is also found in climates without a marked dry season, such as those of the southern Appalachians, southern Japan, or the east coast of Australia. The action of runoff is favoured in this subtropical zone by the easier decomposition of the humus as a result of higher temperature; the soils are thinner than in the maritime and continental zones (especially the A horizon) and their protective role is reduced. If the A horizon is eroded, the raindrops fall directly on a mineral B horizon of illuviation, which they mobilise. By adaptation to the dry seasons, the forest cover of the mediterranean regions is less dense than that of the cool temperate lands; trees are not as tall, leaves are thick, not as abundant, and often replaced by spines. For these reasons, and especially after the intervention of man, who degrades the forest, the screening effect of the vegetation is decreased and splash erosion increased.

The present natural processes, however, only explain a generally rather limited part of the relief of the forested midlatitude zone. Two other factors play a very important role.

The first is the *influence of palaeoclimates* and is reflected in an important number of relict landforms very different from those which the present processes are developing. In Europe and North America the forested midlatitude zone was invaded by the Pleistocene ice sheets and a large part of its area was sculptured as a glacial bed and littered with moraines and proglacial formations. At its maximum extent the ice extended beyond the Great Lakes in North America and came close to the sites of London, Berlin, Warsaw, and Minsk in Europe. On the ice margins a vast region was subject to a periglacial morphogenic system and gentle slopes (5° to 15°) were formed by the process of solifluction. The regions which presently have a subtropical climate were less or differently affected by the cold Pleistocene climate, namely the Mediterranean region, the southeastern United States, the southern half of Japan, Natal, the North Island of New Zealand, and the east coast of Australia. This difference helps to differentiate these regions from the maritime and continental zones.

The second factor affecting the present landforms is the *influence of man*, reflected in the importance of man-induced morphogenic processes. It has been especially marked in the northern hemisphere, particularly the mediterranean regions and the continental zone, because conditions there are more favourable to the development of the mechanical processes. In the Mediterranean region, culturally developed since ancient times, the degradation of the original forest into maquis and garrigue caused by over-grazing has, with only a few exceptions, completely modified the morphogenic system of the non-cultivated land. In California and Western Australia a speculative economic development has resulted in almost comparable modifications in a few decades. In the continental zone man is responsible for the extension of gullying and the retreat of the forest facing the *cultivated steppe* (*Kultursteppe* of German authors), especially in Russia, the Ukraine, and northern China. In less than a century's time a devastating speculative economy in the United States unleashed a powerful wave of erosion and deposition or entrainment of sediments towards the sea which have profoundly modified the natural landscape.

The combination of palaeoclimatic relict landforms, present morphoclimatic processes, and anthropic effects explains the diversity of relief in the lands of the forested midlatitude zone. In view of this complexity one can understand that certain elements such as man-induced gullying and streams fed by springs have arbitrarily been extracted from this combination and regrouped artificially into a system of so-called normal erosion, which is nothing but a product of the mind.

The dry zone

Equatorward from the forested midlatitude zone and towards the interior of the continents drought causes the gradual disappearance of the forest and its replacement by sparser plant formations that form a lesser screen between the lithosphere and the atmosphere. The forest grades into sclerophyllous bush (maquis, garrigue, chaparral) or the wooded steppe which in turn passes into the grass steppe and the desert. Together with increased drought the precipitation becomes more irregular; from year to year annual totals often vary as much as ten or even a hundredfold. Heavy showers play a predominant role in the total. Thus at Cape Juby, in the Spanish Sahara, 57·3 mm (2·25 in) of rain fell in three consecutive days; whereas at Villa Cisneros 88 mm (3·46 in) fell in a single day in September 1944 in a total of 112 mm (4·41 in) for the year. At Port Etienne, Mauritania, 2 mm (0·08 in) fell in 1912 and 301 mm (11·8 in) in 1913, of which 300 mm fell in a single day. Under such conditions the low mean annual totals do not exclude the morphogenic activity of runoff. On the contrary, although rare, runoff is brutal and all the more effective as the plant cover is sparse and the soil thin. Infiltration is therefore minimal and runoff active. But streams are unlike those of humid regions where the flow of water is for a large part fed by springs. The waters which rush down slopes, gashing the hillsides with gullies, disappear rapidly after showers. Bajadas and pediments dominated by dissected inselbergs are thus formed.

The action of water is combined with that of the wind which profits from drought and the lesser density of the plant cover. Alluvial deposits are dry between floods and subject to intense deflation; the coarser fraction of the material is whipped up to form dunes, whereas the fines are blown into dust clouds and may be transported to distant lands. Occasionally the dust colours the rain several thousands of kilometres away, even as far as Paris where Saharan dust has been recognised.

The relative role of runoff and wind on relief varies according to the aridity of the climate and the nature of the plant cover. Three basic environments may be recognised:

1. The *subhumid steppe* (grass steppes and wooded steppes) forms vast expanses to the north and south of the Sahara, in East Africa, around the Kalahari, in Asia Minor, in central Asia, in Australia, on the High Plains of the United States, on the Canadian prairies, on the Mexican Plateau, and in the Argentine Pampas. Under natural conditions the grass cover effectively slows down the mechanical morphogenic processes. Wind deflation occurs locally and is mainly restricted to stream beds and alluvial aprons on the lee of which dune fields may develop. The principal effect of the wind, however, is the accumulation of loess from a local or distant source. The enormous loess deposits of China probably originated in the deserts of central Asia. Runoff on the desiccated steppe soils often takes the form of sheetwash. Where runoff becomes concentrated it may create a drainage net with vertically incised gullies. The eroded materials are often good loams, as under natural conditions the protection of the plant cover is sufficient to permit a soil to form. But drought impedes leaching, and, even in a hot climate like the Sahelian zone of Africa, the soils are neither ferrallitic nor ferritic but brown and chestnut-brown, and even tropical podzolic soils may develop. Where winter cold interrupts the decomposition of humus, very thick A horizons develop as in the *chernozems* or black earths of the Russian steppes or the American Mid-West.

Steppes may support the cultivation of small grains. To herdsmen accustomed to the hard life of the desert they constitute desirable grazing lands. As a result many steppes have been despoiled; unrestricted browsing has caused the retreat of shrubs and the disappearance of wooded steppes and parklands. Overgrazing and browsing impoverishes the plant cover, which has difficulty in reconstituting itself because of the irregular rainfall. The degraded steppe no longer effectively protects the soil, which is then violently attacked by mechanical processes; the effects of deflation, sheetwash, or stream flow become all the more serious because of the vast extent of the friable loamy soils. Cultivation sometimes causes real catastrophes within a few years, especially if the climatic conditions make speculative farming hazardous. In western United States the deflation of the loamy fraction has caused the residual sand to form dunes in many areas formerly covered by wheat.

The steppes have organised drainage networks integrated with stream courses that reach the sea or interior depressions when tectonically active reliefs form topographic barriers that are difficult to overcome (as a number of streams of the Algerian High Plains or in extensive areas of central Asia).

2. The *semiarid regions* are characterised by a discontinuous steppe vegetation. Rains occur in the form of limited showers producing a purely local runoff. Water flows in the canyons of the highlands but is lost at a greater or lesser distance downstream. In contrast to the subhumid steppes there is no integrated drainage or the net has ceased to be active. It is relict from periods when runoff was more important, largely because the climate was colder (Pleistocene pluvials). Thus the Draa, in Morocco, forms a continuous stream course from the High Atlas to the ocean, but the stream bed is practically never used by the same flood over its entire length. Spates occur either in one part or another and become exhausted before reaching the sea. It is under this type of climate that pediments and inselbergs are best developed. It still rains enough (an average of 240 to 300mm per year: 9·6 to 12in) so that showers are not extraordinary events but occur several times per year. Nevertheless, there is not enough rain to produce an integrated drainage net. Finally the plant cover does not provide adequate protection; the regolith is thin and overland flow abundant. Running water is the main morphogenic process; it is concentrated in the hills and spreads in sheets and rills at their foot, producing a distinctive topography. The wind only plays a subordinate role; it reworks most of the loose materials which have first been mechanically fragmented and later spread by water.

Other, usually ancient, semiarid regions are covered by a vegetation of small trees and shrubs, the result of a protracted adaptation to drought. Examples are the bush and scrub of Australia, the caatinga of northeast Brazil, the xerophytic scrub, often with cacti, of southern Mexico, northern Venezuela, and the dry basins of Colombia and northwestern Peru. These plant formations leave the soil bare between the strongly rooted plants. They are adapted to intense rillwash which causes sheet erosion and prevents the formation of humic soil horizons. Conversely, they impede the concentration of runoff, thus favouring the development of pediplains and inselbergs as is the case in a large part of Australia. The vegetation of semiarid regions, unlike that of the subhumid steppes, is unfavourable to aeolian action.

Semiarid regions constitute a margin of intense morphogenic activity around the arid regions. Because of the action of man they extend at the cost of the steppes.

3. The *arid regions* are, etymologically, areas where there is no runoff. Such regions are very few and their existence is related to particular lithologic conditions. They are always sandy or stony, as the Saharan ergs, and too permeable for runoff to take place. On outcrops of solid rocks there is always sporadic overland flow even in the driest regions. In the centre of the Sahara and in the Libyan Desert, certain stations, such as In-Salah, do not receive rain every year; nevertheless rain is not unknown, and when it falls the wadis are in spate. Outside the ergs, or sandy deserts, the Sahara exhibits a more or less dense network of wadis, occasionally occupied by flowing water.

It is true, however, that even in the most arid regions morphologically effective showers are few; they may be totally absent in certain years. Infrequent or not, their action is the same as in semiarid regions. The evolution of relief is therefore very slow. The wind, apparently spectacular, has little effect: once it has removed the fines (which are all it can move) from the surface deposits, it remains powerless and its action on the ergs is limited to a ceaseless reworking of minor forms without importance. The limits of the large ergs have been fixed for thousands of years, and even the details, to within a few metres, are perhaps more stable than is generally believed.

Foggy deserts along coastal regions made arid by a cold marine current, as in Peru and northern Chile, should be considered separately. The disintegration of rocks through hydration caused by condensation is intense, but fluvial runoff is practically nonexistent. The wind is therefore the principal agent of transport; regular and powerful, it shoves the sand towards the top of mountains as much as 1 000 m (3 300 ft) high.

The distinction between semiarid and arid regions is identical to the one German authors make between *Randwüste* (desert margins) and *Kernwüste* or *Vollarid* (central deserts). It is a useful distinction in geomorphology because of the different intensity of the morphogenic processes, which are more active in semiarid regions than in the neighbouring steppes or deserts. In the latter the morphogenic processes are impeded by the infrequency of rain, which is favourable to the preservation of palaeoforms, many of which have formed under more humid climates such as those of the Pleistocene pluvials. This is particularly the case in the Sahara. Reliefs similar to those of the present desert margins were formed in the central Sahara, whose landforms have often created the impression of having been formed by stronger morphogenic processes than presently exist.

Differentiation of the dry zone into steppes, desert margins, and central deserts based on the structure of the plant cover, which reflects the degree of humidity, must be completed by another climatic factor, namely, variations of temperature.

The action of running water and of wind are indeed influenced by the degree of aridity. But they are only processes of transportation and not of fragmentation, which play a primordial role in the morphogenic system. The processes of fragmentation determine the quantity of material removed from the highlands of outcropping rocks.

Variations of temperature above the freezing point produce a process of fragmentation common to all dry regions. Some authors think that they act directly through expansion or differential contraction; others, indirectly by causing a desiccation which splits the rocks. Whatever it is, the sparse plant cover and the usual dryness of the air together permit a large temperature range on the ground surface. This process is thought to be particularly effective on granite, according to numerous authors, and would explain the remarkable development of pediments and inselbergs in areas composed of

247

these rocks. It would perhaps also explain why the crystalline rocks of the Mauritanian Adrar area have suffered a much more intense erosion than the sandstone plateaus: the Inchiri area and the southeast extremity of the Spanish Sahara have been reduced to pediplains at 100 to 200m elevation (330 to 660ft) studded with small inselbergs, which contrast with the plateaus at 400 to 800 m elevation (1 300 to 2 600 ft) of the sandstone Adrar. On other rocks, especially sedimentary rocks such as shales, it seems that temperature variations above the freezing point are not an active process of fragmentation and are a great deal less effective than frost-weathering. There is, therefore, reason to distinguish between dry regions with cold winters and dry regions without winters. Finally, one should also take into account the moisture regime on the rock surface. In foggy deserts with intense condensation there is a frequent alternation of wetting and drying which because of the mechanism of hydration is a powerful agent of fragmentation. Because of this process, foggy deserts tend to resemble deserts with cold winters.

In dry regions with cold winters fragmentation is most intense, snow being a less fruitful source of humidity than rain, which helps frost-weathering. In Mongolia, Tibet, and the Pamirs, slopes are covered with enormous screes and rock escarpments have a serrated aspect. Solifluction lobes may even develop, although they are certainly not active more than a few weeks per year. Slopes create the impression of an intense destruction, which is not common in winterless deserts; no such destruction occurs in the southern Sahara. The activity of fragmentation produces an enormous mass of debris ready to be moved by the agents of transportation in the mid-latitude deserts. This waste contributes to the development of piedmont plains by providing an important load to freshets. In the tropical deserts, outside granitic areas, the slowness of fragmentation retards the evolution of the relief. In the Mauritanian Adrar the sandstone escarpments retreat every time runoff clears the underlying marls and produces overhangs. Uplands composed entirely of sandstones seem to erode extremely slowly, which explains the widespread persistence of old surfaces such as the back-slope of the Grand Dhar, which is Cretaceous or older and has never been fossilised.

Furthermore, on at least the northern fringe of the Sahara, the Pleistocene pluvial phases were characterised by an important lowering of temperature which may have been reflected in the increased role of frost. The magnificent piedmont slopes which survive in this area were then formed, owing to intense fragmentation. Contrary to Birot's opinion, a present overloading of the runoff cannot be imputed to their genesis because at the waning of floods, the waters are clear and tend to incise themselves. For this reason Pleistocene piedmont slopes are more dissected in steppe regions where floods are more frequent than in more arid regions where they are fewer. Because the present intensity of fragmentation is about equal in both types of regions, the floods of the drier regions carry the higher load, permitting them to maintain the morphogenic equilibrium attained during the Pleistocene on the piedmont slopes or, at least, not to dissect them.

The humid tropical zone (forests and savannas)

The low latitudes, like the midlatitudes, are mainly characterised (at least originally) by a forest type of plant cover. The two major forested zones of the earth meet in several places, notably in the Antillian archipelago and especially in southeast Asia. In these areas it is difficult to draw the limits of the morphoclimatic zones because here the two types of forest merge into one another. Moreover, detailed studies which would permit the establishment of specific differences are wanting. Studies of karst topographies show only that the karsts of Jamaica and Cuba are not akin to the karsts of the midlatitudes but belong to a different, tropical, variety.

Morphoclimatic delimitations are clear in the northern half of Africa. The Sahara interposes a clearly defined intermediate zone between well defined life zones to the north and the south. On its southern fringe, close to the seventeenth parallel, the *Sahelian zone* forms the transition between the savannas and the desert. This zone is characterised by a grass steppe wooded with acacias and baobabs. Drought causes runoff to be intermittent and poorly organised; sheet and rill wash play an important role on impermeable rocks, whereas sandy formations are practically areic (i.e. without runoff). The wind is active on alluvial deposits and in areas where the plant cover is degraded. Runoff carves gullies in loamy soils and, because of drought, produces abrupt slopes. Excellent examples of this kind of topography are found in the middle Baoulé Valley of western Mali in a very sparsely populated area. But Pleistocene climates, especially, have imparted to this area a characteristic desert type of relief with a certain number of ergs and magnificent inselbergs and bajadas which present morphogenic processes are unable to erode. In spite of its arborescent plants, the Sahelian zone cannot be classed with the savannas: the intensity of the soil forming processes is minor here; there are no ferrallitic or ferritic soils, and the dominant traits of the relief are those of a desert margin. It is better to study it with the dry zone also because it has problems of applied geomorphology similar to those of other desert margins.

To the south the Sahelian zone passes into the zone of savannas of the alternating wet–dry tropics where the mean annual rainfall is higher than 600 to 800 mm, approximately (24 to 32 in). The rainforest appears with 1 500 mm (60 in) on condition that the dry season is neither too long nor too dry. When the dry season is more pronounced, the wet tropics (with a total rainfall still above 1 500 mm) are occupied by other forests, as those of Vietnam, the Indo-Gangetic plain, or the Brazilian Serra do Mar. The structural difference between the two types of tropical forests does not seem to be appreciable in the present state of our knowledge.

The whole of the humid tropical zone, forests as well as savannas, has in common high temperatures. Because of the lack of frost, rock fragmentation is considerably reduced. Rock escarpments erode extremely slowly; there is no fresh scree at their foot. The boulders that are occasionally found there are coated with patinas and entangled in vegetation as if they had been there for a very long time. At least in the Fouta Djallon they seem to be relics dating back to the last dry climatic phase which here corresponds to the last glaciation (the pluvials appear to have been approximately contemporaneous to the glaciations north of the Sahara, to the interglacials south of the Sahara). Such extremely slow fragmentation probably accounts for the persistence of abrupt and precipitous slopes as on sugarloaves, inselbergs, and tableland escarpments. The lack of frost-weathering makes them almost indestructible. Variations of temperature, which play an appreciable role in dry climates, are much less important in the humid tropics, particularly in the forests. The humidity of the air is greater, the sky more cloudy, and the heating of exceptional rock outcrops during the day is less than in the desert. As to the nights, they are not as cool: whereas temperatures drop to $3°$ or $4°C$ ($37°$ or $39°F$) in the month of January in the south of Mauritania, they only fall to $12°$ or $15°C$ ($54°$ or $59°F$) at Kindia in the Fouta Djallon. The insignificance of mechanical disintegration moreover explains the very frequent existence of flutings on the rock surfaces of sugarloaves and monolithic domes. They are due to chemical corrosion by atmospheric water which flows down the rocky surfaces; such flutings would quickly disappear if fragmentation were intense for their formation is necessarily a very slow process.

High temperatures and high rainfall cause intense chemical action unequalled in other climates. Humus is so rapidly decomposed by bacteria that in spite of the great amount of organic matter supplied by the lush vegetation it does not accumulate, and well drained soils are devoid of it. But the silicates, such as the micas and especially the feldspars, are rapidly attacked and their decomposition, which goes beyond the clayey state, ends in the liberation of silica and alumina. The alumina and the iron hydrates form cuirasses often $10m$ ($30ft$) thick or more; at the very least kaolinitic clays develop. The silica is removed in pseudosolution. The lime is rapidly dissolved because of high rainfall. In total nearly all the original rock minerals are attacked chemically; the most resistant are quartz and a few heavy minerals; for this reason streams only carry a small solid load composed of a little sand and a few pebbles. The evolution of stream beds is therefore very different from that of midlatitude streams. Because of the absence of frost the bedrock is less fissured; it is also worn less by streams carrying a smaller load; rapids are formed as soon as streams pass an unjointed ledge resistant to chemical weathering. Breaks of slope do not wear away; as a result, the streams of the humid tropics have stepped longitudinal profiles. Stream channels are also irregular in size; the waters slip into fracture zones, emptying them of their loose materials, usually previously weathered, without being able to widen them by attacking the unaltered rock. Such characteristics only exist in the tropical forests and savannas; they disappear in the Sahelian zone where chemical activity is reduced by drought. In the Sahel, the Senegal, below Kayes, moves sand banks, and the middle Baoulé cuts magnificent, very regular meanders. An additional reason for excluding the Sahelian zone from the humid tropical zone is that in the Sahelian zone streams are not only fewer but also exert a higher mechanical action.

Let us return to the humid tropical zone. It is necessary to distinguish two realms within it: the savannas and the forests, as each has a distinct morphogenic system:

1. The *tropical savanna* (including cerrados and deciduous seasonal forests). Under natural conditions this zone not only receives less rain than the rainforest but is subject to a very contrasted seasonal regime. Alternations of rainy seasons and marked drought are the rule. The first showers, often violent, fall on a hardened and desiccated ground. Because of a sparser vegetation they produce splash erosion and intensive rillwash. The latter, however, is slowed down by a dense grass cover but may take the form of a sheetflood. Subhorizontal surfaces are unevenly flooded and lack a definitely oriented drainage. Savanna plains are amphibious lands interrupted by pools and are difficult to cross until the waters finally infiltrate or evaporate. Only a fraction of the water finds its way to streams which are unable to incise themselves due to the absence of abrasive tools. This particular type of drainage, emphasised by Dresch (1947), favours colluviation. On topographic bulges the waters become loaded with a certain amount of soil and dissolved products which they deposit in shallow depressions where they stagnate. These dynamics thus contribute to the process of planation, whereas the lack of stream incision prevents such lowlands from being dissected. The combination of these mechanisms probably partly explains the importance of near level surfaces in savanna regions as well as their exceptional uniformity.

There is less chemical activity in savannas than in the rainforest because the quantity of infiltrated water is smaller. The soils being less leached, ferruginous concretions are abundant. Their cementation into cuirasses seems to be more frequent than in forests but results from allogenic influxes of ferric iron or palaeoclimatic oscillations. Whichever the cause, bauxitic or ferruginous cuirasses are the rule on savanna plains in Brazil (where they are called *cangas*) as well as in Africa or southeast Asia. When found in inverted relief, such cuirasses, which are almost immune to fragmentation, help protect plateaus against the attack of regressive erosion while colluviation perfects the planations. Cuirasses thus often form tablelands with abrupt margins incised by short canyons.

The influence of man on the development of savannas is considerable especially in Africa where brush fires play an important role in the cuirassing process and cause the progressive retreat of the forest. The savannas thus have been artificially extended. As cuirasses are laid bare by erosion, runoff is increased but it remains morphologically ineffective because of the lack of physical weathering.

2. The *tropical forest*. By far the most common external dynamic process is *in situ* chemical weathering which produces regolith. Lime is dissolved. Granites, gneisses, and related rocks undergo argillaceous or even allitic weathering and are topped nearly everywhere by a friable, spongy regolith 10m (30ft) thick or more. Most of the rainwater (98 to 99 per cent) infiltrates into it as into a sponge and contributes to the continuance of deep weathering. Unjointed outcrops of the same (above mentioned) rocks, however infrequent, stand in sharp contrast to the regolith, in river banks, on rapids, on monolithic domes or (for limestones) on needle-karst washed by showers preventing seeds from germinating on them. Because no waste is formed and because weathering and dissolution are practically nil, the unweathered rock masses rise above the surrounding hills which are reduced by chemical decomposition under the forest cover.

The slopes of these rather steep and often convex forested hills (forming the *mar do mauros*, a sea of hills, and the *meias laranjas*, or half oranges, of Brazil) are degraded by burrowing animals or falling trees, whether isolated or in groups, which bare the ground and expose it to splash erosion or to earth flowage as, for example, in the Fouta Djallon. Springs and seepages which evacuate large quantities of dissolved products sometimes cause slumps, a local settling of the ground, and even the bursting of mud pockets as, for example, in the dense forests of Colombia or Panama. In some places, as in French Guiana, the soil is covered by a thin layer of protective litter, and runoff does not affect the slopes. In other places the decomposition of vegetal matter by bacteria is so rapid that the soil is bare or almost so, enabling a small proportion of overland flow (1 to 2 per cent in Guinea) and entrainment of suspended matter which will eventually be deposited further down.

Some authors, however, like Rougerie (1960), concede a certain importance to overland flow. It seems to be justified more on shales and micaschists than on granite and under semi-evergreen seasonal forest rather than under evergreen forest.

The geomorphic role of palaeoclimates may be considerable in the forests as well as in the savannas. In Africa, for example, relict Pleistocene landforms are quite common. Coarse alluvial deposits accumulated during the Pleistocene stand witness to a morphogenic system different from the present one. Aeolian sands have even been reported near Kisangani (Stanleyville) on the margin of the rainforest. Recent geomorphological studies, particularly in Africa, attribute an increasingly important role to palaeoclimates in the explanation of the morphology of the presently forested tropics. In Ivory Coast, pediplains of the Sudanese type were developed during dry phases all the way down to the coast. Evidence supporting the former existence of savanna plains has been reported from the area of Kinshasa (Leopoldville).

In Brazil many monolithic domes have been bared by repeated landslides during a climatic phase with a marked dry season which seems to coincide with the end of the postglacial Flandrian transgression. Climatic terraces have been observed on the margins of Lake Maracaibo (Venezuela) beneath the present rainforest, which probably did not exist at the time when the gravels were widely spread as coalescent fans.

Not everything is functional in the topography of the humid tropical zone; only a detailed analysis of the present morphogenic processes combined with careful morphogenic reconstructions will determine the importance of relict phenomena.

Bibliographic orientation

For studies concerning the division of the land masses into morphoclimatic regions the reader is referred to the bibliography of Chapter 3. For particular problems concerning the delimitation of the various regions and their morphoclimatic characteristics, the student is referred to the bibliographies of the volumes of this *Treatise* which deal specifically with each realm.

Conclusion

A certain number of principles on which geomorphology must necessarily be based are derived from the facts mentioned in this work.

1. Relief is in part dependent on climate. The influence of climate is principally felt in an indirect way through the medium of vegetation and the soils which depend on it. In general the plant cover reduces the intensity of mechanical erosion and favours chemical weathering and mechanical accumulation. The plant cover plays the main role in the respective part of the mechanical and the chemical processes of the morphogenic system. This principle is of prime importance to applied geomorphology; it lies at the root of the serious problem of anthropic erosion. It is with the use of plants that accelerated erosion can be checked, as, for example, through the reforestation of torrential basins or through the planting of wooded strips and the sowing of grass to contain aeolian erosion.

2. The considerable role played by the plant cover in climatic geomorphology implies a more thorough climatic differentiation of the relief of the earth's surface than was admitted by the Davisian school which was satisfied with the opposition of 'normal erosion' and glacial and desert 'climatic accidents'. The relief is therefore also sensitive to palaeoclimatic oscillations. The important changes of climate which have affected the Pliocene and the Quaternary have profoundly influenced it. The topography of the greater part of the land areas is now a polygenic relief from the climatic point of view. It is therefore necessary to distinguish carefully between the part played by palaeoclimates and the present climate. The analysis of processes and the study of correlative deposits are the principal means of investigation. Modern geomorphology must lean on them and not relegate them to a secondary role as did the Davisian school. This principle, too, is susceptible to important practical applications: man most often uses, especially in agriculture, palaeosols that are fragile and precious because they do not reconstitute themselves. It is indispensable to know their origin and their age, in order to preserve them and to leave them to future generations whose wealth we will have ceased to despoil.

3. The evolution of relief is much more complex than the progressive reduction towards a peneplain as affirmed by Davis. Our present knowledge of

processes is insufficient to propose definitive generalisations. In the present state of knowledge, however, it can probably be affirmed that slopes do not become increasingly gentle during the course of their evolution, at least under all climates. Do they become increasingly gentle in the forested regions of the midlatitudes? Certainly not in the deserts and in the savannas and forests of the humid tropical zone nor, probably, in the cold zone. In subtropical and intertropical latitudes steep slopes maintain themselves until the very disappearance of the residual reliefs in the form of inselbergs rising above pediplains. Neither do streams seem to evolve towards a regular concave graded profile as do the streams of the midlatitudes, a profile which Davis considered indispensable to peneplanation; lacking abrasives, waterfalls and rapids are preserved and a stepped profile is maintained through the course of geologic time.

4. The considerable role played by vegetation in the development of relief forces the rejection of the cyclic concept. Indeed, considering the length of time involved, the evolution of living organisms, which is more or less continuous in a given region because of climatic oscillations, plays the determinant role. Organisms have the effect, independent of other causes (notably tectonic), of progressively modifying the conditions which affect the morphogenic processes during the course of tens of millions of years which are necessary to reduce a chain of mountains. The morphogenic system which is presently active in the Massif Central is very different from the one which produced the post-Hercynian peneplain. As relief does not seem to evolve uniformly towards a peneplain in all morphoclimatic regions, the relief which will result from the present morphogenic system, supposing it will last long enough, will differ from that of the post-Hercynian peneplain just as the lithologic nature of the rocks which will outcrop at its surface will differ from those which outcropped on the post-Hercynian surface. There is, therefore, no return to a situation identical to that which existed at the starting point. There is no cycle. There is evolution with progressive modifications of the very conditions of this evolution. There are no repeated reversals but a continuous transformation of the landscape, at an unequal rate, it is true. This evolution, it seems, has tended towards a progressive differentiation of the physiognomy of the earth's surface. Are not the present climates more diverse than those of the Cretaceous or the Eocene? Has not the domain of pediplanation shrunk in proportion to the development of the plant cover since the Precambrian? Such evolutionary factors are all recorded in the relief.

Consolidated bibliography

Abbreviations: *IFAN*—Institut Français d'Afrique Noire, now Institut Fondamental d'Afrique Noire. *INEAC*—Institut National d'Etudes Agronomiques du Congo. BRGM: Bureau de Recherche Géologique et Minière. CNRS: Centre National de la Recherche Scientifique.

ACADEMY OF SCIENCE, NEW YORK (1961) 'Solar variations, climatic change and related geophysical problems', *Ann. NY Acad. Sc.* **95**, 1–740.

AHLMANN, H. W., cf. 'Glaciological research on the North Atlantic coasts', *Roy. Geog. Soc. Res. Ser.* **1**, 1948, 83 p.

ALBRECHT, F. (1962) 'Paläoklimatologie und Wärmehaushalt', *Meteorol. Rundsch.* no. 2, pp. 38–44.

ANDERSON, H. and GLEASON, C. (1960) 'Effects of logging and brush removal on snow water runoff', *Int. Ass. Sci. Hydrol.* Helsinki, Surf. Waters Comm. pp. 478–89.

ANDRÉ. J. E. and ANDERSON, H. W. (1961) 'Variation of soil erodibility with geology, geographic zone elevation and vegetation type in northern California wildlands', *J. Geophys. Res.* **66**, pp. 3351–8.

ANTEVS, E. (1952) 'Cenozoic climates of the Great Basin', *Geol. Rundsch.* **40**, no. 1, pp. 94–107.

ARMAND, D. L. (1955) 'Water erosion and its importance for us', *Priroda*, no. 6, pp. 31–41.

AUBERT, G. and CAILLEUX, A. (1950) 'Esquisse d'une étude des sols', *Rev. Gén. Sc.* pp. 28–39.

AUBERT, G. and MONJAUZE, A. (1946) 'Observations sur quelques sols de l'Oranie nord-occidentale. Influence du déboisement, de l'érosion sur leur évolution', *CR Somm. Soc. Biogréogr.* **23**, 44–51.

AUBERT DE LA RUE, E. (1954) *Reconnaissance géologique de la Guyane française méridionale*, Paris, Larose, 128 p.

AUSTIN, R. and BAISINGER, D. (1955) 'Some effects of burning on forest soils of western Oregon and Washington', *J. Forestry*, **3**, 275–80.

BAGNOULS, F. and GAUSSEN, H. (1953) 'Saison sèche et indice xérothermique', *Doc. pour les cartes des prod. végét.*, sér. généralités, **3**, 193–239.

BAGNOULS, F. and GAUSSEN, H. (1957) 'Les climats biologiques et leur classification', *Ann. Géogr.* **66**, 193–220.

BAILEY, R. W. (1939) 'A new epicycle of erosion', *J. Forestry*, **35**, p. 999.

BAILEY, R. W. (1941) 'Land erosion—normal and accelerated in the semiarid West', *Trans. Amer. Geophys. Union*, pp. 240–50.

BALLAL, D. K. and DESHPANDE, R. P. (1960) 'Erodibility studies by a rainfall simulator. Effect on slope, moisture condition and properties of soil on erosion', *J. Soil Water Conserv.* India, **8**, 12–25.

BARAT, C. (1956) 'Les données de la pluviologie dans la zone intertropicale', *Bull. Ass. Géogr. Franc.* **261–2**, 175–84.

256

BARAT, C. (1957) *Pluviologie et aquimétrie dans la zone intertropicale*, Mem. IFAN, Dakar, no. 49, 80 p.

BARNETT, A. P. (1958) 'How intensive rainfall affects run-off and erosion', *Agr. Eng.* (St Joseph), **39**, 703–7 and 711.

BARRELL, J. (1908) 'Relation between climate and terrestrial deposits', *J. Geol.* **16**, 159–90, 225–95 and 363–84.

BASU, J. K. (1952) 'Soil and moisture conservation in the dry regions of the Bombay State', *Empire J. Exp. Agr.* Oct., pp. 326–33.

BATTISTINI, R. (1964) *L'Extrême sud de Madagascar, étude géomorphologique*, Paris, Editions Cujas, 636 p.

BAULIG, H. (1948) 'L'oeuvre de William Morris Davis', *Inf. Géogr.* pp. 101–8 (reproduced in *Essais*, pp. 13–25).

BAULIG, H. (1950) *Essais de géomorphologie*, Strasbourg, Publ. de la Fac. Lettres, no. 114, article entitled: 'La philosophie géomorphologique de James Hutton et de John Playfair', pp. 1–11.

BAULIG, H. (1952) 'Surfaces d'applanissement', *Ann. Géogr.* **61**, 161–83; 245–62.

BAULIG, H. (1956) 'La géomorphologie en France jusqu'en 1940', *La Géographie française au milieu du XXᵉ siècle*, Paris, Baillière, pp. 27–36.

BAULIG, H. (1957) 'Les méthodes de la géomorphologie d'après M. Pierre Birot', second article, *Ann. Géogr.* **66**, 221–35.

BEAUDET, G. (1962) 'Types d'évolution actuelle des versants dans le Rif occidental', *Rev. Géogr. Maroc*, no. 1–2, 41–6.

BEHR, F. M. (1918) 'Über geologisch wichtige Frosterscheinungen in gemässigten Klimaten', *Int. Mitt. Bodenkunde*, **8**, no. 3–4, pp. 50–72.

BEHRMANN, W. (1944) 'Das Klima der Präglazialzeit auf der Erde', *Geol. Rundsch.* **34**, 763–76.

BEIRNAERT, A. (1941) *La Technique culturale sous l'équateur*, Brussels, INEAC.

BELEHRADEK, J. (1935) 'Temperature and living matter', *Protoplasma Monogr.* no. 8, Berlin.

BENCHETRIT, M. (1954) 'L'érosion accélérée dans les chaînes telliennes d'Oranie', *Rev. Géomorph. Dyn.* **5**, pp. 145–67.

BENCHETRIT, M. (1954) 'L'érosion anthropogène: couverture végétale et conséquences du mode d'exploitation du sol', *Inf. Géogr.* pp. 100–8.

BENNETT, H. H. (1955) *Elements of soil conservation*, New York, McGraw-Hill, 358 p.

BENNETT, H. H. (1960) 'Soil erosion in Spain', *Geogr. Rev.* **50**, 59–72.

BERG, H. (1947) *Einführung in die Bioklimatologie*, Bonn, Bouvier, 131 p.

BERG, L. S. (1947) *Climate and Life*, 2nd edn. Moscow, Geografguiz, 356 p.

BERG, L. S. (1950) *The Natural Regions of the USSR*, trans. from Russian by Olga A. Titelbaum under the Russian trans. Project Amer. Council Learned Soc., ed. John A. Morrison and C. C. Nikiforoff, Macmillan, New York, 436 p.

BERGSTEN, K. E. (1950) 'Some characteristics of the dispersion of the annual precipitation in Sweden during the period 1881–1940', *Lund Stud. Geogr.*, ser. A, *Phys. Geogr.* no. 1, 18 p.

BERNARD, E. (1945) *Le Climat écologique de la cuvette congolaise*, Yangambi, Publ. INEAC.

BERNARD, E. A. (1962) 'Théorie astronomique des pluviaux et interpluviaux du Quaternaire africain', *Acad. Roy. Sc. Outre-mer, Cl. Sc. Nat. Medic.* **12**, no. 1, Brussels.

BESAIRIE, H. (1948) 'Deux examples d'érosion accélérée à Madagascar', *Conf. Afr. des sols, Goma*, communic. no. 21.

BÉTREMIEUX, R. (1951) 'Etude expérimentale de l'évolution du fer et du manganèse dans les sols', *Ann. Agron.* pp. 193–5.

BILANCINI, R. (1948) 'Su un tipo di analisi statistica della variabili meteorologiche', *Riv. Meteorol. Aeron.* Rome, **8**, no. 4.

BIRD, E. C. F. (1963) 'The physiography of the Gippsland Lakes, Australia', *Z. Geomorph.* n.s. **7**, 232–45.

BIROT, P. (1949) *Essai sur quelques problèmes de morphologie générale*, Lisbon, Instituto para a Alta Cultura, Centro de Estudios Geográficos, 176 p.

BIROT, P (1960) *Géographie physique générale de la zone intertropicale* (à l'exclusion des déserts), Paris, CDU, 'Cours de Sorbonne' 1960, 1965, 244 p.

BIROT, P. (1965) *Formations végétales du globe*, Paris, SEDES, 508 p.

BIROT, P. and JOLY, F. (1952) 'Observations sur les glacis d'érosion et les reliefs granitiques au Maroc', *Mém. Centre Doc. Cartogr. CNRS*, **3**, 9–55.

BLANCK, E. (1930) *Handbuch der Bodenlehre*, Berlin, Springer, 10 vols.

BLANCK, E. (1949) *Einführung in die genetische Bodenlehre als selbständige Naturwissenschaft und ihre Grundlagen*, Göttingen, 420 p.

BLÖCHLIGER, G. (1932–33) 'Kleinlebewesen und Gesteinsverwitterung', *Z. Geomorph.* **7**. pp. 273–84.

BORCHERT, H. (1961) 'Einfluss der Bodenerosion auf die Bodenstruktur und Methoden ihrer Kennzeichnung', *Geol. Jahrb.* **78**, 439–502.

BORDE, J. (1966) *Les Andes de Santiago et leur avant pays, étude de géomorphologie*, Fac. Lettres, Bordeaux, 559 p.

BORISOV, A. A. (1959) 'Sur les conditions paléoclimatiques de la formation des principaux centres barométriques du climat actuel de la Terre', *Ivz. Vsiess. geogr. Obchtch.* **91**, 259–65 (French trans. by Bur. Rech. Géol. Min. no. 2487).

BORNHARDT, W. (1900) *Zur Oberflächengestaltung und Geologie Deutsch Ostafrikas*, collection *Deutsch Ostafrika*, vol. 7, Berlin, 595 p.

BOTELHO DA COSTA, S. V. and LOBO AZEVEDO, A. 'Aspectos da erosão do solo em Angola', *Agros*, **33**, no. 1–2, 15–22.

BOUDY, P. (1951) 'Considérations sur le problème de la défense et de la restauration des sols (Etats-Unis, Afr. du Nord)', *Soc. Forest. Franche-Comté*, pp. 174–211.

BOUDYKO, M. I. (1956) 'Climatic indices of aridity', *Geogr. Essays, USSR Akad. Sc.* pp. 142–9.

BOYÉ, M. (1950) 'Glaciaire et périglaciaire de l'Ata Sund nord-oriental (Groënland)', *Expéditions Polaires Françaises*, Paris, Herman, vol. 1.

BOYKO, H. (1949) 'On climatic extremes as decisive factors for plant distribution', *Palest. J. Bot. Rehovot*, **7**, 41–52.

BRANDTNER, (1954) *Eiszeitalter und Gegenwart*, **4/5**, 58.

BRAZIER, C. and EBLÉ, L. (1934) 'Introduction à l'étude des températures de l'air et du sol au voisinage de la surface terrestre', *Météorologie*, pp. 97 et seq.

BRAZIER, C. and EBLÉ, L. (1935) 'Sur une particularité de la transmission de la chaleur dans le sol', *C.R. Acad. Agr.*

BROCKMANN, H. (1913) 'Der Einfluss der Klimacharakters auf die Verbreitung der Pflanzen und Pflanzengesellschaften', *Englers Bot. Jahrb.*

BROOKS, C. E. P. (1949) *Climate Through the Ages*, 2nd edn. London, Benn, 395 p.

BROOKS, C. E. P. (1950) 'Selective annotated bibliography on climatic changes', *Amer. Meteorol. Soc., Meteorol. Abstr. Bibliogr.* July, vol. 1, no. 7.

BRUNACKER, K. (1953) 'Die bodenkundlichen Verhältnisse der Würmeiszeitlichen Schotterfluren im Illgebiet', *Geol. Bavarica*, **18**, 113–30.

BRUNET, R. (1957) 'L'érosion accélérée dans le terrefort toulousin (1er examen)', *Rev. Géomorph. Dyn.* **8**, no. 3–4, pp. 33–40.

BRYAN, K. and ALBRITTON, C. C. (1943) 'Soil phenomena as evidence of climatic changes', *Amer. J. Sc.* **241**, 469–90.

BÜDEL, J. (1948) 'Das System der klimatische Morphologie, Beiträge zur Geomorphologie der Klimazonen und Vorzeitklimate', *Deut. Geographentag*, Munich, 1950, pp. 65–100.

BÜDEL, J. (1951) 'Die Klimazonen des Eiszeitalters', *Eiszeitalter und Gegenwart*, Ohringen (Württ.), **1**, 16–26.

BÜDEL, J. (1953) 'Die "periglaziale" morphologische Wirkungen des Eiszeitklimas auf der ganzen Erde', *Erdkunde*, **7**, 249–65; also in Eng. Trans. 'The "periglacial" morphologic effects of the Pleistocene climate over the entire world', Trans. H. E. Wright and D. Alt, *International Geology Review*, **1** (1959), no. 3, pp. 1–16.

BÜDEL, J. (1957a) 'Die "doppelten Einebenungsflächen" in den feuchten Tropen', *Z. Geomorph.* n.s. **1**, 201–28.

BÜDEL, J. (1957b) 'Die Flächenbildung in den feuchten Tropen und die Rolle fossilier solcher Flächen in anderen Klimazonen', *Deut. Geographentag*, Würzburg, pp. 89–121.

BÜDEL, J. (1961) 'Die Morphogenese des Festlandes in Abhängigkeit von Klimazonen', *Die Naturwiss*, no. 9, 313–18.

BUREAU INTERAFRICAIN DES SOLS ET D'ECONOMIE RURALE (1955) *Bull. Bibliogr. Mensuel*, **5**, no. 9, and supplement, June 1958, 3 p.

BUTZER, K. W. (1957) 'The recent climatic fluctuation in lower latitudes and the general circulation of the Pleistocene', *Geogr. Annaler*, **39**, 105–13.

BUTZER, K. W. (1958) 'Russian climate and the hydrological budget of the Caspian Sea', *Rev. Can. Géogr.* **12**, 129–39.

BUTZER, K. W. (1958) 'Die Ursachen des Landschaftswandels der Sahara und Levante seit dem klassischen Altertum', *Akad. Wiss. Litt., Abh. math. Naturwiss*. Kl. no. 1, 2–19.

BUTZER, K. W. (1961) 'Climatic changes in arid regions since the Pliocene', UNESCO, *Arid Zone Res.*

CNRS (1952) *Colloques int. Ecology* 1950, Paris, 350 p.

CNRS (1955) *Colloques int.*, **59**, 12. *Les divisions écologiques du monde*, Paris, 326 p.

CACHAN, P. (1960) *L'Etude des microclimats et de l'écologie de la forêt sempervirente en Côte d'Ivoire*, IDERT–UNESCO, Adiopodioumé (Ivory Coast) mimeographed, 10 p.

CAILLEUX, A. (1950) 'Ecoulements liquides en nappes et applanissements', *Rev. Géomorph. Dyn.* **1**, 243–70.

CAILLEUX, A. (1950) 'Paléoclimats et optimums physiologiques', *C.R. Somm. Sc. Soc.* **230**, Biogéogr. 23–6.

CAILLEUX, A. (1951) 'Reliefs coralliens et paléoclimats', *C.R. Somm. Sc. Soc. Biogéogr.* no. 239, pp. 22–4.

CAILLEUX, A. (1952) *La Géologie*, Paris, A Colin Coll. Que sais-je? 128 p.

CAILLEUX, A. (1953) *Biogéographie mondiale*, A. Colin Coll. Que sais-je? 128 p.

CAILLEUX, A. (1954) 'Le ruissellement en pays tempéré non montagneux', *Ann. Géogr.* **57**, 21–39.

CALEMBERT, L. 'Le sous-sol. Etude de l'influence des facteurs géologiques et miniers sur les déformations du sol de la région liégeoise. Plan d'aménagement de la région liégeoise', I, *l'Enquête*, pp. 57–76.

CAPOT-REY R. (1945) 'Dry and humid morphology on the Western Erg', *Geogr. Rev.* **35**, 391–407.

CARLES, J. (1948) *Géographie botanique*, Coll. Que sais-je? 120 p.

CAVAILLE, A. (1958) 'Sols et séquences de sols en coteaux de Terrefort aquitain', *Ass. Franç. Etude Sol*, no. 94, pp. 4–22.

CAYEUX, LUCIEN (1941) *Causes anciennes et causes actuelles en géologie*, Masson, Paris, 81 p.

CHETELAT, E. DE (1938) 'Le modelé latéritique de l'ouest de la Guinée française', *Rev. Géogr. Phys. Géol. Dyn.* **11**, 5–120.

CHEVALIER, A. (1929) 'Sur la dégradation des sols tropicaux causée par les feux de brousse et sur les formations végétales régressives qui en résultent', *C.R. Acad. Sc.* **188**, 84–6.

CHOLLEY, A. (1950) 'Morphologie structurale et morphologie climatique', *Ann. Géogr.* **59**, 321–35.

CHORLEY, R. J. (1965) 'A re-evaluation of the geomorphologic system of W. M. Davis', in R. J. Chorley and P. Haggett, eds., *Frontiers in Geographical Teaching*, London, Methuen, pp. 21–38.

CHORLEY, R. J., DUNN, A. J., and BECKINSALE, R. P. (1964) *The History of the Study of Landforms or the Development of Geomorphology*, vol. 1, *Geomorphology before Davis*, London, Methuen; New York, Wiley, 678 p.

CLARKE, G. R. (1941) *The Study of the Soil in the field*, London University Press, 228 p.

CLAYTON, H. (1927, 1934) 'World weather records', *Smithonian Misc. Coll.* **79**, 1927, and **90**, 1934, Washington.

CONRAD, V. (1944) *Methods in Climatology*, Harvard University Press, 21 p.

COOK, S. F. (1949) 'Soil Erosion and Population in Central Mexico', *Ibero-Americana*, Berkeley, no. 34, 86 p.

COQUE, R. (1962) *La Tunisie présaharienne, étude géomorphologique*, Paris, Colin, 476 p.

COTTON, C. A. (1948) *Landscape as Developed by the Processes of Normal Erosion*, 2nd edn, Christchurch, New Zealand, Whitcombe & Tombs, 509 p.

COTTON, C. A. (1961) 'The theory of savanna planation', *Geography*, **46**, 89–101.

CRITCHFIELD, H. J. (1960) *General Climatology*, New York, Prentice-Hall, 465 p. 180 fig.

CROSSE, B. (1953) 'Untersuchungen über die Winderosion in Niedersachsen', *Mitt. Inst. Raumforsch.* Bonn, **20**, 137–45.

DABRALL, B., *et al.* (1963) 'Some preliminary investigations on the rainfall interception by leaf litter', *Indian Forester*, **89**, 112–16.

DAMMAN, W. (1948) 'Zur Physiognomie der Niederschläge in Nordwestdeutschland', *Göttinger Geogr. Abhandl.* **1**, 58–69.

DAVIS, W. M. (1899) 'The geographical cycle', *Geogr. J.* **14**, 481–504 (reproduced in *Geographical Essays*, Boston, Ginn, 1909 and New York, Dover, 1954, pp. 249–78).

DAVIS, W. M. (1904) 'Complications of the geographical cycle', *8th Int. Geogr. Congr.* pp. 150–63 (reproduced in *Geographical Essays*, pp. 279–95).

DAVIS, W. M. (1912) *Die Erklärende Beschreibung der Landformen*, Berlin, Teubner, 565 p.

DEEVEY, E. S. (1949) 'Biogeography of the Pleistocene', *Bull. Geol. Soc. Amer.* **60**, 1315–1416.

DEFANT, A. (1916) 'Über die nächtliche Abkühlung der unteren Luftschichten und der Erdoberfläche in Abhängigkeit von Wasserdampfgehalt der Atmosphäre', *Sitzungber. Wien. Akad. Sect.* IIa, pp. 1537 and following.

DELAMARRE, C. and DEBOUTTEVILLE, 'Microfaune du sol', *Actes Scientif. et Industr.* 360 p.

DE LA NOÉ, G. D. and DE MARGERIE, E. (1888) *Les Formes du terrain*, Paris, 205 p.

DEMOLON, A. (1949) *La Génétique des sols*, Paris, PUF, Coll. Que sais-je? 126 p.

DEMONTZEY, P. (1882) *Traité pratique du reboisement et du gazonnement des montagnes*, Paris, Rothschild, vol. 1.

DERRUAU, M. (1960) 'Quel était le climat du Massif Central français pendant la seconde moitié de l'ère tertiaire?' *Rev. d'Auvergne*, **74**, 179–86.

DOIGNON, P. (1946–51) *Le Mésoclimat forestier de Fontainebleau*, Fontainebleau, 3 vols, 142, 128, and 132 p.

DOKUCHAEV, V. V. (1883) *Russian Chernozem*, doctoral diss. St Petersburg University.

DOLLFUS, O. (1959) 'Formes glaciaires et périglaciaires actuelles autour du lac Huampar', *Bull. Ass. Géogr. Franç.* **286/7**, 32–40.

DOLLFUS, O. (1965) *Les Andes centrales du Pérou et leurs piedmonts (entre Lima et le Péréné)*, *étude géomorphologique*, Lima, Inst. Franç. d'Etudes Andines, 404 p.

DOROGANEVSKA, E. A. (1954) 'The question of the hydrothermal coefficient of the period of vegetation of cultivated plants', *Izv. Akad. Nauk. SSSR, Geogr. ser.* no. 6, pp. 51–61.

DRESCH, J. (1941) *Recherches sur l'évolution du relief dans le Massif Central du Grand Atlas, le Haouz et le Sous*, Tours, Arrault, 703 p. 206 fig.

DRESCH, J. (1947) 'Pénéplaines africaines', *Ann. Geogr.* **56**, 125–30.

DREYFUSS, M. (1960) 'Sur quelques problèmes de géomorphologie dynamique', *Ann. Sc. Univ. Besançon*, 2ᵉ sér. Géol. **12**, 93–106.

DUBIEF, J. (1952) 'Le vent et le déplacement du sable au Sahara', *Trav. Inst. Rech. Saharo-ennes*, **8**, 123–64.

DUBOIS, J. and TRICART, J. (1954) 'Esquisse de stratigraphie du Quaternaire du Sénégal et de la Mauritanie du Sud', *C.R. Acad. Sc.* **238**, 2183–5.

DUCHAUFOUR, P. (1965) *Précis de pédologie*, 2nd edn, Paris, Masson, 481 p.

DUCHÉ, J. (1949) *La Biologie des sols*, Paris, PUF, Coll. Que sais-je? 128 p.

DURAND, J. (1953) *Etude géologique, hydrologique et pédologique des croûtes en Algérie*, Dir. Serv. Geol. et Hydr., Serv. des Etudes Sc., Birmandreis, 209 p.

DYLIK, J. (1957) 'Tentative comparison of planation surfaces occurring under warm and under cold semiarid conditions', *Biul. Perygl.* **5**, 175–86.

EAKIN, H. M. (1916) 'The Yukon Keokuk Region (Alaska)', *USGS*, no. 631, 88 p.

EBLÉ, L. (1946) 'Etudes expérimentales sur la propagation de la chaleur dans le sol', *Météorologie*, pp. 269–91.

EKERN, P. C. (1954) 'Rainfall intensity as a measure of storm erosivity', *Proc. Soil Sc. Soc. Amer.* **18**, 212–16.

ELLISON, W. D. (1945) 'Some effects of rain drops and surface flow on soil erosion and infiltration', *Trans. Amer. Geophys. Union*, **26**, 415–29.

ELOUARD, P., FAURE, H., and MICHEL, P. (1967) 'Nouveaux âges absolus (C 14) en Afrique de l'Ouest', *Bull. IFAN*, **29**, ser. A, no. 2, 845–9; see also other articles by Elouard in the same number.

EMBERGER, L. (1955) 'Une classification biogéographique des climats', *Recueil Trav. Lab. Géol. Zool. Univ. Montpellier*, sér. bot. no. 7, pp. 3–43.

EMILIANI, C. (1954) 'Temperature of Pacific bottom waters and polar superficial waters during the Tertiary', *Science*, **119**, 853–5.

EMILIANI, C. (1956) 'Oligocene and Miocene temperatures of the equatorial and subtropical Atlantic Ocean', *J. Geol.* **64**, 281–8.

EMILIANI, C. (1958a) 'Ancient temperatures', *Sc. Amer.* Feb. 11 p.

EMILIANI, C. (1958b) 'Paleotemperature analysis of core 280 and Pleistocene correlations', *J. Geol.* **66**, 264–75.

EMILIANI, C. (1961) 'Cenozoic climatic changes as indicated by the stratigraphy and chronology of deep-sea cores of Globigerina-ooze facies', *Ann. NY Acad. Sc.* **95**, 521–36.

EMILIANI, C. and GEISS, J. (1957) 'On glaciations and their causes', *Geol. Rundsch.* **46**, 576–601.

ERHART, H. (1935) *Traité de pédologie*, Strasbourg, Inst. pédologique, 2 vols. 260 p. (vol. 2, 1938).

ERHART, H. (1955) '"Biostasie" et "rhéxistasie", esquisse d'une théorie sur le rôle de la pédogenèse en tant que phénomène géologique', *C.R. Acad. Sc.* **241**, 1218–20.

ERHART, H. (1956a) *La Genèse des sols en tant que phénomène géologique*, Paris, Masson, Coll. Evolution des Sciences, 90 p.

ERHART, H. (1956b) 'La vie végétale continentale aux époques pré-dévoniennes vue sous l'angle de la théorie bio-rhéxistasique et des dernières découvertes palynologiques', *Bull. Soc. Géol. France*, 6ᵉ ser. **6**, 445–50.

ERHART, H. (1962) 'Témoins pédogénétiques de l'époque permo-carbonifère', *C.R. Somm. Soc. Biogéogr.* **355/7**, 21–53.

ESTIENNE, P. (1952) 'Le problème des variations climatiques en pays tempérés', *Rev. Géogr. Alpine*, **40**, no. 2.

EWING, M. and DONN, W. L. (1958) 'A theory of ice ages II; the theory that certain local terrestrial conditions caused Pleistocene glaciations is discussed further', *Science*, **127**, pp. 1159–62.

EYRE, S. R. (1968) *Vegetation and Soils*, London, Arnold, 328 p.

FAIRBRIDGE, R. F. (1961) 'Convergence of evidence on climatic change and ice ages', *Ann. NY Acad. Sc.* **95**, 542–79.

FAIRBRIDGE, R. F. (1961) 'Radiation solaire et variations cycliques du niveau marin', *Rev. Géogr. Phys. Géol. Dyn.* n.s. **4**, pp. 2–14.

FELT, E. J. (1953) 'Influence of vegetation on soil moisture contents and resulting soil volume changes', *Proc. 3rd Int. Soil Conf.* **1**, pp. 24–7.

FENNEMAN, NEVIN M. (1931) *Physiography of the Western United States*, New York and London, McGraw-Hill, 534 p.

FENNEMAN, NEVIN M. (1938) *Physiography of the Eastern United States*, New York and London, McGraw-Hill, 714 p.

FIRBAS, F. (1950) 'Die Quartäre Vegetationsentwicklung zwischen den Alpen u.d. Nord- und Ostsee', *Erdkunde*, **4**, no. 1, 6–15.

FLEGEL, R. (1958) 'Die Verbreitung der Bodenerosion in der Deutschen Demokratischen Republik', *Bodenkunde und Bodenkultur*, no. 6, 104 p. 1 map.

FLETCHER, J. and BEUTNER, E. (1941) 'Erodibility investigations on some soils of the upper Gila watershed', *U.S. Dept. Agr. Tech. Bull.* no. 794, 31 p.

FLINT, R. F. (1957) *Glacial and Pleistocene Geology*, New York, Wiley, 553 p.

FLOHN, H. (1953) 'Studien über die atmosphärische Zirkulation in der letzten Eiszeit', *Erdkunde*, **7**, 266–75.

FLOHR, E. F. (1962) 'Bodenzerstörungen durch Frühjahrsstarkregen im nordöstlichen Niedersachsen', *Göttinger Geogr. Abh.* no. 28, 119 p. 29 photos, 5 maps.

FOOD AND AGRICULTURE ORGANISATION (1954) 'Soil erosion survey of Latin America', *J. Soil Water Conserv.* no. 4, 158–68.

FOOD AND AGRICULTURE ORGANISATION (1962) *Influences exercées par la forêt sur son milieu*, Etud. Forêts Prod. Forest. no. 15, 341 p. 56 figs.

FOURNIER, F. (1949) 'Les facteurs climatiques de l'érosion du sol', *Bull. Ass. Géogr. Franç.* pp. 97–103.

FOURNIER, F. (1960) *Climat et érosion; la relation entre l'érosion du sol par l'eau et les précipitations atmosphériques*, Paris, PUF, 201 p. 15 graphs.

FREE, G. (1960) 'Erosion characteristics of rainfall', *Agr. Eng.* (St Joseph), **41**, 447–9.

FREISE, F. W. (1933) 'Brasilianische Zuckerhutberge', *Z. Geomorph.* **8**, 49–66.

FRIEDLAND, V. N. (1951) 'Essai de division géographique des sols des systèmes montagneux de l'URSS', *Pédologie*, **9**, pp. 521–53 (French trans. by BRGM).

FRISTRUP, B. (1952–3) 'Wind erosion within the arctic deserts', *Geog. Tidsskr.* **52**, 51–65.

FURON, R. (1941) *La Paléogéographie. Essai sur l'évolution des continents et des océans*, Paris, 530 p.

FURON, R. (1950) 'Les problèmes de paléoclimatologie et de paléobiologie posés par la géologie de l'Arctide', *C.R. Somm. Sc. Soc. Biogéogr.* **230**, 13–23.

FURON, R. (1954) 'Biogéographie et paléogéographie', *Rev. Sc.* **61**, pp. 158–69.

GALEVSKI, M. (1955) 'La corrélation entre les pluies torrentielles et l'intensité de l'érosion', *Ann. Ec. Nat. Eaux Forêts*, **14**, 384–427.

GALON, R. (1954) 'Les principaux paysages morphologiques du monde du point de vue des profils synthétiques qui les caractérisent', *Czasopismo Geogr.* **25**, 26–37.

GARKUSCHA, I. F. (1953) *Bodenkunde* (trans. from Russian), Berlin, Deutscher Bauernverlag, 360 p.

GAUCHER, G. (1948) 'Sur la notion d'optimum climatique d'une formation pédologique', *C.R. Acad. Sc.* **227**, 290–2.

GAUSSEN, H. (1949) 'Projets pour diverses cartes du monde au 1/1 000 000. La carte écologique du tapis végétal', *Ann. Agron.* pp. 78–104.

GAUSSEN, H. (1954) 'Expression des milieux par des formules écologiques. Leur représentation cartographique', *Colloques Int. CNRS, Régions écologiques du globe*, Paris, pp. 13–28.

GAUSSEN, H. (1954) *Géographie des plantes*, Paris, Coll. A. Colin, 223 p.

GAUSSEN, H. (1955) 'Les climats analogues à l'échelle du monde', *C.R. Acad. Agric.* (France), **41**, no. 5, March 9–16 meeting, pp. 234–8.

GEGENWART, W. (1952) 'Die ergiebigen Stark- und Dauerregen im Rhein-Maingebiet und die Gefährung der landwirtschaftlichen Nutzflächen durch die Bodenzerstörung', *Rhein-Main Forsch.* no. 36, 52 p.

GEGENWART, W. and RUPPERT, K. (1955) 'Ein Versuch zur Feststellung der winterlichen Bodenzerstörung', *Peterm. Geogr. Mitt.* **99**, 21–3.

GEIGER, R. (1930) 'Mikroklima und Pflanzenklima', *Handbuch der Klimatologie*, Berlin, Bornträger.

GEIGER, R. (1959) *The Climate Near the Ground*, trans. Steward, Harvard University Press, 494 p.

GEIGER, R. and AMANN, H. (1931) 'Forstmeteorologische Messungen in einem Eichenbestand', *Forstwiss. Zentralbl.*

GEIKIE, A. (1875) *Physical Geography*.

GENTILLI, J. (1949) 'Foundations of the Australian bird geography', *Emu*, **49**, Oct. pp. 85–130 (cf. pp. 106–13).

GENTILLI, J. (1952) 'Southwestern forests of Western Australia, a study in geographical environment', *Queensland Geogr. J.* **54**, 12 p.

GENTILLI, J. (1953) 'Critique de la méthode de Thornthwaite pour la classification des climats', *Ann. Géogr.* **62**, 180–5.

Geol. Rundsch. (Klimaheft) **40** (1952), 200 pp.; also 'Das Klima der Vorzeit als Tagungsthema der Hauptversammlung', 1951 in *Geol. Rundsch.* **37** (1949), 139–40.

GESLIN, H. (1935) 'La température du sol', Rapport à la Commission de l'Association Internationale de la Science du Sol, Versailles, 1934, *Météorologie*.

GÈZE, B. (1959) 'La notion d'âge du sol, son application à quelques exemples régionaux', *Ann. Agron.* pp. 237–55.

GIGOUT, M. (1957) Chronologie du Quaternaire récent marocain. Principes de la corrélation fluvio-marine, *C.R. Acad. Sci.* **244**, 2404–7.

GILBERT, G. K. (1877) *Geology of the Henry Mountains*, Washington, 160 p.

GODARD, M. (1949) 'Microclimats et mésoclimats du point de vue agronomique', *Ann. Agron.* **19**, no. 4.

GOSSELIN, M. (1942) 'La défense des sols cultivés contre l'érosion', *Tunisie agr.*, March.

GREENHALL, A. F. (1951) 'Soil erosion and its control in the south half of the North Island of New Zealand', *Int. Union Geod. Geophys.; Int. Ass. Sc. Hydrol.*, Brussels meeting, vol. 2, 106–14.

GREGORY, S. (1954) 'Climatic classification and climatic change', *Erdkunde*, **8**, 246–52.

GRIGORIEV, A. A. (1942) 'Attempt to characterize the basic types of physical-geographical environment, 3rd part: physical-geographical types of the Arctic zone', *Problems Phys. Geogr., USSR Akad. Sc.*

GRIGORIEV, A. A. (1948) 'Attempt to characterize the basic types of physical-geographical environment, 4th part: basic and general physical-geographical processes of the sub-arctic lands and the temperate zone, and justification for the division of the temperate latitudes into zones', *Problems Phys. Geogr., USSR Akad. Sc.*

GRIGORIEV, A. A. (1954) 'Geographical zonation and some of its applications', *Izv. Akad. Nauk. SSSR, Geogr. Ser.* no. 5, 17–39 and no. 6, 41–50.

GRIGORIEV, A. A. (1956) 'Present state of the theory of zonality in nature', *Geogr. Essays, USSR Akad. Sc.* pp. 365–71.

GRIGORIEV, A. A. (1962) 'Present state of the theory of geographic zonality', *Soviet Geogr.*, Amer. Geogr. Soc. Occas. Public. no. 1, pp. 182–7.

GRITCHUK, B. M. (1952) 'Basic results of a micro-paleontological study of the Quaternary deposits of the Russian plain', *Materials for the Quat. of the USSR*, vol. 3.

GROSSE, B. (1953) 'Untersuchungen über die Winderosion in Niedersachsen', *Mitt. Inst. Raumforsch.* Bonn, no. 20, pp. 137–45.

GROSSE, B. (1955) 'Die Bodenerosion in Westdeutschland, Ergebnisse einiger Kartierungen', *Mitt. Inst. Raumforsch.* Bonn, no. 11, 35 p. 5 sep. maps.

GUERASSIMOV, I. P. (1944) 'La carte mondiale des sols et les lois générales de la géographie du sol', *Ann. Agron.* pp. 448–94.

GUERASSIMOV, I. P. (1954) 'Analogies et différences dans la nature des déserts', *Priroda*, no. 2, pp. 11–22 (French trans. by BRGM, no. 984).

GUILLIEN, Y. (1955) 'La couverture végétale de l'Europe Pleistocène', *Ann. Géogr.* **64**, 241–76.

GUPTA, R., KHYBRI, M., and SINGH, B. (1963) 'Run-off plot studies with different grasses with special reference to conditions prevailing in the Himalayas and Siwalik region', *Indian Forester*, **89**, 128–33.

HAASE, H. (1948) 'Ergebnisse der Abholzung des Harzes auf fliessendes Wasser und Erosion', *Gas- und Wasserwirtschaft*, **89**, no. 9, 265–9.

HARTKE, W. (1954) 'Kartierung von Starkregenzügen auf Grund ihrer bodenzerstörenden Wirkung', *Erdkunde*, **8**, 202–6.

HARTKE, W. and RUPPERT, K. (1959) 'Die ergiebigen Stark- und Dauerregen *in Süddeutschland nördlich der Alpen*, Forsch. 2, Dt. Landeskunde, **115**, 39 p. 32 maps, Bad Godesberg [cf. *Erdkunde*, **15** (1961) p. 246].

HAUDE, W. (1924) 'Temperatur und Austausch der bodenahen Luft über einer Wüste', *Beiträge Phys. frei. Atmos.* **21**, 129 ff.

HEINZELEN, J. DE (1952) *Sols, paléosols et désertifications anciennes dans le secteur nord-oriental du Bassin du Congo*, Brussels, Publ. INEAC, 168 p.

HEINZELIN, J. DE (1953) 'Les stades de récession du glacier Stanley occidental', *Explor. Parc Nat. Albert*, 2nd ser. no. 1, 25 p.

HELIMANN, G. (1925) 'Grenzwerte der Klimaelemente auf Erde', *Naturwiss.* **13**, 845.

HEMPEL, L. (1955) 'Konvergenzen von Oberflächenformen unter dem Einfluss verschiedener klimatischer Kräfte', *Deutsche Geogr. Blätter*, **47**, 188–200.

HEMPEL, L. (1956) 'Über Alter und Herkunfstgebiet von Auelehm im Leinetal', *Eiszeitalter und Gegenwart*, **7**, 35–42.

HEMPEL, L. (1957) 'Soil erosion and water runoff on open ground and underneath wood', *Int. Union Geod. Geophys. Int. Ass. Sc. Hydrol.*, Toronto meeting, vol. 1, 108–14.

HEMPEL, L. (1960) 'Bilanzen zur Reliefgestaltung der Erde', *Geogr. Ber*, **15**, 97–107.

HENIN, S. (1952) 'Quelques remarques concernant l'infiltration de l'eau dans les sols', *Rev. Gén. Hydrol.*, March–April, pp. 77–82.

HENIN, S. (1953) 'Mécanisme de caractère de l'érosion par l'eau', *Technique-Agr.* **24**, 2–11.

HENIN, S. and MONNIER, G. (1956) 'Evaluation de la stabilité structurale du sol', *C.R. VI Congr. Sc. Sol*, Paris.

HENRY, A. J. (1919) 'Increase of precipitation with altitude', *Mon. Weather Rev.* **44**, 33–41.

HERMES, K. (1955) 'Die Lage der oberen Waldgrenze in den Gebirgen der Erde und ihr Abstand zur Schneegrenze', *Kölner Geogr. Arb.* no. 5, 277 p. 4 maps.

HESMER, H. and FELDMANN, A. (1953) 'Der Oberflächenabfluss auf bewaldeten und unbewaldeten Hangflächen des südlichen Sauerlandes', *Forstarchiv.* 11–12, pp. 245–56.

HEYER, E. (1938) 'Uber Frostwechselzahlen in Luft und Boden', *Gerlands Beitr. Geophys.* **52**, 68–122.

HIERNAUX, C. (1955) 'Sur un nouvel indice d'humidité proposé pour l'Afrique occidentale', *Bull. IFAN*, **17**, 1–6.

HOLMES, C. D. (1955) 'Geomorphic development in humid and arid regions: a synthesis', *Amer. J. Sc.* **253**, 377–90.

HOORE, J. D' (1954) 'Essai de classification des zones d'accumulation des sesquioxydes libres sur les bases génétiques', *Sols Africains*, **3**, 66–81.

HORTON, R. E. (1919) 'Rainfall interception', U.S. Weather Bur. *Mon. Weather Rev.* **47**, 603–23.

HURSCH, C. R. (1951) 'Recherches sur les relations entre les forêts et les cours d'eau', *Unasylva*, pp. 3–10.

HUTTON, J. (1795) *Theory of the Earth*, 2 vols, Edinburgh.

INEAC (Yangambi) (1951) *Chutes de pluie au Congo Belge et au Ruanda-Urundi pendant la décade* 1940–1949, Brussels, Bur. climatologique, communication no. 3, 248 p.

IVANOV, N. N. (1948) 'Zones of climatic landscapes of the world', *Mem. Geogr. Soc.* USSR, n.s. no. 1.

JACKSON, M. L. and SHERMANN, G. (1953) 'Chemical weathering of minerals in soils', *Advances in Agron.* **5**, 219–318.

JANSE, A. R. P. and HULSBOS, W. C. (1956) 'Influence de quelque plantes de couverture sur certaines propriétés physiques du sol', *Bull. Bibliogr. Mensuel Bur. Interafricain Sols Econ. Rurale*, no. 11–12, pp. 5–6.

JAUDEL, L. and TRICART, J. (1958) 'Les précurseurs anglo-saxons de la notion davisienne de cycle d'érosion', *Rev. Gén. Sc.* **65**, 237–51.

JENNY, H. (1941) *Factors of Soil Formation*, New York, McGraw-Hill, 270 p.

JESSEN, O. (1936) *Reisen und Forschungen in Angola*, Berlin, Reimer.

JONES, F. O., EMBODY, D. R., and PETERSON, W. L. (1961) 'Landslides along the Columbia River Valley, northeastern Washington', *USGS, Prof. Pap.* 367, 98 p.

JOLY, F. (1962) *Etudes sur le relief du Sud-Est marocain*, Rabat, Institut scientifique chérifien, 578 p.

KANONNIKOV, A. M. (1955) 'Concerning the question of natural zones', *Izv. Vsiess. Geogr. Obchtch.* **87**, 529–34.

KAYSER, B. (1961) *Recherches sur les sols et l'érosion en Italie méridionale, Lucanie*, Paris, SEDES, 127 p.

KELLOGG, C. E. (1950) 'Principal soils', *Trans. 4th Int. Congr. Soil Sc.*, Amsterdam, vol. 1, 266–76.

KENDREW, W. G. (1961) *Climate; a treatise on the principles of weather and climate*, 4th edn, Oxford, Clarendon Press, 329 p.

KENNEDY, A. P. (1952) 'Some factors influencing infiltration into soils under natural and artificial rainfall', *J. Soil Conserv. Serv.* New South Wales, July.

KING, L. C. (1950) 'The study of the world's plainlands: a new approach in geomorphology', *Q.J. Geol. Soc. London*, **106**, 101–31.

KING, L. C. (1953) 'Canons of landscape evolution', *Bull. Geol. Soc. Amer.* **64**, 721–52.

KITTLER, A. G. (1955) 'Merkmale, Verbreitung und Ausmass der Bodenerosion', *Peterm. Geogr. Mitt.* **99**, 269–73.

'Klimaschwankungen und Krustenbewegungen in Afrika, südlich des Aquators von der Kreidezeit bis zum Diluvium', *Geogr. Ges. Hannover*, **3** (1950).

KLUTE, F. (1951) 'Das Klima Europas während das Maximums der Wechsel-Würmeiszeit und die Änderungen bis zur Jetztzeit', *Erdkunde*, **5**, 273–83.

KOCH, P. (1950) 'La violence des orages dans ses relations avec le débit des égouts urbains', *Houille Blanche*, **5**, spec. number B, 679–82.

KOCH, H. G. (1952) 'Bericht über eine Exkursion durch das Nadelwaldgebiet in Lappland', *Polarforsch.* Kiel, p. 193. (Analysis of a study by Regel published in *Experientia*, **8**, 1952, no. 1.)

KÖPPEN, W. (1923) *Die Klimate der Erde; Grundriss der Klimakunde*, Walter de Gruyter Co., Berlin, 369 p.

KÖPPEN, W. (1931) *Grundriss der Klimakunde*, Berlin, de Gruyter.

KRAUS, E. B. (1955) 'Secular changes of tropical rainfall regimes', *Q. J. Roy. Meteorol. Soc.* **81**, 198.

KREBS, N. (1966) *Vergleichende Länderkunde* (3rd edn), Stuttgart, Köhler, 484 p. (pp. 65–77).

KUBIENA, W. L. (1952) *Claves sistemáticos de suelos*, Madrid, 388 p. 26 pl. in colour.

KUCERA, C. L. (1954) 'Some relationships of evaporation rate to vapor pressure deficit and low wind velocity', *Ecology*, **35**, pp. 71–5.

KUHN, W. (1953) 'Hecken, Terrassen und Bodenzerstörung im hohen Vogelsberg', *Rhein-Mainische Forsch.* no. 39.

KUHNHOLZ-LORDAT, G. (1952) 'Le tapis végétal dans ses rapports avec les phénomènes actuels en Basse-Provence', *Encycl. biogéogr. écol.* vol. 9, Paris, Le Chevalier, 208 p.

KURON, H. (1953) 'Berücksichtigung des Bodenschutzes bei Beratung und Umlegung', *Mitt. Raumforsch.* Bonn, **20**, pp. 1–14.

KURON, H. (1954) 'Ergebnisse 15-jährigen Untersuchungen über Bodenerosion durch Wasser in Deutschland', *Int. Union Geod. Geophys., Int. Ass. Sc. Hydrol.* Rome meeting, vol. 1, 220–7.

KURON, H. (1954) 'Landwirtschaft und Bodenerosion Untersuchungen typischer Schadens-gebiete. I, der Rossbacherhof bei Erbach im Odenwald', *Mitt. Inst. Raumforsch.* Bonn, no. 23, 48 p.

KURON, H. and JUNG, L. (1957) 'Über die Erodierbarkeit einiger Böden', *Int. Union Geod. Geophys.* Toronto meeting, vol. 1, pp. 161–5.

KURON, H. and STEINMETZ, H. J. (1957) 'Die Plantschwirkung von regentropfen als ein Faktor der Bodenerosion', *Int. Union Geod. Geophys.* Toronto meeting, vol. 7, pp. 115–21.

LAMB, H. H. (1961) 'Climatic changes with historical time as seen in circulation maps and diagrams', *Ann. NY Acad. Sc.* **95**, 124–61.

LAMB, H. H. and JOHNSON, A. I. (1959) 'Climatic variation and observed changes in the general circulation', *Geogr. Annaler*, **41**, 94–134.

LAMB, H. H. and JOHNSON, A. I. (1961) 'Climatic variation and observed changes in the general circulation', *Geogr. Annaler*, **43**, 363–400.

LANDSBERG, H. (1949) 'Climatology of the Pleistocene', *Bull. Geol. Soc. Amer.* **60**, pp. 1437–42.

LANGLE, P. (1954) 'Sur l'application des études morphologiques pour la prospection des sols', *Soc. Sc. Nat. Maroc*, Trav. Sect. Pédol. **8–9**, 131–4.

LAPPARENT, A. DE (1907) *Leçons de géographie physique*, Paris, Masson, 3rd edn, 728 p.

LAUTENSACH, H. (1950) 'Granitische Abtragungformen auf der iberischen Halbinsel und in Korea, ein Vergleich', *Peterm. Geogr. Mitt.* **94**, pp. 87–196.

LEHMAN, H. (1954) 'Bericht von der Arbeitstagung der internationaler Karstkomision', *Erdkunde*, **8**, 112–21, and the articles which follow this paper, in the same issue, especially those of Corbel, Lasserre, and von Wissmann. An attempt at classification is made by P. Birot.

LEMÉE, G. (1948) 'La méthode de l'analyse pollinique et ses rapports à la connaissance des temps quaternaires', *Année Biol.* **24**, 49–75.

LEMÉE, G. and WEY, R. (1950) 'Observations pédologiques sur les sols actuels de loess aux environs de Strasbourg', *Ann. Agron.* pp. 1–12.

LEOPOLD, L. B. (1951) 'Rainfall frequency: an aspect of climatic variation', *Trans. Amer. Geophys. Union*, **32**, 347–57.

LEPLAE, E. (1937) Comptes rendus et rapports au VIIe Congr. d'Agric. Trop. et Subtrop., Paris.

LEUZINGER, V. R. (1948) *Controversias geomorfológicas*, Rio de Janeiro, 209 p. Excellent critique of Davisian geomorphology.

LISSITCHEK, E. N. (1950) 'Analyse quantitative des processus actuels d'érosion dans les régions semi-sèches du Tian-Chan', *Probl. Phys. Geogr., USSR Akad. Sc.* **16**, 191–2 (French trans. by BRGM, no. 1312).

LOTZE, F. (1938) *Steinsalz und Kalisalze*, Berlin.

LOUIS, H. (1957) 'The Davisian cycle of erosion and climatic geomorphology', *Proc. Int. Geogr. Union, reg. conf. Japan*, pp. 164–6.

LOUIS, H. (1957) 'Rumpfflächen Problem, Erosionszyklus und Klimageomorphologie', *Machatschek Festschrift*, pp. 9–26.

LYELL, C. (1833) *Principles of Geology*, Edinburgh, 3 vols.

LYSGAARD, L. (1952) 'Recent climatic fluctuations', *Folia Geogr. Danica*, vol. 5, Copenhagen.

MAACK, R. (1956) 'Über Waldverwüstung und Bodenerosion im Staate Parana', *Erde*, pp. 191–228.

MACAR, P. (1946) *Principes de géomorphologie normale, étude des formes du terrain des régions à climat humide*, Liège, Masson, 304 p.

MACHATSCHEK, F. (1938–40) *Das Relief der Erde. Versuch einer regionalen Morphologie der Erdoberfläche*, 2 vols. Bornträger, Berlin (2nd edn, 1955).

MCINTYRE, D. S. (1958) 'Soil splash and the formation of surface crusts by raindrop impact', *Soil Sc.* **85**, 261–6.

MAGOMEDOV, A. D. (1950) 'Infiltration des eaux de pluie et de fusion dans les divers types de sols', *Pochvovedenie*, pp. 361–466 (French trans. by BRGM).

MALAURIE, J. (1949) 'Evolution actuelle des pentes sur la côte ouest du Groënland (Baie de Disko)', *Bull. Ass. Géogr. Franç.* pp. 2–8.

MALAURIE, J. (1968) *Thèmes de recherches géomorphologiques dans le Nord-Ouest du Groënland. Mémoires et documents*, CNRS, hors-série, 495 p.

MALKIN, N. P. (1961) 'On the influence of sea transgressions and straits on the Quaternary glaciation', *Izv. Vsiess. Geogr. Obchtch.* pp. 122–35.

MANGENOT, G. (1951) 'Une formule simple permettant de caractériser les climats de l'Afrique intertropicale dans leurs rapports avec la végétation', *Rev. Gén. Bot.* pp. 353–69.

MANLEY, G. (1951) 'Climatic fluctuation, a review', **41**, *Geogr. Rev.* pp. 656–60.

MARKER, M. E. (1959) 'Soil erosion in relation to the development of landforms in the Dundas area of western Victoria, Australia', *Proc. Roy. Soc. Victoria*, **71**, 125–36.

MARKOV, K. K. (1948) *Fundamental Problems of Geomorphology*, USSR Akad. Sc.

MARKOV, K. K. (1956) 'Origin of contemporaneous geographical landscapes', *Geogr. Essays USSR Akad. Sc.* pp. 42–53.

MARKOV, K. K. (1960) *Paleogeography*, 2nd edn, Univ. of Moscow, 267 p.

MARLIÈRE, R. (1951) 'Terrils en marche', *Publ. Ass. Ingénieurs Fac. Polytechn. Mons*, Belgium, Mons.

MARRES, P. (1954) 'Phénomènes actuels de surface et l'équilibre du tapis végétal dans la région méditerranéenne', *Colloq. Int. CNRS*, Régions écologiques du globe, Paris, pp. 117–23.

MARTHELOT, P. (1957) 'L'érosion dans la montagne Kroumir', *Rev. Géogr. Alpine*, **45**, 273–88.

MARTIN, P., SABELS, B., and SHTLER, D. (1961) 'Rampart Cave coprolite and ecology of the Shasta ground sloth', *Amer. J. Sc.* **259**, 102–27.

MARTONNE, E. DE (1913) 'Le climat facteur du relief', *Scientia*, pp. 339–55.

MARTONNE, E. DE (1935) *Traité de géographie physique*, vol. 2, *Le Relief du sol*, 5th edn, Paris, Colin, pp. 499–1057.

MARTONNE, E. DE (1940) 'Problèmes morphologiques du Brésil tropical atlantique', *Ann. Géogr.* **49**, 1–27 and 106–29.

MARTONNE, E. DE (1946) 'Géographie zonale: la zone tropicale', *Ann. Géogr.* **55**, 1–18.

MELTON, M. A. (1958) 'Correlation structure of morphometric properties of drainage systems and their controlling agents', *J. Geol.* **66**, 442–60.

MENSCHING, H. (1951) 'Une accumulation post-glaciaire provoquée par des défrichements', *Rev. Géomorph. Dyn.* **2**, 145–56.

MENSCHING, H. (1957) 'Soil erosion and formation of haugh-loam in Germany', *Int. Union Geod. Geophys., Int. Ass. Sc. Hydrol.* Toronto meeting, vol. 1, 174–80.

MICHEL, P. (1959) 'L'évolution géomorphologique des bassins du Sénégal et de la Haute-Gambie, ses rapports avec la prospection minière', *Rev. Géom. Dyn.* **10**, 117–43.

MILKOV, F. N. (1953) *Influence of Relief on Vegetation and Animal Life*, Moscow, Publ. Geogr. Lit. 164 p.

MILLER, A. (1951) 'Three new climatic maps', *Inst. Brit. Geogr., Trans. Pap.* no. 17, pp. 15–20.

MILLOT, G. and BONIFAS, M. (1955) 'Transformations isovolumétriques dans les phénomènes de latéritisation et de bauxitisation', *Bull. Serv. Carte Géol. Alsace Lorraine*, no. 8, pp. 3–10.

MILOSAVLJEVIC, M. (1950) 'Relation entre la température minimum à 2 m et à 5 cm au dessus du sol à Belgrade', *Bull. Soc. Serbe Géogr.* **30**, no. 1, pp. 11–25.

MINTIER, T. (1941) 'Soil erosion in China', *Geogr. Rev.*, **31**, pp. 570–90.

MITCHELL, J. M. (1965) 'Theoretical paleoclimatology', in *The Quaternary of the United States* (a review volume for the VIIth Congress of the Int. Assoc. Quatern. Res.), ed. H. E. Wright, Jr and David G. Frey, pp. 881–901.

MISZCZAK, A. (1960) 'Amalgamation of landholdings—an agent of increasing soil erosion', *Czasopismo Geogr.* **31**, 179–90.

MONOD, T. (1957) *Les grandes divisions chronologiques de l'Afrique*, London, Publ. CSA, 147 p. one map.

MORTENSEN, H. (1950) *Das Gesetz der Wüstenbildung*, Universitas, Stuttgart, vol. 5, no. 7, pp. 801–14.

MORTENSEN, H. (1957) 'Temperaturgradient und Eiszeitklima am Beispiel der Pleistozänen Schneegrenzdepression in den Rand- und Subtropen', *Z. Geomorph.* n.s. **1**, 44–56.

Consolidated bibliography

MULLER-MINY, H. (1954) 'Bodenabtragung und Erosion im südbergischen Bergland. Ein Beitrag zur Frage der Bodenzerstörung und zur quantitativen Morphologie', *Ber. Deut. Landeskunde*, Remagen, **12**, 277–92.

MURZAIEV, E. M. (1953) 'Outline for the regional physical-geographical subdivision of Central Asia', *Izv. Akad. Nauk. SSSR, Geogr. Ser.* no. 6, 17–30.

MUSGRAVE, G. and NORTON, R. (1937) 'Soil and water conservation investigations at the soil experiment station, Missouri loess valley region, Clarinda, Iowa', *US Dept. Agr. Tech. Bull.* no. 558, 181 p.

NANSEN, F. (1922) *The Strandflat and Isostacy*, Christiania, 313 p.

NICOD, J (1951) 'Sur le rôle de l'homme dans la dégradation des sols et du tapis végétal en Basse-Provence calcaire', *Rev. Géogr. Alpine*, **39**, no. 4. pp. 709–48.

NYE, P. H. (1954) 'Some soil forming processes in the humid tropics, I: a field study of a catena in the West African forest', *J. Soil Sc.* **5**, 7–21.

ÖPIK, E. J. (1950) 'Secular changes of stellar structures and the ice ages', *Mon. Not. Roy. Astron. Soc.* London, **110**, 49–68.

OSBORN, B. (1954a) 'Soil splash by raindrop impact on bare soils', *J. Soil Water Conserv.* US, **9**, 33–8, 43 and 49.

OSBORN, B. (1954b) 'Effectiveness of cover in reducing soil splash by raindrop impact', *J. Soil Water Conserv.* US, **9**, 70–6.

OVINGTON, J. D. (1954) 'A comparison of rainfall in different woodlands', *Forestry*, **27**, 41–53.

PACKER, P. E. (1953) 'Effect of trampling disturbance on watershed condition, runoff and erosion', *J. Forestry*, **51**, 28–31.

PARKER, G. (1963) 'Piping, a geomorphic agent in landform development of the drylands', *Int. Union Geod. Geophys.* Publ. 65, pp. 103–13.

PASSARGE, S. (1904) *Die Kalahari*, Berlin, 822 p.

PASSARGE, S. (1924) *Vergleichende Landschaftskunde*, Heft 4: *Der heisse Gürtel*, Berlin.

PASSARGE, S. (1926) 'Morphologie der Klimazonen oder Morphologie der Landschaftsgürtel?' *Peterm. Geogr. Mitt.*

PEATTIE, R. (1936) *Mountain Geography: a critique and field study*, Harvard University Press, 257 p.

PEGUY, C. P. (1953) 'Hautes altitudes et hautes latitudes', *Cahiers Inf. Géogr.* no. 2, 58–66.

PEGUY, C. P. (1961) *Précis de climatologie*, Paris, Masson, p. 250.

PELLETIER, J. (1953) 'La Bordure orientale du Massif Central de Vienne à Tournon', *Revue de Géographie de Lyon*, **4**.

PENCK, A. (1894) *Morphologie der Erdoberfläche*, Stuttgart, 2 vols.

PENCK, A. (1905) 'Climatic features in the land surface', *Amer. J. Sc.* **169**, 165–74.

PENCK, A. (1913) 'Versuch einer Klimaklassifikation auf physiographischer Grundlage', *Sitzungber. Preuss. Akad. Wiss.* Berlin, 1st sem. pp. 77–97.

PERELMAN, A. I. (1954) 'The natural landscapes of the European part of the U.S.S.R. and their geochemical properties', *Priroda*, no. 3, pp. 38–47, one map.

PHILIPPSON, A. (1886) 'Ein Beitrag zur Erosionstheorie', *Pet. Geog. Mitt.* vol. 32, 67–79.

PHILLIPS, J. (1959) *Agriculture and Ecology in Africa*, London, Faber, 424 p. one separate map.

PIOTROVSKII, M. (1948) 'Concerning the theory of the fluvial cycle of erosion', *Proc. Inst. Geogr. USSR Akad. Sc.* **39**, 119–33.

PITOT, A. and MASSON, H. (1951) 'Quelques données sur la température au cours des feux de brousse aux environs de Dakar', *Bull. IFAN*, **13**, no. 3.

PLAISANCE, G. (1953) 'Les chaînes de sol', *Rev. Forest. Franc.* pp. 565–77.

PLAISANCE, G. (1958) *Lexique pédologique trilingue*, Paris, CDU, 357 p.

PLAYFAIR, J. (1802) *Illustrations of the Huttonian Theory of the Earth*, Edinburgh.

POLOVITSKAIA, M. E. (1953) 'Soil erosion in the United States', *Priroda*, no. 6, pp. 38–43.

POLYNOV, B. B. (1948, 1951) 'Modern ideas of soil formation and development' (trans. from the Russian in *Pochvovedenie*, 1948, pp. 3–13), *Soils Fertilizers*, **14**, 1951, no. 2, 7 p.

PONCET, J. (1963) 'Défense des sols et restauration', *La Pensée*, **111**, 43–54.

PONCET, J. (no date) 'Les rapports entre les modes d'exploitation agricole et l'érosion des sols en Tunisie', *Secr. Agric., Et. et Mém.* no. 2, Tunis, 173 p. 23 photogr. pl.

POUQUET, J. (1951) *L'Erosion*, Paris, PUF, Coll. Que sais-je?, 128 p.

POUQUET, J. (1956) 'Le Plateau du Labé (Guinée française, AOF). Remarques sur le caractère dramatique des phénomèmes d'érosion des sols et sur les remèdes proposés', *Bull. IFAN*, **18**, sér. A, pp. 1–16.

POUQUET, J. (1966) *Les Sols et la géographie; initiation géopédologique*, SEDES, Paris, 267 p.

POWELL, J. W. (1875) *Exploration of the Colorado River of the West and its Tributaries*, Washington; reprint by Univ. of Chicago and Cambridge University Press, 1957.

PRASSOLOV, L. I. (1944) 'Géographie et aires occupées par les différents types de sols', *Ann. Agron.* pp. 495–9.

PRENANT, M. *Géographie des animaux*, Paris, Coll. A. Colin.

QUANTIN, P. and COMBEAU, A. (1962) 'Relation entre érosion et stabilité structurale du sol', *C.R. Acad. Sc.* **254**, 1855–7.

RAMSAY, A. C. (1863) *The Physical Geology and Geography of Great Britain*, London, 145 p.

RAYNAL, R. (1957) 'Bodenerosion in Marokko', *Wiss. Z.*, Martin Luther University, Halle, Math-Nat. Kl. **6**, 885–94.

RAYNAL, R. (1961) *Plaines et piedmonts du Bassin de la Moulouya (Maroc oriental)*, *étude géomorphologique*, Rabat, Imframar, 619 p.

REE, W. and PALMER, V. (1949) 'Flow of water in channels protected by vegetative linings', *U.S. Dept. Agr., Tech. Bull.* no. 967, 115 p. A great number of numerical data.

REICHE, P. (1947) 'A study of weathering processes and products', *Univ. New Mexico, Geol. Public.* no. 1.

REMPP, G. (1937) 'Sur les frontières et les relations entre le macroclimat, le mésoclimat et le microclimat et entre le climat physique et le bioclimat', *Météorologie*, pp. 265–74 and 379–91.

RENAULT, P. (1959) 'Processus morphogénétiques dans les karsts équatoriaux', *Bull. Assoc. Géog. Franç.* 15–22.

'Research studies on soil erosion in Poland', *Wiadomosci Inst. Melioracji i Uz Ziel*, **1** (1960), 11–26.

REYA, O. (1948–49) 'The intensity of precipitation on the Slovenian coast', *Geogr. Vestnik*, **20–21**.

RICHTHOFEN, F. VON (1886) *Führer für Forschungsreisende*, Hannover, 734 p.

RIQUIER, J. (1958) 'Les "lavaka" de Madagascar', *Bull. Soc. Géogr. Aix-Marseille*, **69**, 181–91.

ROBINSON, G. W. (1952) *Soils, their Origin, Constitution and Classification*, London, Murby, 573 p.

ROSE, C. W. (1960) 'Soil detachment caused by rainfall', *Soil Sc.* **89**, 28–35.

ROUGERIE, G. (1960) *Le Façonnement actuel des modelés en Côte d'Ivoire forestière*, Mém. IFAN, no. 58, Dakar, 542 p.

ROUSSEAU, J. (1952) 'Les zones biologiques de la péninsule Quebec-Labrador et l'hémiarctique', *Can. J. Bot.* **30**, 436–74.

ROZOV, N. N. (1954) 'Development up to the present time of Dokuchaev's doctrines on zonal soils', *Izv. Akad. Nauk. SSSR, Geogr. Ser.* no. 4, pp. 3–17.

RUDAKOV, B. E. (1952) 'Méthode d'utilisation des anneaux annuels des arbres pour la mise en évidence des variations du climat d'après leur épaisseur', *Dokl. Akad. Nauk. SSSR*, **84**, no. 1, pp. 169–71 (French trans. by CEDP, no. 787).

RUELLAN, F. (1953) 'O papel das anxurradas no modelo do relêvo brasileiro', *Bol. Paulista Geogr.* no. 13, pp. 5–18 and no. 14, pp. 3–25.

RUHE, R. V. (1954) *Erosion Surfaces of Central African Interior High Plateaus*, Publ. INEAC (Yangambi), Sér. Sc. no. 59, 40 p.

RUHE, R. V. (1956) *Landscape Evolution in the High Ituri, Belgian Congo*, Publ. INEAC, Sér. Sc. no. 66, 91 p.

RUHE, R. V. and DANIELS, R. B. (1958) 'Soils, paleosols and soil-horizon nomenclature', *Proc. Soil Sc. Soc. Amer.* **20**, 66–9.

RUTIMEYER, P. (1869) *Über Thal- und Seebildung*, Basel, 95 p.

RUXTON, B. and BERRY, L. (1957) 'Weathering of granite and associated erosional features in Hong-Kong', *Bull. Geol. Soc. Amer.* **68**, 1263–92.

SABAN (1951) 'Thermomètres géologiques', *Bull. Trim. d'Inf. CEDP*, **3**, no. 12, 3–13.

SALOMON, W. (1916) 'Die Bedeutung der Solifluktion für die Erklärung der deutscher Landschafts- und Bodenformen', *Geol. Rundsch.* **7**, 30–41.

SAPPER, K. (1914) 'Über Abtragungsvorgänge in den regenfeuchten Tropen', *Geogr. Z.* pp. 5–18 and 81–92.

SAPPER, K. (1935) *Geomorphologie der feuchten Tropen*, Geogr. Schriften, herausgegeben von A. Hettner, Leipzig, Teubner, 150 p.

SAPPER, K. and GEIGER, R. (1934) 'Die dauernd frostfreien Räume der Erde und ihre Begrenzung', *Meteorol. Z.* pp. 465–8.

SCAETTA, H., SCHOEP, A., and MEURICE, R. (1937) *La Genèse climatique des sols montagnards d'Afrique Centrale*, Brussels, Inst. Roy. Congo Belge, 351 p.

SCHEIDEGGER, A. E. (1961) *Theoretical Geomorphology*, Berlin, Springer, 333 p.

SCHMID, J. (1955) *Der Bodenfrost als morphologischer Faktor*, Heidelberg, Hüthig, 144 p. 27 figs.

SCHMIDT, W. (1952) 'Art und Entwicklung der Bodenerosion in Südrussland', *Mitt. Inst. Raumforsch.* Bonn, no. 13, 50 p.

SCHMITT, O. (1952) 'Grundlagen und Verbreitung der Bodenzerstörung im Rhein-Mainischen Gebiet mit einer Untersuchung über Bodenzerstörung durch starkregen im Vorspessart', *Rhein-Mainische Forsch.* no. 33, 130 p.

SCHMITTHENNER, H. (1956) 'Die Entstehung der Geomorphologie als geographische Disziplin', *Petermanns Geogr. Mitt.* **100**, pp. 257–68.

SCHNELL, R. (1948) 'Observations sur l'instabilité de certaines forêts de la Haute-Guinée française en rapport avec le modelé et la nature du sol', *Bull. Agron. Congo Belge*, **40**, 671–6.

SCHOELLER, H. (1962) *Les Eaux souterraines*, Paris, Masson, 627 p.

SCHÖNHALS, E. (1951a) 'Fossile gleiartige Böden des Pleistozäns im Usinnger Becken und am Rand des Vogelsberges', *Notizbl. Hessischen Landesamt Bodenforsch.* **6**, 160–83.

SCHÖNHALS, E. (1951b) 'Über einige wichtige Lössprofile und begrabene Böden im Rheingau', *Notizbl. Hessischen Landesamt Bodenforsch.* **6**, 243–59.

SCHÖNHALS, E. (1951c) 'Uber fossile Böden im nicht-vereisten Gebiet', *Eiszeitalter Gegenwart*, **1**, 109–30.

SCHOVE, D. J. (1954) 'Summer temperatures and tree-rings in north Scandinavia, A.D. 1461–1950', *Geogr. Annaler*, **36**, 40–80.

SCHULTZE, J. (1952) 'Bodenerosion in Thüringen, Wessen, Stärke, Abwehrmöglichkeiten', *Peterm. Geogr. Mitt. Ergänzungsheft*, no. 247, 186 p.

SCHULTZE, J. (1953) 'Neuere theoretische und praktische Ergebnisse der Bodenerosions-forschung in Deutschland', *Forsch. und Fortschritte*, **27**, no. 1, pp. 1–7.

SCHWARZBACH, M. (1949) 'Fossile Korallenriffe und Wegeners Drifthypothese', *Naturw.* **36**, no. 8. pp. 229–33.

SCHWARZBACH, M. (1963) *The Climate of the Past; an Introduction to Paleoclimatology*, trans. and ed. by Richard O. Muir, London and Princeton, Van Nostrand, 328 p.

SEEYLE, C. *The Frequency of Heavy Rainfalls in New Zealand*, Wellington, N.Z. Inst. Eng.

SEGALEN, P. (1948) 'L'érosion des sols à Madagascar', *Conf. Afr. des sols, Goma*, pp. 1127–37.

SELIVANOV, A. P. (1960) 'Water-stability of the structure of different soil groups and its dependence on agricultural practices', *Pochvovedenie*, pp. 280–6.

SELTZER, P. (1935) 'Etudes micrométéorologiques en Alsace', doct. diss. Fac. Sc. Univ. Strasbourg, 58 p.

SERRA, L. (1951) 'Interprétation des mesures pluviométriques, lois de la pluviosité', *Int. Union Geod. Geophys.* Brussels meeting.

SHALEM, N. (1952) 'La stabilité du climat en Palestine', *Rev. Biblique*, **58**, 54–74. Reproduced with modifications in *Desert Research*, UNESCO, pp. 153–75.

SHAW, ROBERT H. (1967) *Ground Level Climatology*, Amer. Assoc. Adv. Sc. 408 p.

SMITH, D. D. (1957) 'Factors affecting rainfall erosion and their evaluation', *Int. Union Geod. Geophys., Int. Ass. Sc. Hydrol.* Toronto meeting, vol. 1, pp. 97–107.

SOKOLOV, A. A. (1955) 'Decrease in the durability of ice in relation with the climatic amelioration', *Priroda*, no. 7, pp. 96–8.

SOPPER, W. E. and LULL, H. W. (1967) *International Symposium on Forest Hydrology*, New York, Pergamon, 813 p.

SOUBIES, L., GADET, R., and MAURY, P. (1960) 'Le climat de la région toulousaine et son influence sur les récoltes de blé et de maïs', *C.R. Acad. Agric.* (France), pp. 185–95.

STALLINGS, J. H. (1953) 'Continuous plant cover, the key to soil and water conservation', *J. Soil Water Conserv.* US, **8**, 37–43.

STALLINGS, J. H. (1957) *Soil Conservation*, New York, Prentice-Hall, 575 p.

STEINMETZ, H. J. (1956) *Ausmass des Bodenabtrages in einem Teilgebiet der Wetteran und Vorschläge zur Verhütung der durch ihn entstandenen Schäden*, Hessischen Ministerpräsidenten, Wiesbaden, 105 p. 10 sep. maps.

STRAHLER, A. N. (1956) 'The nature of induced erosion and aggradation', in: Thomas, W. L. Jr. (Ed.) *Man's Role in Changing the Face of the Earth*, University of Chicago Press.

STRETTA, E. (1958) 'Délimitation des zones arides et semi-arides en Turquie', *Rev. Géomorph. Dyn.* **9**, 97–102.

STRZEMSKI, M. (1956) 'A schema of soil distribution over the surface of the globe according to climatic zones', *Przeglad Geogr.* **28**, 131–42.

SURELL, A. (1841) *Etudes sur les torrents des Hautes-Alpes*, Paris.

TAMHANE, R. V., BISWAS., T. D., DAS, B., and NASKAR, G. C. (1959) 'Effect of intensity of rainfall on the soil loss and run off', *J. Indian Soc. Soil Sc.* **7**, no. 4, 231–8.

TANNER, W. P. (1961) 'An alternate approach to morphogenetic climates', *Southeastern Geol.* **2**, 251–7.

TERMIER, H. and TERMIER, G. (1952) *Traité de géologie*, vol. 1: *Histoire géologique de la biosphère (La vie et les sédiments dans les géographies successives)*, Paris, Masson, 721 p. 35 maps in colour.

THAMBYAHPILLAI, G. (1958) 'Rainfall fluctuations in Ceylon', *Ceylon Geogr.* **12**, 51–74.

THEOBALD, N. (1952) 'Les climats de l'Europe occidentale au cours des temps tertiaires d'après l'étude des insectes fossiles', *Geol. Rundsch.* **40**, 89–92.

THORARINSSON, S. (1962) 'L'érosion éolienne en Islande à la lumière des études téphrochronologiques', *Rev. Géomorph. Dyn.* **13**, 107–24.

THORBECKE, F. (1927) 'Morphologie der Klimazonen, herausgegeben von ...', Selection of conferences by various German authors, Breslau, Düsseldorfer Vorträge, 100 p.

THORNTHWAITE, C. (1933) 'The climates of the earth', *Geogr. Rev.* **23**, 433–40.

THORNTHWAITE, C. (1943) 'Problems in the classification of climates', *Geogr. Rev.* **33**, 233–55; with map.

THORNTHWAITE, C. (1948) 'An approach toward a rational classification of climate', *Geogr. Rev.* **38**, 55–94.

TIKHONOV, A. V. (1959) 'Complex problems of soil conservation in the Volga heights', *Izv. Akad. Nauk. SSSR, Geogr. ser.* no. 4, pp. 55–66.

TIVY, J. (1957) 'Influence des facteurs biologiques sur l'érosion dans les Southern Uplands écossais', *Rev. Géomorph. Dyn.* **12**, 9–19.

TRICART, J. (1949, 1952) *La Partie orientale du Bassin de Paris, étude morphologique*, Paris, SEDES, vols. 1 and 2, 474 p.

TRICART, J. (1952) 'Climat, végétation, sols et morphologie', *Jubilaire 50ᵉ Anniv. Labo. Géogr. Rennes*, pp. 225–39.

TRICART, J. (1953) 'Climat et géomorphologie', *Cahiers Inf. Géogr.* no. 2, pp. 39–51.

TRICART, J. (1953) 'Erosion naturelle et érosion anthropogène à Madagascar', *Rev. Géomorph. Dyn.* **4**, 225–30.

TRICART, J. (1953) 'La géomorphologie et les hommes', *Rev. Géomorph. Dyn.* **4**, 153–6.

TRICART, J. (1955) 'Types de fleuves et systèmes morphogénétiques en Afrique occidentale', *Bull. Sect. Géogr., Com. Trav. Hist. Sc.* **68**, 303–45, 21 pl.

TRICART, J. (1956) 'Aspects géomorphologiques du delta du Sénégal', *Rev. Géom. Dyn.* **7**, May–June, pp. 65–86.

TRICART, J. (1957a) 'Application du concept de zonalité à la géomorphologie', *Tijdschr. Kon. Ned. Aardrijksk. Gen.* Festschrift Jacoba Hol, **74**, no. 3, 422–34, with map showing the morphoclimatic divisions of the earth.

TRICART, J. (1957b) 'L'évolution des versants', *Inf. Géogr.* **21**, 109–10.

TRICART, J. (1958) 'Division morphoclimatique du Brésil atlantique central', *Rev. Géomorph. Dyn.* **9**, 1–22.

TRICART, J. (1958b) 'Les variations Quaternaires du niveau marin', *Inf. Géogr.* **22**, 100–4.

TRICART, J. (1959) 'L'évolution du lit du Guil au cours de la crue de juin 1957'. *Bull. Comité Trav. Hist. Scient., Sect. Géogr.* (Ministère de l'Education Nationale), **72** (1960) 169–403; summary in Tricart, J. (1959), 'Les modalités de la morphogenèse dans le lit du Guil au cours de la crue de la mi-juin 1957', *Int. Ass. Sci. Hydrol. Publ.* no. 53, Commission d'Erosion Continentale, pp. 65–73.

TRICART, J. (1961) 'Les caractéristiques fondamentales des systèmes morphogénétiques des pays tropicaux humides', *Inf. Géogr.* **25**, pp. 155–69.

TRICART, J. (1961) 'Note explicative de la carte géomorphologique du delta du Sénégal', *Mém. Bur. Rech. Geol. Min.* no. 8, 137 p., 9 plates, 3 coloured maps, scale 1 : 100 000.

TRICART, J. (1962) *L'Epiderme de la terre, esquisse d'une géomorphologie appliquée*, Paris, Masson, Coll. Evolution des Sciences, 167 p.

TRICART, J. (1963) *Géomorphologie des régions froides* trans. by Edward Watson, 1970, as *Geomorphology of Cold Environments*, London, Macmillan, Paris, PUF, 289 p. 320 pp. (abbreviated edition of vols. 2 and 3 of the *Treatise*).

TRICART, J. (1965) *Principes et méthodes de la géomorphologie*, Paris, Masson, 496 p.

TRICART, J. and CAILLEUX, A. (1962–) [titles translated]; *Treatise of Geomorphology*, Paris, 12 volumes; SEDES.

Part I: EXTERNAL DYNAMICS
1 *Introduction to Climatic Geomorphology* (1955) 1965—English trans. by KdeJ, Longman, 1972.
2 *Periglacial Geomorphology* (1950) 1967
3 *Glacial Geomorphology* (1955) 1962
4 *The Landforms of Dry Regions* (1960–61) 1969
5 *The Landforms of the tropics, Forests and Savannas* (1965)—English trans. by KdeJ, Longman, 1972.
6 *The Action of Running Water*
7 *Littoral Geomorphology*
8 *The Development of Slopes*
9 *The Action of the Wind*
Part II: STRUCTURAL GEOMORPHOLOGY
10 *General Characteristics of the Earth*
11 *Orogenic Belts* (1954)
12 *The Continental Plates* (1957)

The dates without brackets refer to the definitive edition published by the Société d'Edition d'Enseignement Supérieur, Paris. The dates in brackets refer to the preliminary editions published in mimeographed form by CDU as 'Cours de Géomorphologie'. Volumes 6–10 have yet to be published in either form.

TRICART, J. and CAILLEUX, A. (1962) *Le Modelé glaciaire et nival*, Paris, SEDES, 508 p. vol. III of the *Traité de Géomorphologie*.

272

TRICART, J. and CAILLEUX, A. (1965) *Introduction à la Géomorphologie climatique*, Paris, SEDES, 306 p. vol. I of the *Traité de Géomorphologie*.

TRICART, J. (1965) *Le Modelé des régions Chaudes, forêts et savanes*, Paris, SEDES, 322 p. vol. V of the *Traité de Géomorphologie*.

TRICART, J. (1967) *Le Modelé des régions périglaciaires*, Paris, SEDES, 512 p. vol. II of the *Traité de Géomorphologie*.

TRICART, J. (1969) *Le Modelé des régions séches*, Paris, SEDES, 472 p. vol. IV of the *Traité de Géomorphologie*.

TRICART, J. (1968) *Précis de géomorphologie*, vol. 1 : *Géomorphologie structurale*, Paris, SEDES, 322 pp.

TRICART, J. and CAILLEUX, A. (1952) 'Causes actuelles et causes anciennes dans la genèse des pénéplaines', *Proc. Int. Geogr. Congr., Washington*, pp. 396–9.

TRICART, J., CAILLEUX, A., and RAYNAL, R. (1962) *Les particularités de la morphogenèse dans les régions de montagnes*, Paris, CDU, 136 p. mimeographed 'Cours de l'Université de Strasbourg.'

TROLL, C. (1944) 'Diluvial-Geologie und Klima', Klimaheft der geologischen Rundschau Einführung, *Geol. Rundsch.* **34**, no. 7–8, 307–25.

TROLL, C. (1944) 'Strukturböden, Solifluktion und Frostklimate der Erde', *Geol. Rundsch.* **34**, 545–694.

TROLL, C. (1955) 'Der jahreszeitliche Ablauf des Naturgeschehens in den verschiedenen Klimagürteln der Erde', *Studium Generale*, **8**, no. 3, 713–33.

TROLL, C. (1955) 'Forschungen in Zentralmexiko 1954. Die Stellung des Landes in dreidimensionalen Landschaftaufbau der Erde', *Deutschen Geographentag*, Hamburg, pp. 191–213.

TROLL, C. (1959) 'Die tropischen Gebirge. Ihre dreidimensionale klimatische und pflanzengeographische Zonierung', *Bonner Geogr. Abh.* Heft 25, 93 p.

TYCZYNSKA, M. (1957) 'Climate in Poland during the Tertiary and the Quaternary', *Czasopismo Geogr.* **28**, 131–70.

UNO (1954) 'Soil erosion survey in Latin America, II: South America', *J. Soil Water Conserv.* **9**, 214–19, 223–9, and 237.

UNESCO (1952) Consultative committee on arid zones. *Eastern Hemisphere Western Hemisphere, Distribution of arid homoclimates*, Paris, 2 maps.

VEILLARD, J. A. (1963) *Civilisations des Andes; Évolution des populations du haut plateau bolivien*, Paris, Gallimard.

VERYARD, R. G. (1963) 'A review of studies on climatic fluctuations during the period of the meteorological record', UNESCO, *Arid Zone Res.*

VIDAL, J. M. and POTAU, M. (1951) *Intensidad de las lluvias en Barcelona*, Serv. Meteorol. Nac. 14 p.

VISHER, STEPHEN S. (1941) 'Climate and geomorphology', *J. Geomorph.* **4**, 54–64.

VOGT, J. (1953) 'Erosion des sols et techniques de culture en climat tempéré maritime de transition (France et Allemagne)', *Rev. Géomorph. Dyn.* **4**, 157–83.

VOGT, J. (1956) 'Culture sur brûlis et érosion des sols', *Bull. Sect. Géogr., Comité Trav. Hist. et Sc.* **69**, 339–42.

VOGT, J. (1957a) 'La dégradation des terroirs lorrains au milieu du XVIIIᵉ siècle', *Bull. Comité Trav. Hist. et Sc. Sect. Géogr.* **70**, 111–16.

VOGT, J. (1957b) 'Protection des sols et modes de tenure dans l'agriculture ancienne', *Bull. Comité Trav. Hist. et Sc. Sect. Géogr.* **70**, 117–29.

VOLOBUIEF, V. P. (1953) *Soils and Climate*, Baku, Azerbaijan Acad. Sc. 319 p.

WALLWORK, K. L. (1956) 'Subsidence in the Mid-Cheshire industrial area', *Geogr. J.* **122**, 40–53.

WANDEL, G. and MÜCKENHAUSEN, E. (1951) 'Neue vergleichende Untersuchungen über den Bodenabtrag an bewaldeten und unbewaldeten Hangflächen in Nordrheinland' *Geol. Jahrbuch*, **65**, 1949, pp. 507–50.

WARMING, E. (1909) *Oecology of Plants; an introduction to the study of plant communities*, English ed. by Percy Groom and I. B. Balfour, Oxford, Clarendon Press, 422 p.

WENTWORTH, C. K. and DICKEY, R. I. (1935) 'Ventifact localities in the United States', *J. Geol.* **43**, 97–104.

WEST, R. G. (1968) Pleistocene geology and biology. London, Longman, 377 p.

WETZEL, W. (1952) 'Sediment und Boden eine Grenzbestimmung', *Z. Pflanzenernährung, Düngung, Bodenkunde.*

WILLETT, H. C. (1949) 'Long period fluctuations of the general circulation of the atmosphere', *J. Meteorol.* **6**, no. 1, 34–50.

WILLETT, H. C. (1953) *Atmospheric and Oceanic Circulation as Factors in Glacial–Interglacial Changes of Climates*, Cambridge, Harlow-Sharphey.

WISSMANN, H. VON (1951) *Über seitliche Erosion. Beiträge zu ihrer Beobachtung, Theorie und Systematik im Gesamtaushalt fluviatiler Formenbildung*, Bonn, Dümmler, 71 p.

WOLDSTEDT, P. (1954) 'Die Klimakurve des Tertiärs und Quartärs in Mitteleuropa', *Eiszeitalter und Gegenwart*, **4–5**, 5–9.

WOLDSTEDT, P. (1954) *Das Eiszeitalter. Grundlinien einer Geologie des Quartärs*, vol. i, *Die allgemeine Erscheinungen des Eiszeitalters*, 2nd edn, Stuttgart, Enke, 374 p.

WOLFF, W. (1950–51) 'Bodenerosion in Deutschland', *Erde*, no. 3–4 pp. 215–28.

WRIGHT, H. E. and FREY, D. G., eds. (1965) *The Quaternary of the United States* (a review volume for the 7th INQUA Congress), Princeton University Press, 922 p.

WUNDT, W. (1948) 'Eiszeiten und Warmzeiten in der Erdgeschichte', *Deut. Geographentag*, Munich, Amt für Landeskunde, **27**, 114–19.

ZVONKOVA, T. V. (1959) *Etude du relief à des fins pratiques*, Moscow, Geografguiz, 304 p. (French trans. by SIG, Bur. Rech Géol. Min., no. 3188).

Biographical sketch of the author

Jean Tricart was born in Montmorency, France, in 1920. In 1943 he finished his university studies and obtained the title of *agrégé*. He received his geography education at the Sorbonne under the direction of the great masters of French geography, Emm. de Martonne, Albert Demangeon, and Max Sorre. From 1945 to 1948 he was the assistant of A. Cholley, under whose direction he wrote his dissertation, *La Partie orientale du Bassin de Paris*, a geomorphological study in two volumes (1949, 1952). At the same time he studied geology and sedimentology under the direction of L. Lutaud and J. Bourcart.

After the retirement of H. Baulig from the University of Strasbourg, Tricart took over the latter's classes and later became director of the Institut de Géographie of that university. In 1949 he created the Laboratoire de Géographie Physique, and in 1956 the Centre de Géographie Appliquée of which he became the director. Also, in 1948, he joined the service charged with the geologic mapping of France and later became principal collaborator, entrusted to direct the mapping of Quaternary formations.

Within the framework of the Laboratoire and later of the Centre de Géographie Appliquée, Tricart developed several new approaches to the study of landforms, including methods of sedimentological investigation in connection with morphogenic processes, the use of palaeoclimatic data in geomorphic cartography (as exemplified in the map of the Senegal delta at the scale of 1 : 50 000, finished in 1956), the study of the Quaternary and of climatic oscillations, particularly in France, Africa, and South America. His main effort, however, went into applied geography which was thereby considerably stimulated. At the date of its foundation in 1956 he became secretary of the Commission of Applied Geography of the International Geographical Union and in 1960 its president. He helped to develop an international geomorphic cartography within the framework of the Commission and later founded a team charged with the establishment of a detailed geomorphic map of France with the help of the Centre National de la Recherche Scientifique. Lastly, in conjunction with the development of the Centre de Géographie Appliquée, he increasingly occupied himself with the study of the interactions between the various elements of the physical geographical environment (geology, geomorphology, climate, vegetation, and soils) in order to understand better the laws which control them and thus to provide a surer foundation to the solution of problems of land improvement and economic development.

As consultant for UNESCO, FAO, the Council of Europe, the Ministerio de Obras Públicas of Venezuela, the Institut de Recherches Agronomiques Tropicales and with teams of the Center of Applied Geography, Professor Tricart has conducted various low cost projects concerned with problems of land classification, agricultural development and management, including soil and hydrologic surveys with detailed maps, whether in France, Senegal, Mali, Argentina, Chile, Peru, Venezuela or Algeria. The Center of Applied Geography of the University of Strasbourg therefore not only provides a unique service among institutes of geography but ensures part of its own income, which is equally unique and may well serve as a model.

Books by J. Tricart

La Partie Orientale du Bassin de Paris: Etude morphologique, 2 vols., Paris, SEDES, 1949, 1952, 474 p.
Le Modelé des chaînes plissées, Paris, CDU, 'Cours de l'Université de Strasbourg', 1954, 1963, 330 p.
Les Massifs anciens, Paris, CDU, 'Cours de l'Université de Strasbourg', 1957, 1963, 252 p.
Les Particularités de la morphogenèse dans les régions de montagnes (with A. Cailleux and R. Raynal), Paris, CDU, 'Cours de l'Université de Strasbourg', 1962, 136 p.
L'Epiderme de la terre. Esquisse d'une géomorphologie appliquée, Paris, Masson, 1962, 167 p.
Le Modelé glaciaire et nival (with A. Cailleux), Paris, SEDES, 1962, 508 p.
Géomorphologie des régions froides, Paris, P.U.F., Collection 'Orbis' directed by A. Cholley, 1963, 289 p. English translation by Edward Watson: *Geomorphology of Cold Environments*, Macmillan and St. Martin's Press, 1969, 320 p.
Le Relief des côtes (cuestas), Paris, CDU, 'Cours de l'Université de Strasbourg', 1963, 137 p.
Le Type de bordures de massifs anciens, Paris, CDU, 'Cours de l'Université de Strasbourg', 1963, 118 p.
Introduction à la géomorphologie climatique (with A. Cailleux), Paris, SEDES, 1965, 306 p. English translation by Conrad J. Kiewiet de Jonge: *Introduction to Climatic Geomorphology*, London, Longman, 1972, 324 p.
Le Modelé des régions chaudes, forêts et savanes, Paris, SEDES, 1965, 322 p. English translation by Conrad J. Kiewiet de Jonge: *The Landforms of the Tropics, Forests and Savannas*, London, Longman, 1972, 304 p.

Index of authors

Index of authors

H

Heinzelin, J. de., 85, 173
Hiernaux, C., 137
Horton, R. E., 71
Hutton, J., foreword, 10, 11

J

Joly, F., 182, 183
Jones, F. O. (Embody, D. R. and Peterson, W. L.), 203

K

King, L. C., 15, 41, 194, 195
Klüte, F., 172
Köppen, W., 14, 134–5
Krebs, N., 241
Kuhnholtz-Lordat, G., 86

L

Lapparent, A. de, 8–9, 14
Lautensach, H., 241
Lemée, G., 112
Leonardo da Vinci, foreword
Lotze, F., 190
Lyell, Ch., foreword, 11, 55, 191

M

Macar, P., 12
Magomedov, A. D., 76, 104
Malaurie, J., 54, 146
Margerie, E. de, foreword, 11
Martonne, E. de, 8, 9, 13, 14–15, 64, 161, 162
Meynier, A., 202
Michel, P., 101, 182
Millot, G. and Bonifas, M., 97

N

Nansen, F., 55
Noë, de la, 11

P

Palissy, B., foreword
Passarge, S., 12
Pelletier, J., 195
Penck, A., 116
Peterson, W. L., 203
Philippson, A., 161
Playfair, J., 10, 11
Powell, J. W., 10, 11

R

Raynal, R., 100, 152, 153, 182, 183
Ree, W. and Palmer, V., 79
Renault, P., 95
Richthofen, F. von, 14
Rougerie, G., 71, 101, 252
Rousseau, J., 240
Ruxton, B. and Berry, L., 97

S

Salomon, W., 12
Sapper, K., 12
Schönhals, E., 117
Schwarzbach, M., 188
Soil Conservation Service, US, 61
Surell, A., foreword, 10, 11, 161

T

Thornthwaite, C., 134, 135–7, 232, 233
Tournefort, J. Pitton de, 147
Tricart, J., 15, 54, 57, 76, 85, 99, 101, 106, 116, 152, 173, 174, 195, 196, 204, 232
Troll, C., 53, 138, 152, 237

V

Van 't Hof, 58
Victor, P.-E., 128
Vogt, J., 101, 182, 222
Voltaire, 9

W

Woldstedt, P., 191

Index of place names

281

Subject index

A

ablation, 152

accelerated erosion, 165, 180, 199, 200; *see also under* anthropic and soil erosion

'accidents': climatic, 6, 7, 15, 47, 64, 193, 254; volcanic, 7

action: reaction, 4; effects, influence of man, 84, 129, 131, 145, 165, 198, 243, 244, 245, 246

active layer, 54, 128, 239

adret, 152

aeolian: deposits, 58, 62, 239; processes, 2, 62, 105, 141, 159, 185, 238, *see also under* wind

aerial photos, 112

afforestation, 72

aggradation, 53

agreste, 175

agricultural: development, 63, 169, 178, 180, 200; lands (cultivated regions), 179, 198, 206–8; (cultivation) techniques, 63, 206–8

agronomists, 180

airports, 203

air temperature in forests, 74

algal reefs, 185

alkaline deposits, 190

alluvial: aprons, 35, 182, 245; bars, 63; fans, 85, 214; outskirts, 183

alluviation, inland, 101

alpine (mountain) meadows, 79, 151, 152, 238

alps (*alpes, alpages*), 209

altiplanation, 12, 35; terraces, 239

altitudinal zonation, 147–54; *see also under* vertical zonation

alumina, 101

American geomorphology, 15

ancient massifs, 3, 31, 35, 114, 180, 184

angiosperms, 189

anthropic erosion, 13, 63, 64, 84, 108, 109, 112, 160, 180, **198–223**, 254; *see also under* accelerated and soil erosion

applied geomorphology, 204, 254 biographical sketch

aquifers, exploitation of, 201

areic, areism, 143, 249

argillaceous rocks, 149

arid: climates, 36, 62, 174; phases, 173; regions, 30, 55, 93, 131, 143, 148; subtropics, 173; tropics, 73; zone, 4, 116

arkoses, 194

Australian bush, 246

autocatalysis, thermal, 170

avalanche chutes, 239

azonal, 53, 138, 139–43; phenomena, 2, 4, 5, 54, 55, 56, 57, 165

B

Bacillus exturquens, 93

bacteria (cf. micro-organisms), 92, 93, 98, 250, 252

badlands, 3, 63, 64, 66, 145, 146, 217

bajadas, 49, 100, 182, 183, 244, 249

barchan, 165

barren: grounds (cf. frost deserts), 62, 233, 238, 240; lands, 130

basalts, 94, 96, 104, 115, 212

base-level concept, 10, 25

'basic climate', 149

bauxites, 185, 193

bauxitic terraces, 114

beaches, 63, 138, 187, 201, 202

bedload, 16, 54, 133

biblical Flood, 10

biochores, 83, 160

bioclimatic processes, 4

biogeographers, 170, 174

bioherms, 185

biological deserts, 17